Mathematics and Astronomy in the Ancient World

Daniel Patrick Morgan • Tang Quan
Editors

Mathematics and Astronomy in the Ancient World

An East-Asian Perspective

Springer

Editors
Daniel Patrick Morgan
CRCAO - UMR8155
PARIS, France

Tang Quan
Northwest University
Shaanxi, China

ISBN 978-3-031-93808-5 ISBN 978-3-031-93809-2 (eBook)
https://doi.org/10.1007/978-3-031-93809-2

This Springer imprint is published by the registered company Springer Nature Switzerland AG
The registered company address is: Gewerbestrasse 11, 6330 Cham, Switzerland

Preface

Different civilizations have different mathematical and astronomical traditions. To study the mathematics and astronomy of ancient civilizations is of great significance to reveal the diversity of ancient scientific traditions. So the history of ancient mathematics and astronomy has long been attractive to historians of science in the world.

From July 31 to August 3, 2005, chaired by Michio Yano and myself, the *International Conference on History of Exact Sciences Along the Silk Road* was held in Xi'an. This conference was supported by Wu Wen-Tsun's Silk Road Foundation. One theme of this conference was the transmission and transformation of the exact sciences, and the other was mathematical methods in astronomy. In addition, the conference also focused on research related to the history of mathematics and astronomy along the Silk Road.

From August 23 to 29, 2015, chaired by Karine Chemla and myself, the *International Conference on History of Ancient Mathematics and Astronomy* was held in Xi'an. This conference was held in memory of the late Professor Li Jimin (1938–1993), who founded the history of science and technology program at Northwest University in Xi'an. In tribute to Professor Li's interest in the history of mathematics in ancient China, the theme of the conference was algorithms in the mathematical sciences in ancient world.

Whether it be pure mathematics or the astral sciences, mathematical knowledge in the ancient world was often expressed in the form of procedures (in modern terms, "algorithms"). The 2015 conference was focused on procedures, from a variety of perspectives, and participants were invited to explore how procedures were written, the knowledge involved in formulating mathematics in this way, the variety of ways that procedures were used, and the ways in which mathematical knowledge developed within the framework of procedures.

In 2016, Northwest University then founded the Institute for Advanced Study in History of Science (IAS-HS), and the IAS-HS would collaborate with some of the foremost institutes in the field to form the International Joint Laboratory on Chinese Science and Civilization. On December 2–8, 2018, the Joint Laboratory organized the *Second International Conference on History of Mathematics and Astronomy* to celebrate its establishment, to inaugurate a new series of international conferences, and to discuss how the Institute and Laboratory might function as a new center of international exchange in the history of science.

Held in Xi'an, with a strong contingent of Chinese historians, the second ICHAMA saw contributions by forty-eight invited speakers on the history of mathematics and astronomy in China, Japan, Korea, India, Oceana, and the Islamic and Mediterranean worlds. Befitting the conference's goals and wide range of international speakers, its themes were the diversity of scientific cultures within individual civilizations, scientific exchange between cultures, and the interaction between science and society. Lastly, the talks were divided into eight sessions: (1) astronomy and mathematics in India and the Arabic world; (2) other sciences in the ancient world; (3) astronomy in ancient China; (4) the sciences in Greece, Mesopotamia, and Europe; (5) Chinese mathematical astronomy; (6) exchange and comparison of astronomy and mathematics between civilizations; (7) mathematics in ancient China; and (8) mathematics and astronomy in ancient east Asia.

Of the forty-eight speakers, fourteen expressed interest in reworking their talks over 2019 to contribute to an edited volume representative of the conference's international scope and of the highlights of the state of our respective fields. The preparation of this publication and of future ICHAMA meetings was regrettably delayed due, in part, to the global COVID-19 pandemic. However, the extra time permitted the contributors to refine and update their papers, and they and the editors decided to finish what they had begun as a symbol of the culture of international exchange that brought us together in Xi'an in 2018 and which will hopefully bring us back together there soon.

Since the 2000s, the history of mathematics and astronomy, particularly in China, has seen an important turn toward the philosophy of mathematical practice. "Concerned with the way that mathematics is done, evaluated, and applied," this approach is consistent with the paradigm, advocated by Professor Wu Wen-Tsun since the 1970s, of "recovering mathematical procedures with their original thought" 古证复原. Since the first ICHAMA in 2015, the history of ancient mathematics and astronomy has seen new progress and new advances, some of which are reflected in the present volume. We are greatly indebted to the authors of these papers for their contributions and their patience faced with the many delays.

A study on the history of ancient mathematics and astronomy is still a fascinating topic, and we plan to continue the series of international conferences on this topic in next years. We expect more and more scholars to participate in this series.

Last but not least, thanks to Dr. Daniel Morgan and Professor Tang Quan for their hard work in editing the papers in this edited volume.

Shaanxi, China — Qu Anjing
May 8, 2022

Contents

Part I
China

“Modeling Heaven and Earth”: Application of the Constellation Yi and the Nine Palaces in the Planning of the Western Zhou Capital at Luoyi

Xu Bin

Abstract This paper takes a look at historical literature and archeological evidence concerning the city planning of the Western Zhou’s 西周 (1070–771 BCE) eastern capital at Luoyi 雒邑 in the tradition of “modeling heaven and earth.” After providing a survey of relevant pre-imperial sources, we will examine evidence concerning the theory behind the planning, the choice and location of the site, the timeline of events, and the timing and implications of “positioning” (*gongwei* 攻位). By way of conclusion, this paper will argue that a coincidence in timing between the start of “positioning” and the spring equinox in 1062 BCE suggests rereading a passage from the *Yi Zhou shu* 逸周書 to the effect that the city’s north–south axis was aligned to the constellation Yi 繹 (also known as Tiantian 天田, ο–τ Vir) and that its layout was modeled on the Nine Palaces (*jiugong* 九宮).

Keywords Archeoastronomy · Capital city planning · Luoyang (Luoyi) · Western Zhou · Modeling heaven and earth

1 Introduction

As recorded in the *Book of Changes* (*Zhouyi* 周易), the spirit of “looking up (in the sky) and looking down (at the ground)” 仰觀俯察[1] in ancient Chinese philosophy, informed a unique tradition of “modeling heaven and earth” 象天法地 in the area of ancient city planning, which includes modeling a city’s layout on the configuration of stars at some important moment (such as at dusk, at the beginning of the year).[2] Given the high status of astronomy in ancient politics and culture, there are

[1] *Zhouyi yizhu*, p. 402.

[2] See Pankenier (2013).

Xu Bin (✉)
The Palace Museum, Beijing, People’s Republic of China
e-mail: xubin@dpm.org.cn

D. P. Morgan, T. Quan (eds.), *Mathematics and Astronomy in the Ancient World*,
https://doi.org/10.1007/978-3-031-93809-2_1

numerous historical records of this practice. In brief, the idea of "modeling heaven and earth" arose in the Shang 商 (1600–1070 BCE) and Zhou 周 (1070–256 BCE), prospered in the Qin 秦 (221–207 BCE) and Han 漢 (206 BCE–9 CE, 25–220 CE), matured in the Sui 隋 (581–618) and Tang 唐 (618–907), and reached fruition in the Yuan 元 (1279–1368), Ming 明 (1368–1644), and Qing 清 (1644–1912). This chapter goes back to the early Western Zhou 西周 (1070–771 BCE) to explore the formation of the theory and method of "modeling heaven and earth" city planning and its application in the construction of the eastern capital, Luoyi 雒邑.

More specifically, we will be discussing four key issues concerning the planning of Luoyi. The first is the core theory behind the planning: to determine the location of the capital by following the Tianshi 天室 ("Heavenly Residence"). Was Tianshi an asterism (i.e., the constellation Tianshi) or, as Lin Yun (1993) has suggested, a geographical place (i.e., Mount Tianshi or Songshan 嵩山)? Next is the location of the site, which we will attempt to triangulate using descriptions at three different geographic scales. Third is the time of the planning, concerning which we will establish a timeline of events, determine the number of days dedicated to "positioning," and ask if there is a reason that planning occurred on these specific days? Lastly, we will turn to the layout of the city, the subject most closely related to today's city planning, inquiring about the specific entailments of "positioning."

2 Sources

Also referred to as Luoyi 洛邑 (the City of Luo), Luoshi 洛師 (Luo Headquarters), Dayi 大邑 (the Great City), Xinyi 新邑 (the New City), Xinjing 新京 (the New Capital), Xinyi Luo 新邑洛 (New Luo City), Chengzhou 成周 (the Accomplishment of Zhou), or simply Luo 洛, the planning and construction of the Zhou capital at Luoyi is widely recorded in received literature and excavated bronze inscriptions.[3] One of the earliest references to the construction of a new, centrally-located capital comes in the inscription of the "He *zun*" 何尊 tureen (Fig. 1), which dates to the reign of King Cheng 周成王 (r. 1068–1045 BCE):[4]

> 唯武王既克大邑商，則廷告於天，曰:「余其宅茲中國，自之乂民。」
> When King Wu (r. 1070–1069 BCE) conquered the great city of Shang, he held a ceremony at court to report to Heaven, saying "I will make my home (i.e., build a new capital) at the center of the kingdom from which to rule the people."[5]

Turning to the earliest received literature from and about the period, one finds additional information about the background and theory of the planning. Consider as the annals of the Zhou in the *Shiji* 史記, "Zhou benji" 周本紀 (91 BCE):

[3] Zhong Chunhui (2010).

[4] Note that this reign includes the seven-year regency of the Duke of Zhou (r. 1068–1062 BCE).

[5] Li Xueqin (1981: 35).

Fig. 1 He *zun* inscription. Source: Duan Dexin (2005: 31)

武王徵九牧之君，登豳之阜，以望商邑。武王至于周，自夜不寐。周公旦即王所，曰:「曷為不寐?」王曰:「……我未定天保，何暇寐!」王曰:「定天保，依天室……自洛汭延于伊汭，居易毋固，其有夏之居。我南望三塗，北望嶽鄙，顧詹有河，粵詹雒、伊，毋遠天室。」…
King Wu summoned the lords of the nine lands. He climbed the hill at Bin to gaze towards the City of Shang. When King Wu returned to Zhou, he was not able to sleep at night. Dan, the Duke of Zhou, went to the king's place and said, "Why can't you sleep?" The king said, "… I have not yet secured Heaven's protection, how could I find time to sleep!" The king [also said], "To secure Heaven's protection, we must rely on a Heavenly Residence (Tianshi 天室). … From the north shore of the Luo 洛 [River] to the north shore of the Yi 伊 [River], the land is easily settled and without obstruction. Here was the settlement of the Xia 夏. If I might gaze south to the Santu 三塗 [Mountains] and north to the edge of the Sacred Mountains (Taihang Mountains), if I might look back to the He 河 (Yellow River) and look out to the Luo 雒 and the Yi, this would not be far from a Heavenly Residence." …

成王在豐，使召公複營洛邑，如武王之意。周公複卜申視，卒營築，居九鼎焉。曰:「此天下之中，四方入貢道裡均。」.
King Cheng stayed at [the western capital,] Feng 豐, and had the Duke of Shao revive construction of the City of Luo, as King Wu intended. The Duke of Zhou again consulted the oracle and [went there] to inspect. Finally, when they finished construction, he put the Nine Tripods there. He said, "This is the center of the world. When people from the four quarters pay tribute, the distance will be equal for them."[6]

[6] *Shiji*, 4.128–133; translation modified from Nienhauser et al. (1994: 63–65).

In the biography of the Duke of Zhou in the *Shiji*, "Lu Zhougong shijia" 魯周公世家, we read:

> 成王七年二月乙未，王朝步自周，至豐，使太保召公先之雒，相土。其三月，周公往營成周雒邑，卜居焉，曰:「吉」，遂國之。
> On King Cheng year 7 month II day *yiwei*$_{32/60}$, the King in the morning walked from Zhou and reached Feng, He sent the Grand Guardian, the Duke of Shao, to go first to Luo to survey the land. In month III, the Duke of Zhou went there and constructed the Accomplishment of Zhou (Chengzhou), the City of Luo (Luoyi). He divined [with shells and bones] on whether to reside there. [The result] read "It would be auspicious." Thus it was made the capital.[7]

In similar terms, the earlier *Yi Zhou shu* 逸周書, "Du yi jie" 度邑解, tells us:

> 王曰:「我圖夷茲殷，其惟依天室，其有憲命，求茲無遠慮。天有求繹，相我不難。自洛汭延于伊汭，居陽無固，其有夏之居。我南望過於三途，我北望過於有岳，顧瞻過於河，宛瞻于伊洛，無遠天室。」
> King Wu said: "If I intend to pacify the Yin 殷 (i.e., the Shang), then I must rely on a Heavenly Residence to receive the [Heavenly] Mandate so that there will be no further worries. If I sought the Yi in Heaven (the sky), it will not be difficult to get help from it (from Heaven). From the north shore of the Luo [River] to the north shore of the Yi [River], the land is easily settled and without obstruction. Here was the settlement of the Xia. If I might gaze south to the Santu [Mountains] and north to the slopes of the Sacred Mountains, if I might look back to the He (Yellow River) and look out to the Luo and the Yi, this would not be far from a Heavenly Residence."[8]

The *Shangshu* 尚書 breaks the planning down day by day. In the chapter "Shao gao" 召誥, we read:

> 惟二月既望。越六日乙未，王朝步自周，則至於豐。惟太保先周公相宅。
> It was month II, the sixth day after full moon (day 21), [sexagenary day] *yiwei*$_{32/60}$,[9] when the king in the morning walked from Zhou to Feng. The grand guardian (the Duke of Shao) proceeded the Duke of Zhou to inspect the site.
>
> 越若來三月，惟丙午朏。
> When it came to month III, the first shine of the moon (= day 3) fell on *bingwu*$_{43/60}$.
>
> 越三日戊申，太保朝至於洛，卜宅。厥既得蔔，則經營。
> On the third day (two days later), *wushen*$_{45/60}$, the grand guardian in the morning arrived at Luo and took tortoise oracle about the site. When he had obtained the oracle, he planned and laid out [the city].
>
> 越三日庚戌，太保乃以庶殷攻位於洛汭。
> On the third day (two days later), *gengxu*$_{47/60}$, the grand guardian with all the Yin people started work on positioning [the site] on the north shore of the [River] Luo.
>
> 越五日甲寅，位成。
> On the fifth day (four days later), *jiayin*$_{51/60}$, the positioning was (achieved =) determined.

---

[7] *Shiji*, 33.1519; translation modified from Nienhauser et al. (2006: 136).

[8] *Yi Zhou shu huijiao jizhu*, 5.479–483; translation modified from Huang Huaixin (2006: 219–220).

[9] I follow Liu Qiyi (1979) in reading the lunar notation *jiwang* 既望 ("already/since/after full moon") as referring to the day after full moon, i.e. day 16 as counted from the beginning of the month, which places the sixth day (i.e., 5 days later) on day 21. The notation #/60 indicates a sexagenary date and its place in the sixty-day cycle, counting from *jiazi*$_{01/60}$ to *guihai*$_{60/60}$.

若翼日乙卯，周公朝至於洛，則達觀于新邑營。
The next day, *yimao*$_{52/60}$, the Duke of Zhou in the morning arrived at Luo and all over he inspected the alignment of the new city.

越三日丁巳，用牲于郊，牛二。
On the third day (two days later), *dingsi*$_{54/60}$, he sacrificed victims on the suburban altar, namely two oxen.

越翼日戊午，乃社於新邑，牛一，羊一，豕一。
On the next day, *wuwu*$_{55/60}$, he sacrificed to the God of the Soil in the New City, namely one ox, one sheep, one pig.

越七日甲子，周公乃朝用書命庶殷侯甸男邦伯。厥既命殷庶，庶殷丕作。
On the seventh day (six days later), *jiazi*$_{01/60}$, the Duke of Zhou in the morning by written documents gave charges (about the labor service to be rendered) to all the rulers of states of the *hou*, *dian*, *nan* and *bangbo* zones in the Yin [realm]. When orders had been given to the Yin multitude, all the Yin grandly started work.[10]

As concerns the site's location, the "Luo gao" 洛誥 chapter tells us that:

周公拜手稽首曰:「……予惟乙卯朝至于洛師。我卜河朔黎水，我乃卜澗水東、瀍水西，惟洛食；我又卜瀍水東，亦惟洛食。伻來以圖及獻蔔。」
The Duke of Zhou saluted and bowed down the head and said: "… On the day *yimao*$_{52/60}$, in the morning I came to (the intended) headquarters at Luo. I prognosticated about (the region of) the River Li, north of the He; I then prognosticated about (the region) east of the River Jian, and west of the River Chan; and it was (the region of) Luo that was ordered (i.e., by the oracle). Again I prognosticated about (the region) east of the River Chan; and again, it was (the region of) Luo that was ordered. I have sent a messenger to come (to the king) and to bring a map and to present the oracles."[11]

Lastly, as concerns the spatial arrangement of Luoyi, the *Yi Zhou shu*, "Zuo Luo jie" 作雒解 states:

周公敬念於後，曰:「……乃作大邑成周於土中。立城方千七百二十丈，郛方(七十二)〔二十七〕里。南系于雒水，北因於郟山……大縣立城，方王城三之一，小縣立城，方王城九之一……乃位五宮:太廟、宗宮、考宮、路寢、明堂。」
The Duke of Zhou respectfully reflected for posterity, saying: "… then we built the Great City (Dayi) of the Accomplishment of Zhou (Chengzhou) at the center of the earth. We erected a square inner wall 1720 *zhang* to a side and an outer wall (72) [27] *li* to a side. To the south, it linked to the River Luo, and to the north, it rested against Mount Jia. … The large county [seats] erected city walls one third [the length] of the sides of the royal city, and the small county [seats] erected walls one ninth [the length] of the sides of the royal city. … Then we placed the Five Palaces: the Taimiao 太廟, Zonggong 宗宮, Kaogong 考宮, Luqin 路寢, and Mingtang 明堂.[12]

To summarize, the aforementioned sources name four principal characters as being involved in the planning of Luoyi. The first is King Wu, who came up with the idea after his conquest of the Yin-Shang and determined the principles of the planning and the range of the city. The second is King Cheng, son of King Wu, who put his father's plan into action. Third is the Duke of Shao, an important

[10] *Shangshu yizhu*, 285; translation modified from Karlgren (1950: 48).

[11] *Shangshu yizhu*, 294; translation modified from Karlgren (1950: 51).

[12] *Yi Zhou shu huijiao jizhu*, 5.524–537.

official serving Kings Wu and Cheng in the capacity of Grand Guardian (*taibao* 太保). His job was to observe and to divine the site, to determine the location of the central axis, and to arrange the layout of the city (*gong wei* 攻位, "start work on positioning"). Lastly, there is the Duke of Zhou, the brother of King Wu and regent to King Cheng. The Duke of Zhou's job was to perform the sacrifices to Heaven and Earth at the appropriate alters, to supervise the construction, and to move people into the new capital.[13]

3 Location of the Site

Though the exact location and layout of Luoyi are still the subject of debate, recent studies in historical geography and archaeology in the Luoyang area have revealed the ancient courses of the aforementioned rivers,[14] the spatial evolution of the relevant city walls in different dynasties,[15] and the overall distribution of Western Zhou relics.[16] From this, we can narrow the site of the Western Zhou's eastern capital to the area illustrated in Fig. 2.

Concerning the selection of the site of Luoyi, as decided by King Wu and the Duke of Shao, we may distinguish three scales of geographical references. At the largest scale, the *Yi Zhou shu*, "Zuo Luo jie," places the capital "at the center of the earth," where, as the *Shiji*, "Zhou Benji," explains, "when people from the four quarters pay tribute, the distance will be equal for them." One level down, as the "Zhou Benji" and *Yi Zhoushu*, "Du yi jie," tell us, King Wu situates the city in relation to a series of mountains and rivers: "If I might gaze south to the Santu [Mountains] and north to the slopes of the Sacred Mountains, if I might look back to the He (Yellow River) and look out to the Luo and the Yi, this would not be far from a Heavenly Residence." Lastly, as concerns the boundaries of the city, we read that the city was to be built "east of the River Jian, and west of the River Chan" and/or "east of the River Chan" (*Shangshu*, "Luo gao"), "on the north shore of the [River] Luo" (*Shangshu*, "Shao gao"), where "to the south, it linked to the River Luo, and to the north, it rested against Mount Jia" (*Yi Zhou shu*, "Zuo Luo jie").

Geographically, the site of Luoyi was chosen south of Mount Jia and close to the confluence of the rivers Luo, Jian and Chan (Fig. 3). This comports with the map of the overall distribution of Western Zhou relics in the Luoyang area. Also, one notes

[13] Yan Haiwen and Hu Chunli (2008).

[14] Ye Wansong et al. (1991).

[15] Li Jiuchang (2007).

[16] Iijima (2003).

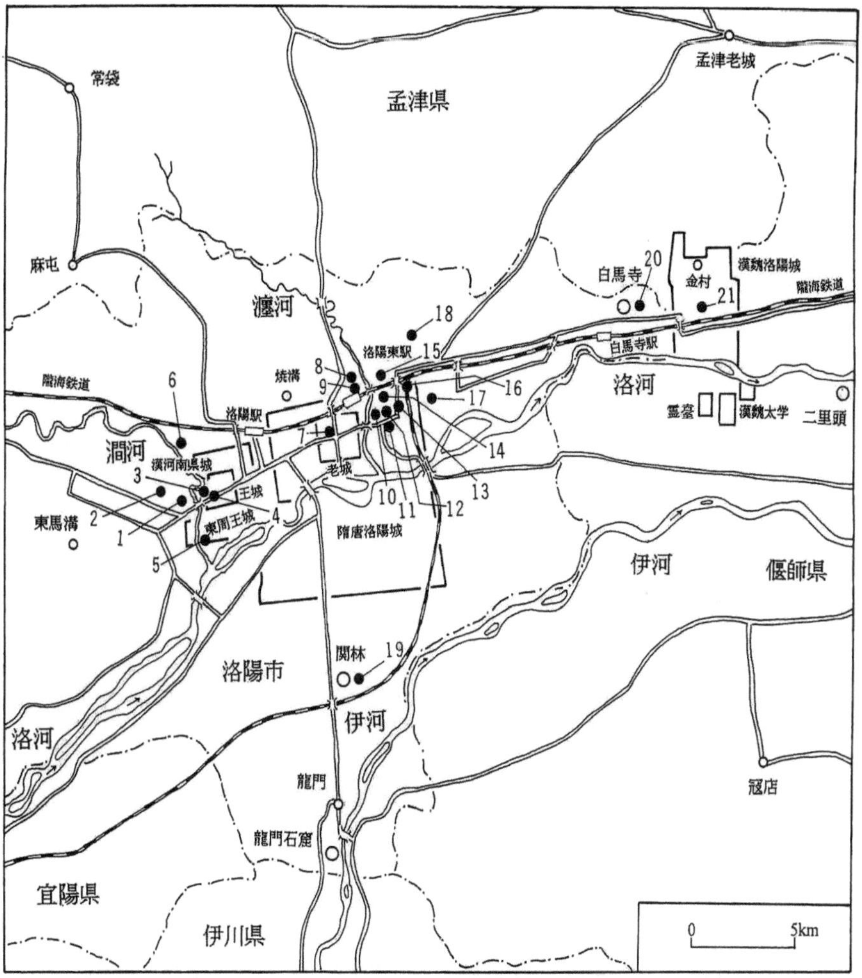

1. 西干沟遗址 2. 涧滨西周墓 3. T203-M3号墓 4. 中州路西周墓 5. 瞿家屯遗址 6. 五女冢西周墓
7. 老城西周车马坑 8. 庞家沟西周墓地 9. 北窑铸铜遗址・西周墓 10. 西周窑址 11. 泰山庙遗址
12. 东大寺遗址・林业学校车马坑 13. 东关遗址 14. 下瑶村遗址 15. 摆驾路口遗址
16. 焦枝线甲骨出土遗址 17. 东郊M13号墓 18. 交通部西周墓地 19. 关林遗址 20. 白马寺西周墓
21. 汉魏洛阳城西周城郭

Fig. 2 Distribution of Western Zhou relics in the Luoyang area. Source: Iijima (2003: 563)

that the site is at the exact center of the circle drawn around three points: the Santu Mountains, the Sacred (Taihang) Mountains, and Mount Tianshi (Songshan), suggesting itself as the "center of the world."

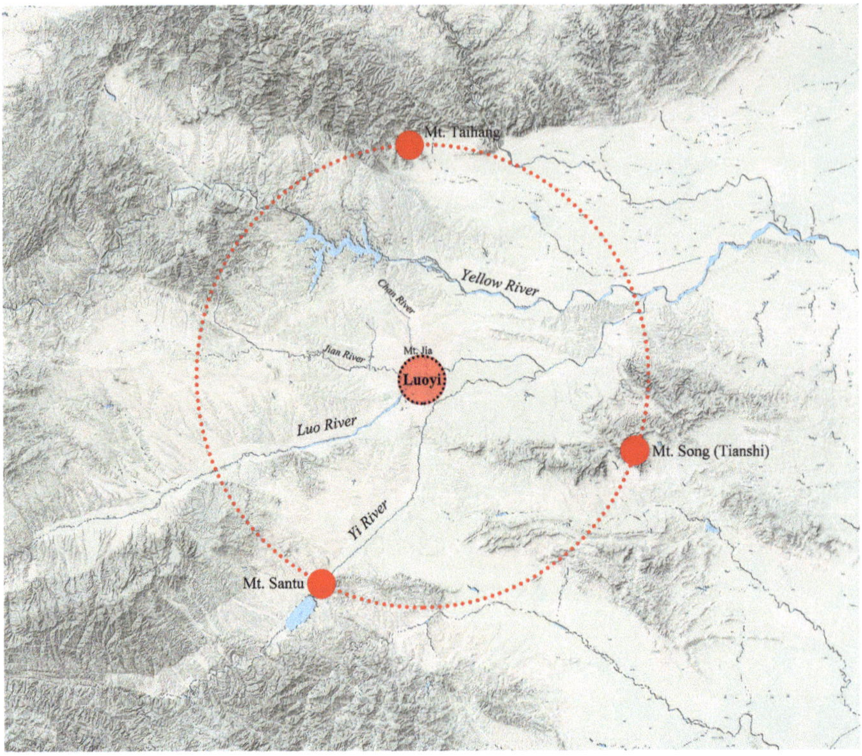

Fig. 3 Situating Luoyi on the basis of the mountains and rivers mentioned in early received sources

4 Time of the Planning

Moving on from the selection of the site, let us consider what the "Shao gao" chapter of the *Shangshu* tells us about *gong wei* or "starting work on positioning" the Heavenly Residence at Luoyi. Specifically, the "Shao gao" details the timing and duration of this process, and therein lies a clue about the city's layout.

We read that the planning of Luoyi took from month II day $yiwei_{32/60}$ to month III day $jiazi_{01/60}$, covering a total of 28 days, i.e. a full lunar month. Within this time, the process of *gong wei* took 5 days, from month III day $gengxu_{47/60}$ to day $jiayin_{51/60}$. What did this entail? The term *gong wei* is somewhat vague, but what happens next gives us an important clue: the Duke of Zhou came to "inspect the new city's *ying*" 觀于新邑營. As Pankenier (2013, 126) explains, in the context of early city planning *ying* refers to "the use of stretched cords to establish the orientation, alignment, and *enceinte* of the pounded-earth footings for the walls and buildings to come." The object of the city's alignment, Pankenier argues, was the constellation Ding 定, also known as Yingshi 營室 ("Align-the-Hall," i.e. α–β Pegasi), and "the optimal time for such alignment activity in late Shang and Western Zhou would have been in early evening in late autumn" (p. 138).

That may be, but the 5 days of *gong wei*, in month III, took place in *spring*, when Yingshi was beneath the horizon on the opposite side of the sky. When exactly in spring? According to Liu Qiyi's 劉啟益 (2002) chronology of the Western Zhou, the seventh year of King Cheng corresponds to 1062 BCE, and the lunar phases and sexagenary and ordinal dates given in the "Shao gao" are a perfect match for the month beginning on day *yihai*$_{12/60}$ in the astronomical tables of Zhang Peiyu 張培瑜 (1997), namely February 24, 1062 BCE. According to those same tables, one notes, the spring equinox of that year was at 23:20, month III day *gengxu*$_{47/60}$ (March 31), the very day that "the grand guardian with all the Yin people started work on positioning the site (*gong wei*)." Admittedly, the exact calendar of the Western Zhou is still the subject of debate, but let us consider what this coincidence might explain if Liu Qiyi's chronology is correct.[17]

So, why was the day of spring equinox chosen to "start work on positioning" (*gong wei*)? Spring equinox was obviously an important moment in their lives. It was the day when the sun directly shone at the equator, and that day and night shared the same length. It was not only important in astronomy, it also marked the beginning of the farming season in most parts of ancient China. The Zhou people valued farming and, moreover, claimed Houji 后稷, the god of soil and "Master of Farming" to Emperor Yao 帝堯, as their mythical ancestor.[18] Another answer to this question, in my estimation, lies in the curious phrase from the *Yi Zhou shu* "Du yi jie" in which King Wu announces "If I sought the Yi in Heaven (the sky), it will not be difficult to get help from it (from Heaven)" 天有求繹，相我不難. Whereas most scholars read *yi* 繹 (lit. "unraveled silk," "extend") as a loan for *yi* 懌 (happy), there is evidence to suggest that this *yi* might refer to a constellation.[19]

Specifically, in the preface to the ode "Silk Clothing" (Siyi 絲衣) in the *Shijing* 詩經 we read that "Silk clothing is [for] the impersonator of the extended guest (*yibin* 繹賓) [ceremony]; Master Gao says that it is [for] the impersonator of [the sacrifice to] the Numinous Asterism (Lingxing 靈星)."[20] In his subcommentary, Kong Yingda 孔穎達 (574–648) does not know what to make of this line, but he quotes Zhang Yan 張晏 (third cent.) to the effect that "the Numinous Asterism" is

---

[17] Pankenier (2013: 124), for example, places the founding of Chengzhou (Luoyi) in 1036 BCE, and one notes that the "Shang gao" dates also fit 2 months prior in Zhang Peiyu's tables, in the lunar month beginning on December 26, 1062 BCE (also day *yihai*$_{12/60}$). Nevertheless, the coincidence of the "positioning" and "alignment" of Luoyi with the exact date of spring equinox 1062 BCE would seem to provide supporting evidence for Liu Qiyi's chronology.

[18] *Shiji*, 4.112.

[19] Scholars have long had difficultly parsing this particular line, the most common reading being "... if I strive for this then [Heaven] will not be far. I think that if Heaven is happy, then it will not be difficult for it to help me" 求茲無遠。慮天有(求)繹，相我不難 (tr. modified from Huang Huaixin 2006: 219–220). I think it wise that we attempt to make sense of this passage exactly as written.

[20] 絲衣繹賓尸也。高子曰:靈星之尸也 (*Maoshi zhengyi*, 19.750b).

"the left horn of the Dragon Asterism, called the Celestial Field (Tiantian 天田; ο–τ Vir), which is lucky for agriculture and is given sacrifice upon its [first] morning appearance."[21] In the *Xu Hanshu* 續漢書, "Jisi zhi" 祭祀志, Sima Biao 司馬彪 (*c.*240–*c.*306) explains:

> 漢興八年，有言周興而邑立后稷之祀，於是高帝令天下立靈星祠。言祠后稷而謂之靈星者，以后稷又配食星也。舊說，星謂天田星也。一曰，龍左角為天田官，主穀。
>
> In the eighth year after the rise of the Han (199 BCE), it was said that upon the rise of the Zhou, cities established the sacrifice to Houji 后稷, and thus it was that the High Emperor (r. 206–195 BCE) ordered the establishment of temples to the Numinous Asterism throughout the subcelestial realm. [He/they] spoke of temples to Houji but referred to him as "the Numinous Asterism," because Houji was paired with the asterism in receiving sacrifice. An old theory is that "the Asterism" refers to the Celestial Field asterism. Another states that the left horn of the Dragon (ζ Vir) is the office (constellation) of the Celestial Field, which governs grain [production].[22]

Houji, the god of soil, was the mythical founder of the Zhou royal line, and it makes sense that he would receive sacrifice and indeed become interchangeable with an agriculture-related asterism, since we are dealing with a culture in which such figures frequently travelled to and found their home among the stars (e.g., Fu Yue 傅說, who became the star G Sco). It also makes sense that, generations later, King Wu would look to him for guidance in founding the Zhou kingdom and its "Heavenly Residence." As it happens, at 23:20, March 31, 1062 BCE, the moment of spring equinox and first day of "positioning" the new capital, the Celestial Field culminated on the meridian of Luoyi, forming a straight north-south line with the North Celestial Pole (Fig. 4). And since "the Numinous Star" and Yi would likewise appear to be interchangeable, it could well be that this is what King Wu designated should be "sought in Heaven" if Luoyi was to win its blessing.

The theory and method of aligning the central axis of a city to a constellation can be seen in other early sources such as the *Shijing* and *Guoyu* 國語.[23] For example, the ode "When Ding Had Just Culminated" (Ding zhi fang zhong 定之方中) describes how Duke Wen of Wei 衛文公 (r.?–635 BCE) began construction of the Chu Palace (Chugong 楚宮) when Ding (i.e., Align-the-Hall (Yingshi)) had just culminated, providing a similar north-south axis through the celestial pole.[24] The officials in charge of such affairs are recorded in the *Zhouli* 周禮 as the "Astronomical Observer" (Pingxiang shi 馮相氏), "Royal astrologer" (Baozhang shi 保章氏), and "Surveyor" (Tufang shi 土方氏).[25]

---

[21] 龍星左角曰天田，則農祥也。晨見而祭之。史傳之說靈星，唯有此耳 (ibid.).

[22] *Hou Hanshu*, *zhi* 9, 3204.

[23] *Shijing yizhu*, 84–85; *Guoyu*, 2.31.

[24] See Pankenier (2013: 127–130).

[25] *Zhouli yizhu*, 378–379, 487.

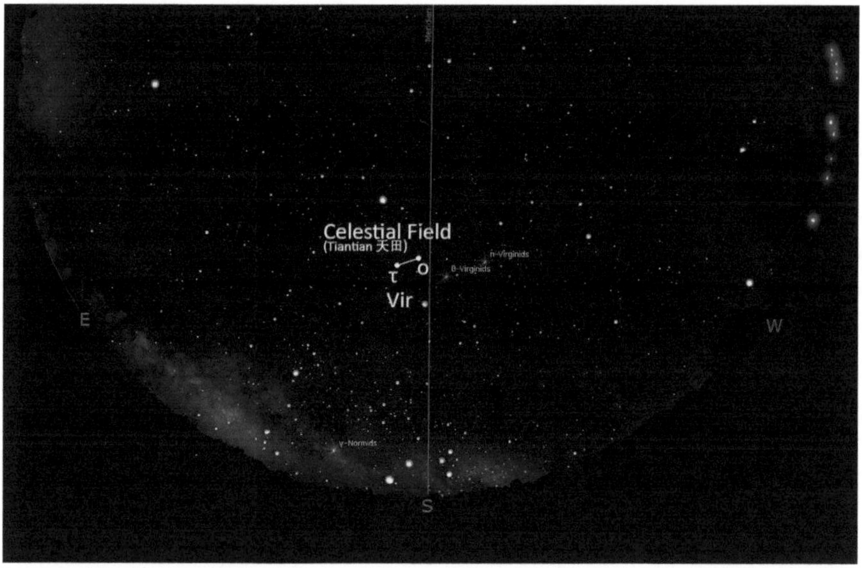

Fig. 4 Night sky at Luoyi (34°41′N, 112°27′E) at spring equinox, 23:20, March 31, 1062 BCE, on the first day of "positioning" (*gong wei*). Note that the asterism Celestial Field (or Yi) is several degrees east of the meridian. Source: Stellarium

5 Layout of the City

Once the central axis was set, there were still 4 days left in the process of "positioning" (*gongwei*) the new capital. What did they do in these 4 days? Generally, this would involve arranging the layout of the city. This work is thoroughly recorded in the "Zuo Luo jie" chapter of the *Yi Zhou shu*, covering four scales of spatial activity. First, at the scale of the kingdom, they decided the fields of the suburbs, divided the land into counties (*xian* 縣) and commanderies (*jun*郡), and settled the population. Second, at the scale of the region, they built city walls around the various seats. Third, at the scale of city, they "set up the Altar of Heaven" 設丘兆, "constructed the Altar of Earth" 建大社, and "positioned the Five Palaces" 位五宮. Fourth, at the scale of individual buildings, they decided the architecture.

It is worth noting that a single pattern was applied to the layouts of spaces at these different scales. The well-known spatial pattern of the capital city is the "Nine Palaces" (*jiugong* 九宮), as described in the "Kaogong ji" 考工記 chapter of the *Zhouli*:

> 匠人營國，方九里，旁三門。國中九經九緯，經塗九軌，左祖右社，面朝後市，市朝一夫。
>
> The artisans plan the layout of the capital, which forms a square nine *li* to a side with three gates on each side. Within the capital, there are nine roads and nine cross streets, and the

path of the former have nine carriage lanes. To the left is the Ancestral Hall, and to the right the Altar of the Soil; in front (south) is the audience [chamber], and in back (north) is the market. The market and the audience chamber are [both] one *fu* [in area].[26]

With the capital city (*guo* 國) divided equally into nine parts, it can be inferred that the Palace is at the center, its area occupying one ninth of the city.

The *Yi Zhou shu*, "Zuo Luo jie," ascribes Luoyi with "an outer wall 72 *li* to a side" (above), which appears as 70, 700, 72 and 17 *li* in different versions. Of these, Sun Yirang 孫詒讓 (1848–1908) argues that 27 *li* makes the most sense:

竊疑當為方二十七里，乃三城方之數也。……王宮方九百步，即三里，城內包宮垣，距城面各三里，並之為城方九里。郛內包城，城距郛面亦各九里，並之為郛方二十七里也。
I suspect that it should be a square 27 *li* to a side, since that is three times of the length of each wall of the inner city… The royal palace is a square 900 *bu* (double-paces) to a side, i.e. 3 *li*. The city wraps around the palace which is 3 *li* from the face of the city wall, which, combined, makes the city wall a square 9 *li* to a side. Likewise, the outer wall wraps around the city, whose walls are 9 *li* from the face of the outer wall, which, combined, makes the outer wall a square 27 *li* to a side.[27]

The "Zuo Luo jie" further specifies that the walls of the large county seats were "one third [the length] of the sides of the royal city," with the small county seats at "one ninth" (above). With the royal city at 9 *li* to a side (81 li^2), that makes the large county seats 3 *li* to a side (9 li^2) and the small county seats 1 *li* to a side (1 li^2). As shown in Fig. 5, the result is a very neat system, with a clear spatial hierarchy, repeatedly applying the pattern of Nine Palaces at different scales.

This indeed applied all the way down to the scale of architecture, as seen most clearly in the building of the Bright Hall (Mingtang 明堂).[28] The scale and layout of this ritual site are recorded in the "Mingtang jie" 明堂解 chapter of the *Yi Zhou shu*:

明堂方百一十二尺，高四尺，階廣六尺三寸。室居中，方百尺。室中方六十尺。戶高八尺，廣四尺。東應門，南庫門，西皋門，北雉門。東方曰青陽，南方曰明堂，西方曰總章，北方曰玄堂。中央曰太廟，左為左介，右為右介。
The Bright Hall is a square 112 *chi* to a side, 4 *chi* in height, and with steps 6 *chi* 3 *cun* in width. The halls are in the center [of the complex], forming a square 100 *chi* to a side. The interior of the halls is a square 60 *chi* to a side. The windows are 8 *chi* in height and 4 *chi* in width. [There is] the Ying Gate to the east, the Ku Gate to the south, the Gao Gate to the west, and the Zhi Gate to the north. [As to the rooms, there is] Qingyang on the east, the Mingtang on the south, Zongzhang on the west, the Xuantang on the north, and the Taimiao in the center, the left of which is the Zuojie, and right Youjie.[29]

The record indicates that it is describing the Bright Hall of Zongzhou 宗周, the capital Feng and Hao of the Western Zhou, which was built earlier than that at Luoyi, making it a reliable reference to understand the layout of the latter. As

---

[26] *Zhouli yizhu*, 665.

[27] *Yi Zhou shu huijiao jizhu*, 5.526–528 (comm.).

[28] On the Bright Hall and the application of the Nine Palaces scheme to architecture, see Pankenier (2013: 343–349) and Tseng (2011: 58–88).

[29] *Yi Zhou shu huijiao jizhu*, 6.710–716.

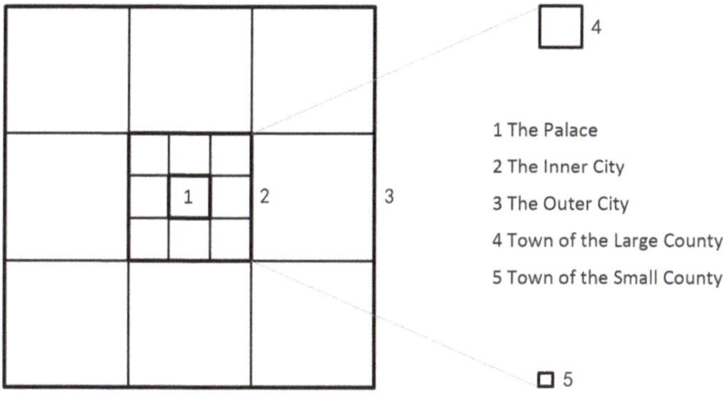

Fig. 5 City-town system of the Western Zhou as recorded in the *Yi Zhou shu* and "Kaogong ji"

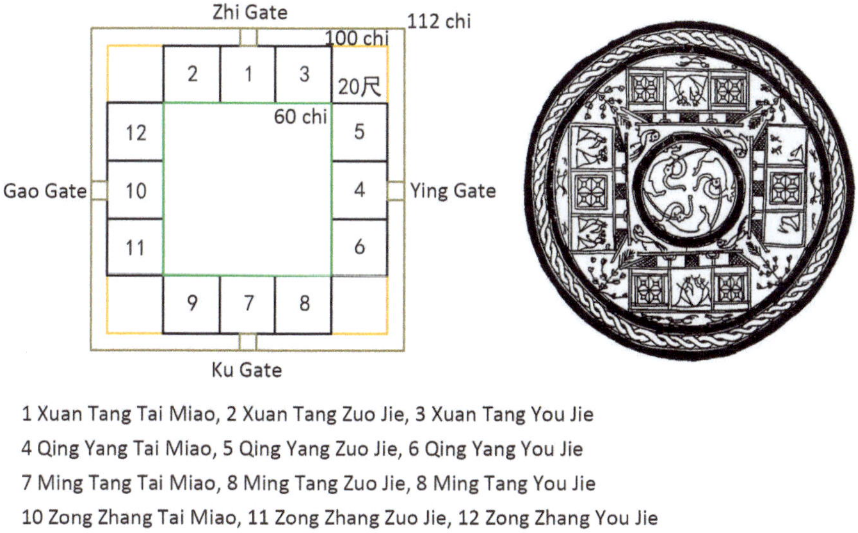

Fig. 6 Reconstructed floorplan of the Bright Hall (Mingtang) as recorded in the *Yi Zhou shu* (left) compared to an illustration of its layout on a piece of Eastern Zhou lacquerware (right). Source: the image on the right is from Cao Chunping (1996: 53)

illustrated in Fig. 6, the floorplan of the Zongzhou Bright Hall that one is able to reconstruct from this passage closely resembles a lacquer painting excavated from the Eastern Zhou (771–256 BCE). This is just one more example of the prevalence of the Nine Palaces scheme at every scale of spatial organization.[30]

In sum, the Nine Palaces scheme was widely used in planning and design at different scales, from that of the kingdom all the way down to architecture. The *Zhouli* calls this method "Principle of Form" (*xingti zhi fa* 形體之灋) and identifies the

[30] Cao Chunping (1996) and Yang Hongxun (1998).

officials in charge of this as the Supervisor of Exterior Districts (Suiren 遂人) and Supervisor of Territories (Xingfang shi 形方氏).[31] Later, in the *Hanshu* 漢書 "Yiwen zhi" 藝文志, this was renamed *xingfa* 形法.[32]

As described in the "Yiwen zhi," the origin of the Principle of Form, or *xingfa*, came from "the disposition of the Nine Regions" 九州之勢. Therefore, the origin of the pattern of Nine Palaces can be traced back to the method of land division, a very early spatial planning activity, invented by Yu the Great 大禹. As preserved in the *Shangshu zhengyi* 尚書正義, Kong Anguo's 孔安國 commentary establishes a relationship between Yu the Great and the *Luoshu* 洛書.[33] Coincidentally, the place that the *Luoshu* appeared was the site selected to build the capital at Luoyi: the north shore of the River Luo.[34] As the very birthplace of the Nine Palaces, we can see that the planning of Luoyi was be deeply rooted in local tradition and that this pattern was widely used to arrange the layout of spaces at different scales.

6 The Reconstruction of Luoyi

At long last, a rough reconstruction of the planning of Luoyi can be made based on the analyses of this study. To summarize, the capital Luoyi should have been located near the intersection of the rivers Luo, Jian, and Chan. Its inner wall was 9 *li* long (81 li^2), and its outer wall 27 *li* (729 li^2). The central axis runs roughly north–south, aligned with two peaks in the south, named the Gate of Yi (Yique 伊闕). According to this reconstruction, the inner walls of Luoyi would have overlapped the Palace City (*gongcheng* 宮城) and Imperial City (*huangcheng* 皇城) of Sui-Tang Luoyang, and the southern portion of their respective outer walls would have likewise overlapped (Fig. 7). Still, this hypothesis remains to be proven by solid evidence from further archaeological excavation.

7 Conclusion

In this paper, we have explored the theory and method behind the construction of the Western Zhou capital at Luoyi based on historical literature and archaeological evidence. To summarize, early sources describe the choice and location of the site at three different geographical scales from which we can triangulate to place it exactly where the highest concentration of contemporary relics in the Luoyang area

[31] *Zhouli yizhu*, 223, 489.
[32] *Hanshu*, 30.1775.
[33] *Shangshu zhengyi*, 165.
[34] *Shuijing zhu jiao*, 15.486.

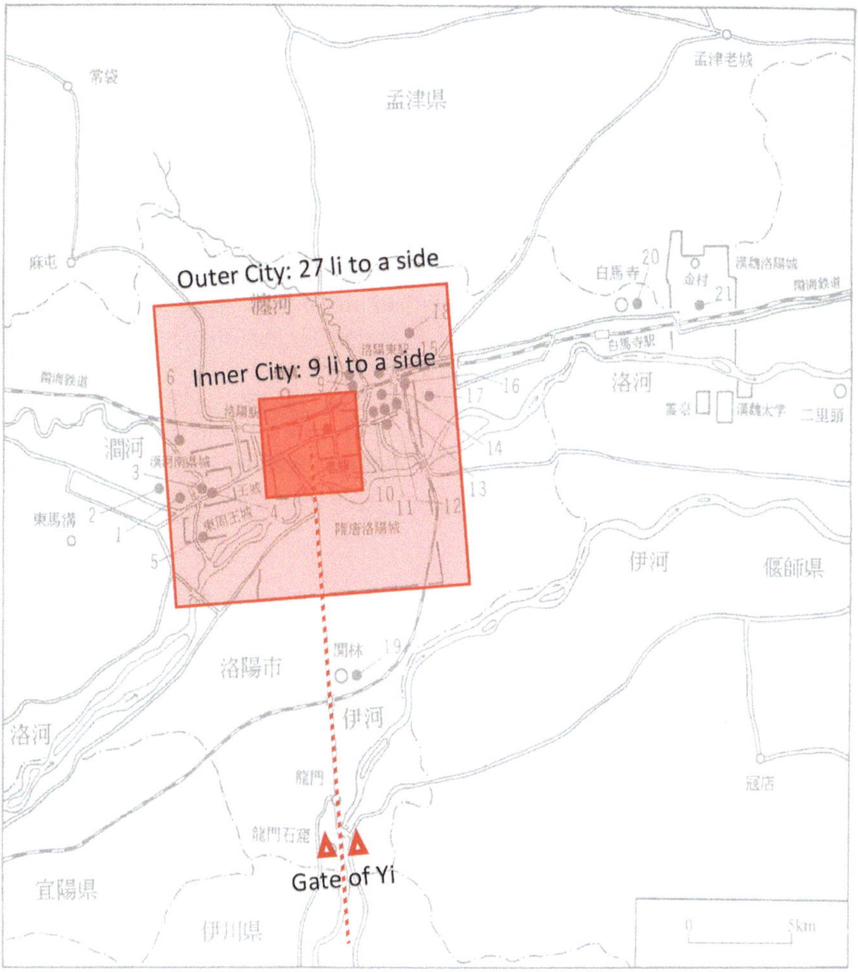

Fig. 7 Reconstruction of the location, central axis, and general layout of the capital Luoyi

appear. Turning to the sequence of events in the city's planning, we note that the coincidence of the first day of "positioning" (*gongwei*) with the moment of spring equinox and culmination of the dynasty's star suggests rereading a passage of the *Yi Zhou shu*, "Du yi jie," to refer to aligning the city's central axis to the constellation Yi. Lastly, from the scale of the kingdom all the way down to individual buildings, we have seen that everything was modeled upon the Nine Palaces grid.

As such, it would appear that the planning of Luoyi was one of the earliest attempts at "modeling heaven and earth," imitating the patterns of the sky and the land to establish its central axis and arrange the layout of the city. Indeed, this can be regarded as the founding of said tradition in ancient Chinese city planning. This ideal went on to play an important role in the planning of the Qin capital at

Xianyang 咸陽 the Han capitals at Chang'an 長安 and Luoyang, the Sui capital at Daxing 大興, the Sui-Tang capital at Luoyang, the Yuan capital at Dadu 大都, and the Ming-Qing capital at Beijing 北京.

Further research into Luoyi may shed new light on the capital cities of the Shang and Western Han. The "Yao dian" 堯典 chapter of the *Shangshu* records the observation of the four constellations Niao 鳥, Huo 火, Xu 虛, and Mao 昴 in early times,[35] while the *Zuo zhuan* 左傳 notes that "Chen is the star of the Shang" 辰為商星,[36] and the *Guoyu* 國語 that "the Shang rule [the star] Dahuo" 商主大火."[37] One notes that the three Shang cities of Zhengzhou 鄭州, Yanshi 偃師, and Huanbei 洹北, occupied in 1600–1300 BCE, were all laid out in rectangular shape with their central axes pointing to the southwest.[38] Is it possible that they applied the same planning method we see at Luoyi but with their axes aligned to the star of the Shang? Turning to the Han, one notes that Zhang Heng 張衡 (78–139) describes the Western Han capital at Chang'an as having "followed the principles of the Qin" 覽秦制 and "surpassed the model of the Zhou" 跨周法.[39] It may well have surpassed it on different principles, but how much did the planning of Chang'an inherit from the planning of Luoyi?

Bibliography

Primary Sources

Guoyu 國語, pre-Qin (< 221 BCE), commentary by Wei Zhao 韋昭 (204–273). Shanghai: Shanghai guji chubanshe, 2008.

Hanshu 漢書, Ban Gu 班固 (32–92) et al., commentary by Yan Shigu 顏師古 (581–645), Beijing: Zhonghua shuju, 2013.

Hou Hanshu 後漢書, Fan Ye 范曄 (398–445) et al., commentary by Li Xian 李賢 (654–684), Beijing: Zhonghua shuju, 1965.

Maoshi zhengyi 毛詩正義, pre-Qin (< 221 BCE), commentary by Zheng Xuan 鄭玄 (127–200), subcommentary by Kong Yingda 孔穎達 (574–648 CE), Shanghai: Shanghai guji chubanshe, 1990.

Shangshu yizhu 尚書譯注, translation and exegesis by Li Min 李民 and Wang Jian 王健, Shanghai: Shanghai guji chubanshe, 2000.

Shiji 史記, Sima Qian 司馬遷 (*c.*146–*c.*86 BCE) et al., commentary by Pei Yin 裴駰 (fifth century CE), Zhang Shoujie 張守節 (Tang), and Sima Zhen 司馬貞 (eighth century CE), Beijing: Zhonghua shuju, 2013.

Shijing yizhu 詩經譯注, translation and exegesis by Cheng Junying 程俊英, Shanghai: Shanghai guji chubanshe, 2016.

---

[35] *Shangshu yizhu*, 3–4.

[36] *Zuo zhuan*, 702.

[37] *Guoyu*, 10.157–158.

[38] Xu Hong (2017).

[39] "Xijing fu" 西京賦, cited in *Wenxuan*, 2.52.

Shujing zhu jiao 水經注校, commentary by Li Daoyuan 酈道元 (466/472–527), critical edition of Wang Guowei 王國維, Shanghai: Shanghai renmin chubanshe, 1984.
Tongdian 通典, Du You 杜佑 (735–812); Changsha: Yuelu shushe, 1995.
Wenxuan 文選, compiled by Xiao Tong 蕭統 (501–531), commentary by Li Shan 李善 (630–689), Shanghai: Shanghai guji chubanshe, 2011.
Yi Zhou shu huijiao jizhu 逸周書匯校集注, commentary by Huang Huaixin 黃懷信, Zhang Maorong 張懋镕, and Tian Xudong 田旭東, Shanghai: Shanghai guji chubanshe, 2007.
Zhouli yizhu 周禮譯注, pre-Qin (< 221 BCE), translation and exegesis by Yang Tianyu 楊天宇, Shanghai: Shanghai guji chubanshe, 2004.
Zhouyi yizhu 周易譯注, pre-Qin (< 221 BCE), translation and exegesis by Huang Shouqi 黃壽祺 and Zhang Shanwen 張善文, Shanghai: Shanghai guji chubanshe, 2007.
Zuozhuan 左傳, Zuo Qiuming 左丘明 (502–422 BCE), commentary by Du Yu 杜預 (222–285 CE), Shanghai: Shanghai guji chubanshe, 2016.

Secondary Sources

Cao Chunping 曹春萍: 1996. "Si'e chongwu" tankao 「四阿重屋」探考. *Huazhong jianzhu* 華中建築 1996.1: 50–55.
Duan Dexin 段德新. 2005. Guobao He zun yu "Zhongguo" 國寶何尊與「中國」. *Wenbo* 文博 2005.6: 30–31.
Huang Huaixin 黃懷信. 2006. *Yi Zhou shu jiaobu zhuyi* 逸周書校補注譯. Xi'an: Sanqin chubanshe.
Iijima Taketsugu 飯島武次. 2003. Luoyang Xizhou shidai de yizhi yu Chengzhou 洛陽西周時代的遺址與成周, translated by Hu Mingming 胡明明. In: *Kaoguxue yanjiu (5)* 考古學研究(5), ed. Beijing daxue kaogu wenbo xueyuan 北京大学考古文博学院, 557–571. Beijing: Kexue chubanshe.
Karlgren, Bernhard. 1950. The Book of Documents. *Bulletin of the Museum of Far Eastern Antiquities* 22: 1–81.
Li Jiuchang 李久昌. 2007. *Guojia, kongjian yu shehui: gudai Luoyang ducheng kongjian yanbian yanjiu* 國家、空間與社會:古代洛陽都城空間演變研究. Xi'an: San Qin chubanshe.
Li Xueqin 李學勤. 1981. He zun xinshi 何尊新釋. *Zhongyuan wenwu* 中原文物1981.1: 37–41, 47.
Lin Yun 林澐. 1993. Tian Wang gui "Wang si yu Tianshi" xinjie 天亡簋「王祀於天室」新解. *Shixue jikan* 史學集刊 1993.3: 24–29.
Liu Qiyi 劉啟益. 1979. Xizhou jinwen zhong yuexiang ciyu de jieshi 西周金文中月相詞語的解釋. Lishi jiaoxue 歷史教學 1979.6: 21–26.
Liu Qiyi 劉啟益. 2002. Xi Zhou jinian 西周紀年. Guangzhou: Guangdong jiaoyu chubanshe.
Nienhauser, William H., Jr., et al. 1994. *The Grand Scribe's Records. Volume I: the Basic Annals of Pre-Han China*. Bloomington: Indiana University Press.
Nienhauser, William H., Jr., et al. 2006. *The Grand Scribe's Records. Volume V.1: The Hereditary Houses of Pre-Han China, Part I*. Bloomington. Indiana University Press.
Pankenier, David W. 2013. *Astrology and Cosmology in Early China: Conforming Earth to Heaven*. Cambridge: Cambridge University Press.
Tseng, Lillian Lan-ying. 2011. *Picturing Heaven in Early China*. Cambridge, MA: Harvard University Asia Center.
Wen Renjun 聞人軍. 2008. *Kaogong ji yizhu* 考工記譯注. Shanghai: Shanghai guji chubanshe.
Xu Hong 許宏. 2017. *Xian Qin chengyi kaogu* 先秦城邑考古. Beijing: Xiyuan chubanshe.
Yan Haiwen 閻海文 and Hu Chunli 胡春麗. 2008. "Dingbao Tianshi": Zhou chu "Dongdu Luo" zhi zai kaocha 「定保天室」:周初「東都洛」之再考察. *Lanzhou xuekan* 蘭州學刊 5: 140–143.

Yang Hongxun 楊鴻勛. 1998. Mingtang fanlun: Mingtang de kaoguxue yanjiu 明堂泛論:明堂的考古學研究. In: *Yingzao diyi ji* 營造第一輯, ed. Zhongguo jianzhu xuehui jianzhu shixue fenhui 中國建築學會建築史學分會 and Qinghua daxue Jianzhu lishi yu wenwu jianzhu baohu yanjiusuo 清華大學建築歷史與文物建築保護研究所, 3–96. Beijing: Wenjin chubanshe,1998: 3–96

Ye Wansong 葉萬松, Zhang Jian 張劍, and Li Defang 李德方. 1991. Xi Zhou Luoyi chengzhi kao 西周洛邑城址考. *Hua Xia Kaogu* 華夏考古 1991:2: 70–76.

Zhang Peiyu 張培瑜. 1997. *Sanqianwubai nian liri tianxiang* 三千五百年曆日天象 (Calendar and Astronomical Phenomenon over 3500 Years). Zhengzhou: Daxiang chubanshe.

Zhong Chunhui 鐘春暉. 2010. Tianxia guannian yu Zhou chu zhi jiandu Luoyi 天下觀念與周初之建都雒邑. In: *Guoji qingnian xueshu huiyi ji di er jie Qinghua qingnian shixue luntan lunwen ji zhong* 國際青年學術會議暨第二屆清華青年史學論壇論文集中, 4–14. Beijing: Qinghua daxue lishi xi.

A Statistical Approach to Numerical Sequences on Ancient Chinese Artifacts

Andrea Bréard

Abstract In the absence of contemporary literary sources providing context to numerical inscriptions on ancient Chinese artifacts (twelfth to fourth century BCE), this chapter proposes a statistical approach. It allows to make some hypothesis about possible practices in which sequences of numbers (ranging from 4 or 5 to 8 or 9) devoid of explanations might have emerged. Structural similarities to sets of three or six numbers in later trigram and hexagram divination techniques as described in the *Book of Changes* (*Yijing* 易經), as well as archaeological finds of dice of six, fourteen, and eighteen sides in relation to divinatory practices, are taken into consideration in this new approach. Several theoretical models are proposed and tested against the relative frequencies of numbers as they appear on ancient artifacts and in manuscripts. Future archaeological finds of more inscriptions of numerical sequences will allow to support or refute the conclusions drawn here, in particular (1) the result that before canonization of the *Dayan* procedure described in a commentary to the *Yijing*, a variety of similar practices for counting off was in circulation much earlier than expected, and (2) that certain techniques involving dice might have played a role in producing the observed numerical sequences.

Keywords Numerical inscriptions · *Book of changes* · Hexagram divination · Irregular dice · Six-sided die

The statistical approach proposed in this paper has been developed in the framework of my ongoing collaboration with Constance A. Cook. Preliminary results from our collaboration are published in (Bréard and Cook 2020) and (Cook and Bréard 2022) and provide much more archaeological background and historic context than this chapter, which focuses on methodology.

A. Bréard (✉)
Friedrich-Alexander-Universität Erlangen-Nürnberg, Erlangen, Germany
e-mail: andrea.breard@fau.de

D. P. Morgan, T. Quan (eds.), *Mathematics and Astronomy in the Ancient World*,
https://doi.org/10.1007/978-3-031-93809-2_2

1 Introduction

Recent excavations of ancient tombs in China have brought to light numerous oracle bones, bronze, jade, stone and pottery objects, and bamboo manuscripts dating from the twelfth to the fourth century BCE. Among these artifacts are a number of records of numerical sequences, for which no textual or ritual explanation has been found of how they were produced. Structural links to the *Book of Changes* (*Yijing* 易經), a divination manual that entered the Confucian canon, are evident; yet, the algorithm described therein produces sequences of numbers from six to nine only and dates to the slightly later second to first century BCE. By combining historical with numerical evidence, this article proposes a statistical approach that tests and identifies possible models for how numerical sequences were generated during different historical periods. In these models, the use of dice and divination stalks is assumed, based on archaeological finds that confirm their possible link to the number inscriptions. Both dice and stalks, appear in fact to have been underlying the rather stable numerical patterns in ancient China at different times all the way back to the early Zhou dynasty. For the late Shang dynasty (1300–1046 BCE), none of the tested models could be confirmed statistically.

In the following, the various theoretical models are presented, along with their historical legitimacy, and conclusions are drawn from the statistical tests conducted. In order to keep the technical aspects apart from the main discussion of my contribution, the underlying data and the applied methods are described in the supplementary material in the Appendix.

2 Modeling Dice and Stalks Procedures

In order to pursue the kind of statistical approach proposed here, it is necessary to have a sufficiently large sample size of numerical inscriptions at hand. We do now possess enough data since the massive infrastructure projects undertaken in China over the last 40 years have led to the discovery of many archeological sites and have brought to light a considerable number of artifacts and manuscripts. Pits filled with inscribed oracle bones and tombs with entire libraries of high officials, as well as objects of everyday life, shed new light on our understanding of life in ancient China but also raise new questions. Among the most enigmatic issues is the presence of hundreds of sequences of numbers, mostly grouped into sets of six, and only occasionally accompanied with omen texts. Their relation to divinatory practices is undoubted, but how these sequences were produced remained a puzzle. One of the reasons is that the numbers involved vary: while we find 1s, 5s, 6s, 7s, 8s, and 9s from the Shang (twelfth century BCE), no 9s are found on early to mid-Western Zhou objects (11th–ninth cent. BCE); only 1s, 5s, 6s, 8s and 9s appear during the late Western and Eastern Zhou (8th–fifth cent. BCE); in addition, the number 4, which along with 1s, 2s, and 3s appears in non-divinatory contexts since the Shang, occurs in a fourth century BCE divination manuscript.

The total number of excavated examples available today thus allows the problem to be approached from a statistical point of view. Distinguishing between four different time periods, from twelfth to fourth century BCE, my statistical analysis suggests that divination procedures involving both dice throwing and yarrow stalk sorting can account for most of the relative frequencies of the different numbers among 1594 available numerical records.[1] The stability of the pattern—the grand majority of numbers recorded are 6s and 1s (or 7s—see explanation below), followed by 8s, few 5s and eventually a very few 9s and 4s (Table 1)—allows us to assume some technical continuity across the nine centuries.

In spite of such a stable discrete distribution, it has to be taken into consideration that the nature of the available evidence is nuanced by the material context in which the numbers are embedded. While pre-fourth-century BCE number sequences are scattered among singular bones, ritual bronze, jade, and stone objects or pieces of pottery, fourth-century BCE evidence rests on manuscripts whose texts are constructed upon editorial principles: the recorded number sequences are thus less likely to reflect a random sample of outcomes of divinatory procedures.

Three historical assumptions are tested in the following reconstruction. Let us recall that, in the canonical form of Chinese divination techniques, the lines of a hexagram are found through manipulation of milfoil stalks as described in detail in the *Great Commentary* to the *Book of Changes* from the 1st cent. BCE (Peterson 1982). This amounts to the following general algorithm, henceforth noted A_n:

> After elimination of one stalk from the initial bunch one starts the following subprocedure: the stalks are randomly divided into two bunches. One then first takes away one stalk from either bunch, resulting in s_1 and s_2 stalks, divides by 4 the cardinality of either pack, s_1 and s_2, subtracts from each of them a number of stalks equal to the remainders, and finally reassembles the residual stalks. The subprocedure is iterated n times. Note that if $s_i \equiv 0 \pmod 4$, the remainder was considered as equal to 4, not to zero.

In the "Great Commentary" to the *Book of Changes*, 50 initial stalks are given, yielding after $n = 3$ iterations a total of 24, 28, 32 or 36 stalks (Smith 2008: 24–30). Division by four results in the numbers 6, 7, 8 or 9 (see Fig. 1 for an illustration of the procedure A_3 with $n = 50$); one easily calculates the respective probabilities of these outcomes as 5.89%, 30.30%, 44.12% and 19.68%.[2] Although this has been the canonical procedure for two millennia, recent excavations reveal a certain degree of variability: the number of stalks varies from 30 to 60. For example, in the *Tricks of Jing*, a newly discovered manuscript from first century BCE, 30 stalks are divided in three packs, consisting of s_1, s_2, and s_3 stalks respectively. The number of stalks in each of them is divided by 4, and the three remainders r_1, r_2, and r_3 directly taken as the mantic number triple (Fig. 2).[3]

---

[1] Our database was constructed on the basis of (Bian 2016) with minor corrections and inclusions. See **2. Materials** in the Appendix.

[2] Probabilities based on a uniform model of dividing the stalks. See **3. Methods**: **Modeling the division of the stalks** in the Appendix.

[3] All 16 possible outcomes of number triples (r_1 r_2 r_3) are listed in the manuscript. It is therefore a complete manual and not a record of actual random results of the procedure (*Beijing daxue* 2014).

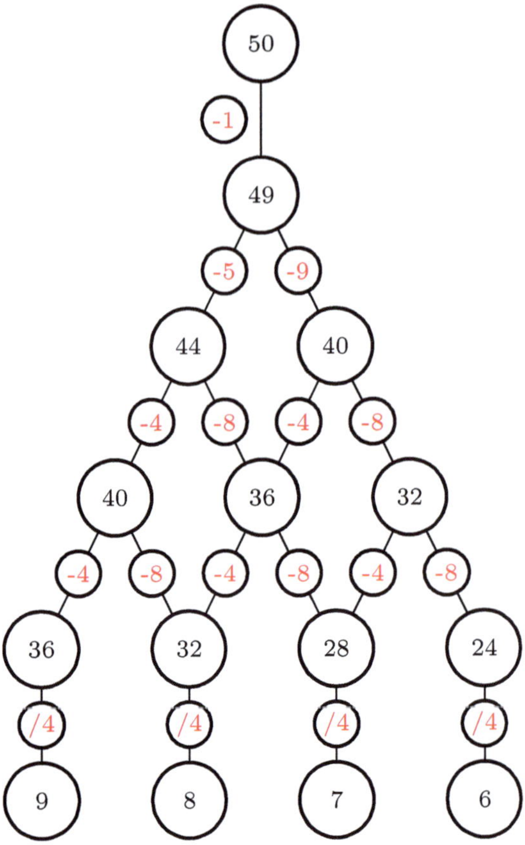

Fig. 1 Possible paths of the total number of stalks at each step of the canonical procedure A_3 with an initial 50 stalks and $n = 3$ iterations resulting in the numbers 6, 7, 8, and 9. The same numerical results would be obtained with 49 or 48 stalks and A_3, or with 55 stalks and A_4, albeit each with different probabilities (see Table 2)

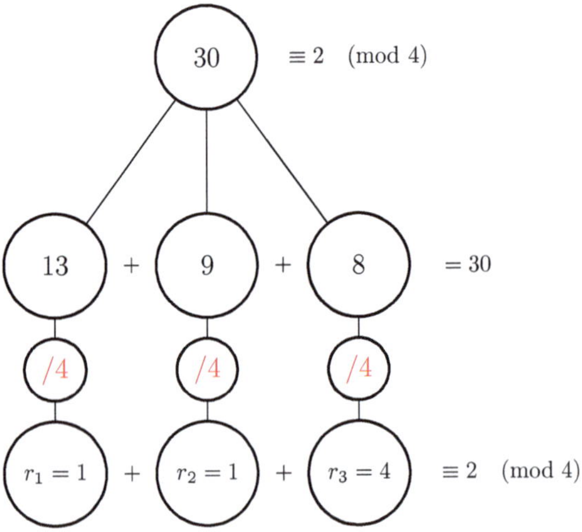

Fig. 2 Procedure as described in the *Tricks of Jing*. With an initial 30 stalks and $n = 1$ iteration, it results in the remainders 1, 2, 3, or 4 where $r_1 + r_2 + r_3$ is either 6 or 10

One can thus reasonably assume, and this is my first assumption, that the number of stalks was not bound to 50, in particular since we are dealing here with material that predates the amount of stalks fixed by the Confucian textual tradition. A variety of non-standardized procedures probably pre-dated the canonized version given in the *Book of Changes*.

Another important point is the confusion of 1s and 7s, now generally accepted by historians and attested not only graphically (Ma 2014) but also conceptually. This is shown for example in two symmetrical number series in the *Shifa* 筮法, a fourth-century BCE divination manual (Cook and Zhao 2017), that gives the sequence 4-5-6-1-8-9 and its reverse 9-8-1-6-5-4 both twice. We thus read, and this is our second assumption, any occurrence of 1 as equivalent to a 7.[4] Thirdly and finally, the use of dice is assumed. 23 regular six-sided wooden dice accompanied the burial of divination material including a manual on bamboo-strips from the fourth century BCE, the *Guicang* 歸藏 (Lewis 2002; Shaughnessy 2014). It is crucial to note that two of the dice had two blank sides with the remaining four sides marked with 1s and 6s in opposition. The other cubical dice were identical to the now conventional die numbered 1 to 6 (Jingzhou 1995; cf. Fig. 3). Moreover, 18-sided dice with the numbers 1 to 16 and two non-numerical or blank sides, for example, were found together with stalks in third to first century BCE sets for a popular divination game (Xiong 1979; Röllicke 1999; cf. Fig. 6 in the Appendix).[5] Another die, 14-sided, from a late Eastern Zhou dynasty tomb, with the numbers 1 to 6 twice on four sides and two blanks (Shandong 2014; cf. Fig. 5 in the Appendix). Both are irregular polyhedra and seem to be intended to model, with one single die, the throw of three and two dice.

Given the many 1s (or 7s), and 6s and the skewed distributions for all but the late Western Zhou number series, and to a lesser extent the *Shifa* records (see Table 1), we took the following as our model B_n^m:

> A m-sided die carrying two blanks was thrown: the numbers 1 or 6 were recorded if an odd or even number came up, with 1 possibly noted symbolically like a 7. If the die fell on a blank side, the diviner used stalks instead to obtain a number through the general procedure A_n described above.

In order to obtain, for example in the Shang dynasty, the numbers 5, 6, 7, 8 and 9, $n = 4$ iterations in A_n were necessary, starting either with 54, 53, or 52 stalks. The theoretical probabilities of model B_4^6 clearly provide a better approximation of the frequencies observed on Shang oracle bones than A_4 for 54 and 53 stalks (see Table 5

[4] In (Pischedda and Bréard forthcoming) I propose some models for the case where a distinction between 1s and 7s is assumed. It turns out that only for the Shang a reasonable fit can be obtained for a mixed dice-stalk procedure, where 1s and 6s resulting from the throw of a 14-sided die or a 6-sided special die are directly recorded as a result, and 52 stalks are used to generate numbers from 6 to 9 in case the die falls on a blank side.

[5] According to Xiong (1979, 37) the 18-sided die from Mawangdui carries in addition to the numbers 1 to 12, the characters *jiao* 驕 and 𫘝, the latter read *wei* or *gui* according to Röllicke (1999, 31).

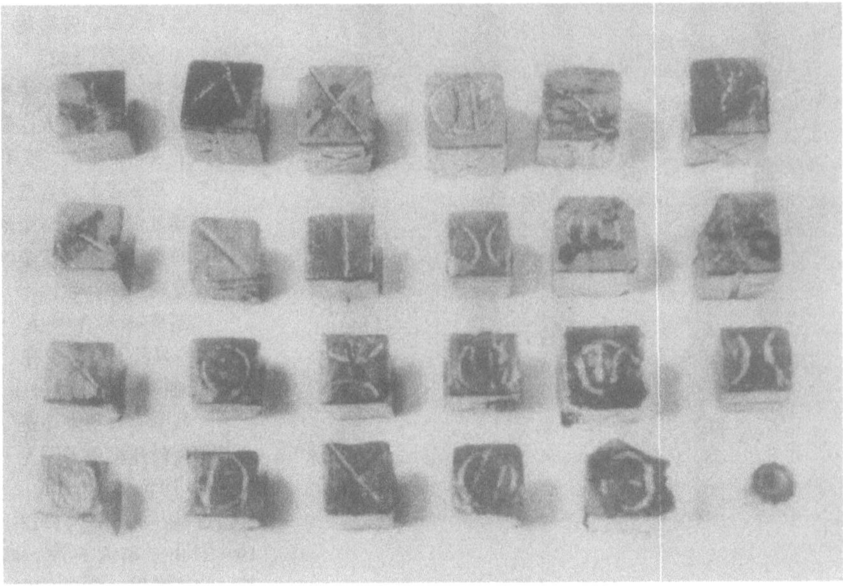

Fig. 3 23 wooden dice from Qin dynasty, unearthed in Jiangling Wangjiatai, Hubei Province, in 1993, unfortunately already rotten by now. Source: (Jingzhou 1995, 42, fig. 14)

in the Appendix), but none of the models provides a convincing fit. All models B_5^m of one special die with $m = 6$, 14 or 18 sides, combined with 58 (57, or 56) stalks and A_5, fit better with the data from the *Shifa* manual (see Table 9 in the Appendix), although it is unclear to what extent the text recorded actual outcomes of random experiments. Some of the number sequences seem in fact to be constructed so as to serve as exemplary number sets for the interpretative framework, but the overall majority might well be based upon a B_5^m model. A related problem occurs with the samples taken from three further manuscripts from the fourth century BCE. They include bamboo records of the attempts by diviners over the course of 3 years to find the spirit cursing a particular tomb occupant. Although the texts from Tianxingguan, Baoshan, and Xincai Geling, with only 96, 72, and 143 numbers recorded each, show similar patterns of distribution, they are compilations of records and as such can *a priori* not be considered as random and independent from each other as in our database of singular oracle bones, scattered pottery pieces and other objects from the earlier periods. This can explain why the fourth-century BCE data do provide a less accurate fit with the combined model B_n^m of stalks and dice (Table 8). Nevertheless, the hypothesis that only stalks were used according to procedure A_n (Liu Bin 2014) can certainly be discarded for both fourth-century BCE data sets.

As for the sequences containing 5s, 6s, 7s, and 8s found on early to mid-Western Zhou dynasty material, a model A_3 with stalks only, in this case 46 (Fig. 4), produces an excellent fit with the actually observed distribution, whereas the involvement of a die does not (Table 6).

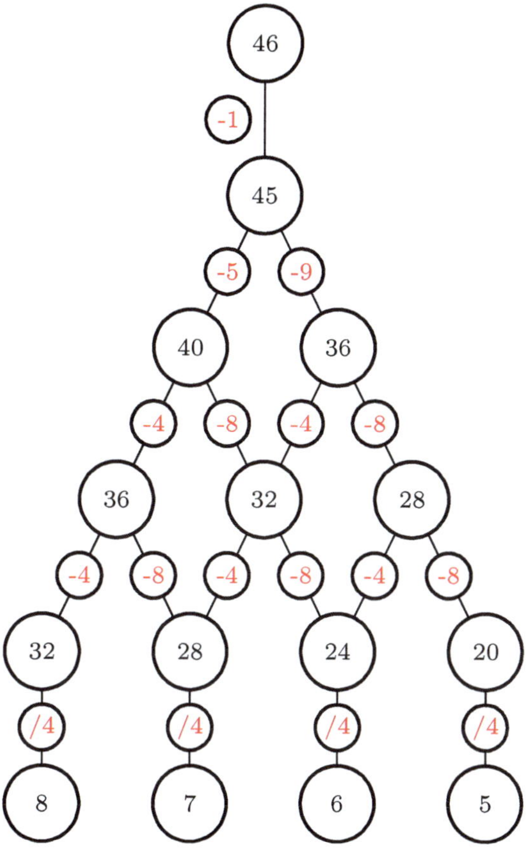

Fig. 4 Possible paths of the total number of stalks at each step of procedure A_3. With an initial 46 stalks and $n = 3$ iterations, it results in the numbers 5, 6, 7, and 8. The same numerical results would be obtained with 45, 51, or 44 stalks, albeit each with different probabilities (see Table 6)

For the highly symmetric distribution of the relative frequencies of the numbers from the sequences of 5s, 6s, 7s, 8s, and 9s during the late Western and early Eastern Zhou, a model *C* involving only dice throwing fits perfectly the relative frequencies (Table 3):

> The sum S of the outcomes of throwing two regular six-sided dice numbered 1 to 6 is taken. The lowest and highest possible outcome S = 2 and S = 12 are recorded as the highest and lowest numbers in the sequence, 5 and 9 respectively, whereas the numbers 6, 7, and 8 in the sequence are the tokens of the outcomes $S = \{3,6\}$, $S = \{4,5,7,9,10\}$, and $S = \{8,11\}$ respectively.

Although no early textual evidence exists for such a technique, the importance of number 12 in this model places our reconstruction not only into the context of ancient Chinese cyclic calendric numerology, but also reflects the grouping of numbers into sequences of six or even twelve as in the *Shifa*. Note also that the 14-sided die, showing two blank sides and four sides with the numbers from 1 to 6 twice, might attest to a practice of throwing two dice, and in fact amount to modeling it by means of one die only (Fig. 5).

In a similar vein, a model *D* with three dice can be proposed for the numbers in the *Shifa* manuscript (Table 10):

> The sum S of the outcomes of throwing three regular six-sided dice numbered 1 to 6 is taken. The lowest and highest possible outcome S = {3,4} and S = {17,18} are recorded as the highest and lowest numbers in the sequence, 4 and 9 respectively, the sums of 5 and 16 are recorded as 5 and 8 respectively, whereas the numbers 6 and 7 are the tokens of the even outcomes S = {6,8,10,12,14} and of the odd outcomes S = {7,9,11,13,15} respectively.

Again, the 18-sided die, showing the numbers from 1 to 16 and two characters or two blanks, might attest to a practice of throwing three dice (note that a sum of outcomes $S = 1$ or 2 cannot occur in throwing three dice, so that only 16 values of S are possible), and in fact amount to modeling it by means of one die only (Fig. 6).

3 Conclusion

Given the large amount of available records of number sequences on ancient Chinese artifacts, and in the absence of contemporary textual explanations of how these were produced, we have proposed a statistical approach as a new methodology to suggest new hypothesis in an admittedly already fairly hypothetical field of research. This methodology is applied to data from China's high antiquity, a time that predates the canonization and standardization of a divination procedure given in the "Great Commentary" of the *Book of Changes*. That various techniques were indeed in circulation earlier is attested by the available sources from the fourth century BCE, where five methods are named that scholars associate with stalk methods (Beijing 2014; Cook 2006; Kalinowski 2009). Although only three involved the recording of twelve numbers in groups of three or six, one can reasonably assume variations of techniques in space and time. Yet, given a similar distribution pattern of number occurrences over a long period of time (Table 1), there seems to be a historical correlation and continuity between twelfth and fourth century BCE material, that requires exploration, in the present case, statistical exploration to account for the observed continuity.

So far, quantitative approaches to the understanding of the origins of ancient numerical sequences have been limited to the numbers found in the fourth-century BCE *Shifa* strips. Liu Bin (2014), Jia Lianxiang (2014), and Cheng Hao (2014) only proposed a modified A_n-style procedure without statistically testing the fit of the model. The present study goes back further in time and includes older records all the way back to the Shang dynasty (twelfth century BCE). Applying for the first time various goodness-of-fit methods that are standard tests in other disciplines—Chi-squared in physics, Bray-Curtis for population ecology and the G-test in linguistics—allowed us to test different models and identify possible techniques and to discard others on a sound archaeological and statistical basis.

By taking into account the presence of various forms of dice in combination with stalks in fourth and third century BCE burial grounds, we developed models that included random throws of dice (models *B*, *C*, and *D*). Our analysis showed that dice might indeed have gained importance for certain types of numerical sequences in particular historical contexts. The more symmetrical the frequencies are distributed between the possible numbers, the more likely is the involvement of dice in their production. In particular for the late Zhou, model *C* involving the throw of two dice showed an excellent fit (Table 7), whereas for fourth-century BCE records, one should not reject the hypothesis of mixed techniques involving both, dice and stalks (Tables 8 and 9). The odd-even dichotomy, although not explicitly spelled out, might underlie the encoding of the results of the numerical sum in dice throws into symbols for 1 (or 7) and 6 at least as early as the late Western Zhou dynasty.

From the point of view of the history of stalks divination, the most striking result coming out of a statistical approach is the perfect fit of a pure stalks technique using 46 (instead of the canonical 50) stalks in case of the records on early Zhou artefacts (Table 6). This strongly suggests that the procedure described in the commentary to the *Book of Changes* goes back to much earlier stalk divination traditions from Zhou dynasty.

Another historical conclusion that can be drawn from the present study relates to the nature of fourth-century BCE manuscripts, in which samples of numerical sequences are not expected to be as randomly assembled as in a loose collection of unrelated artifacts. We found that the observed occurrences of numbers therein nicely correlate to the theoretically expected numbers of certain models. That is particularly the case for numbers in the *Shifa*,[6] in which artificially constructed numerical sequences of exemplary divinations have only a marginal impact upon the overall distribution of number occurrences. As the distribution is similar to other, older records, the *Shifa* can be read as a compilation of actual outcomes of divination procedures.

Altogether, statistical reconstructions suggest that stalk and dice use traces back to the early Zhou and that while the actors and specific ritual contexts may have changed over time, their divinatory techniques retained a certain continuity and are much older than previously assumed. In the face of continuous new discoveries of artefacts in China, we can expect that a statistical approach will be based on an ever growing database, and that our conclusions drawn from the kind of research methodology proposed here, will be further sharpened and become statistically more significant and historically more relevant.

[6] The *Shifa* bamboo strips are dated to ca. 300 BCE according to carbon-14 analysis and preserved at Tsinghua University in Beijing, with a total of 228 number triples recorded, see (Li Xueqin 2013).

Appendices

Tables

Table 1 Relative frequencies of numbers as recorded on artifacts and in fourth-century manuscripts. Numbers shown are the compounded values for 1 and 7. *N* indicates the total number of counts. For uncompounded values, see Table 4

	4	**5**	**6**	**7 (1)**	**8**	**9**	***N***
Shang		6.00%	41.50%	37.00%	14.00%	1.50%	**200**
early-mid Western Zhou		6.27%	33.73%	40.00%	20.00%		**255**
late Western-early Eastern Zhou		3.47%	19.44%	54.86%	19.44%	2.78%	**144**
Shifa	1.02%	1.90%	46.20%	45.61%	1.61%	3.65%	**684**
4th cent. BCE (other than Shifa)		2.25%	50.80%	40.51%	5.47%	0.96%	**311**

Table 2 Theoretical probabilities for obtaining the numbers 6, 7, 8, and 9 with a procedure A_n. It is noteworthy that A_4 applied to 55 stalks leads to the same numerical sequence as A_3 applied to 50, 49, or 48 stalks (see Appendix 3. Methods: Modeling the stalk procedure A_n)

	6	**7**	**8**	**9**
55 stalks (A_4)	11,54%	36,61%	38,46%	13,40%
50 stalks (A_3)	5,89%	30,30%	44,12%	19,68%
49 stalks (A_3)	11,54%	36,61%	38,46%	13,40%
48 stalks (A_3)	17,42%	43,18%	32,55%	6,84%

Table 3 Compounded probabilities of the sums *S* of two regular six-sided dice numbered 1 to 6 in model *C*, compared to the relative frequencies found on late Western Zhou to early Eastern Zhou dynasty artifacts

S	2	3	4	5	6	7	8	9	10	11	12
***P*(*X* = *S*)**	1/36	2/36	3/36	4/36	5/36	6/36	5/36	4/36	3/36	2/36	1/36
	2.78%	**5.56%**	8.33%	11.11%	**13.89%**	16.67%	**13.89%**	11.11%	8.33%	**5.56%**	2.78%
$P_{compounded}$	2.78%			**19.44%**		55.56%		**19.44%**			2.78%
for *S*	2			3, 6		4,5,7,9,10		8,11			12
	↓			↓		↓		↓			↓
	5			**6**		**7(1)**		**8**			**9**
late Western - early Eastern Zhou	3.47%			**19.44%**		54.86%		**19.44%**			2.78%

Materials

The data underlying the present statistical analysis is based upon a list of oracular inscriptions of numbers in Bian Xiaobing (2016: 5–30).

Items 1 to 38 therein were included in our category of Shang dynasty number sequences; items 39 to 83 in the early to mid-Western Zhou group; items 84 to 109 constitute the late Western to early Eastern Zhou set. Items 108 and 109—inscriptions on a bone arrowhead from Zhangjiapo containing the two number triples 511 and 161—date from the early Western Zhou and were thus included in this category, as well as items 110–114 from the Zhouyuan Museum (Li 2006), corresponding to the numerical sequences 1-1-1-1-8-1, 1-1-1-1-8-1, 6-6-6, 1-1-8, and 1-6-6. Item 115 has not been included into our database. It is a sequence of three numbers, 8-7-7, on the hilt of a Qin sword from the eastern section of the Qin Shihuang Mausoleum. Whether or not it is a product of divination cannot be said. The numbers on the sword would have been hidden by the wrappings used to make up the handle. So they likely had to do with the manufacture of the sword and might be considered a craftsman's mark. That by the Qin period the only inscriptions made on bronze vessels seem to be functional rather than religious supports our choice. Items 116 to 141 constitute the sequences in the three fourth-century BCE bamboo manuscripts, items 142 to 198 are the contemporary sequences from the *Shifa* manual.

The erroneous transcription of item 84 (p. 16) has been corrected to 1-6-1-1-1-6. Two occurrences of "9" in items 51 & 52 are dubious due to their unusual graphic variance. The first has been proposed to be read as an 8 (Cao 2002), but the same is also possible for the second occurrence in the sequence 9-1-1-1-6-5. We have tested two options: (1) reading them as 8s due to the graphic ambiguity and (2) excluding them from the early to mid-Western Zhou category because the tomb in Qijiacun where the inscribed bovine shoulder bones were found contained earlier and later temporal layers. In both cases, a statistical distribution of numbers from the early to mid-Western Zhou without 9s fits perfectly a model A_3 with 46 stalks that does not produce 9s (see Table 6, calculated with items 51 & 52 read as 8s).

Finally, items 71 and 72 are the same number sequences because they are bronze copies of single casting events. The same is the case for items 80 to 82, which document the same divination event. They were thus each counted only once in our database.

Methods

Modeling the stalk procedure A_n. In calculating the theoretical probabilities of the outcomes of procedures A_n and B_n^m we included all possibilities to randomly divide a bunch of stalks into two bunches. Although we assume a tendency to split the bunch in half, in absence of any textual evidence all cases have been taken uniformly into account. From a probabilistic point of view there would be no significant difference between considering all or only a central portion of all possibilities (see below: **Modeling the division of the stalks**).

Table 11 gives as an example all possibilities to divide an initial bunch of $Y = 1 + s_1 + s_2 = 58$ stalks into two bunches of s_1 and s_2 stalks. From one bunch a stalk is subtracted and then from each bunch the remainder from division by 4 is subtracted. In case of $s_i \equiv 0(mod\ 4)$, 4 stalks are subtracted. For example, if 57 stalks are split into 24 and 33 stalks, a total of either 5 or 9 stalks are deduced if the single stalk is taken away from the 24 or from the 33 stalks. In general, when $Y - 1 \equiv 1(mod\ 4)$, the probability to reduce $Y - 1$ by 5 or 9 stalks is approximately 0.75 and 0.25, respectively (precise probabilities for all values underlying our calculations are given in Table 12). If $Y - 1 \equiv 0(mod\ 4)$, the probabilities to reduce $Y - 1$ by 4 or 8 stalks are approximately 0.5 each, for $Y - 1 \equiv 3(mod\ 4)$, the probabilities to reduce $Y - 1$ by 3 or 7 stalks are approximately 0.25 and 0.75 and, in case of $Y - 1 \equiv 2(mod\ 4)$, the probability of subtracting 5 stalks from $Y - 1$ is 1. This is the reason why fewer numbers result from a procedure A_n if $Y - 1 \equiv 2(mod\ 4)$. For example, A_4 with 55 stalks leads to the numbers 6, 7, 8 or 9, whereas A_4 with 54 stalks results in a 5, 6, 7, 8 or 9. It also explains why, if $Y - 1 \equiv 2(mod\ 4)$, the same results with the same probabilities are obtained with a procedure A_n as with a procedure A_{n-1} and $Y - 6$.

For example, A_4 with 51 stalks leads to the same results with the same probabilities as A_3 with 45 stalks. More prominently and with wider exegetical consequences is the fact that A_3 with 50 stalks, the canonical *Dayan* procedure as it was understood by the Song Neo-Confucianist Zhu Xi (1130–1200) and described in the "Great Commentary" to the *Book of Changes*, leads to the same results as A_4 with 55 stalks (for the respective probabilities cf. Table 2). This fact has, according to my knowledge, gone unnoticed so far. It would nevertheless reinforce an alternative reading with 55 stalks of the *Dayan* passage in the "Great Commentary", based on purely numerological arguments (Chen and Guo 1998). Probabilistic considerations of alternative stalk procedures can thus contribute to the controversies over the exegesis of the canonical *Dayan* procedure, and the question if the initial number of stalks should be 50 or 55. This point will be developed in detail in a future publication.

Modeling dice in procedures B_n^m, *C*, and *D*. It is well known that no 14 or 18-sided die can be regular. Nevertheless, as can be seen from Figs. 5 and 6, the kind of dice unearthed in China are fairly equilibrated polyhedra. We have therefore assumed in all our calculations an equal probability for all sides of each *m*-sided die. The same holds for any of the six-sided cubical dice (Fig. 3).

In combination with the stalk procedure, it is evident that the 14 and 18-sided die reduces significantly the overall influence of stalks on the outcome, since blanks, which lead to the use of stalks in model B_n^m, only appear with a probability of 2/14 and 2/18 respectively.

Models *C* and *D* are admittedly *a posteriori* constructions, grouping together certain results to obtain the best fits with the observed frequencies (Tables 7 and 9). Other combinations of outcomes are possible, but we have limited ourselves to those solutions here that reflect best symmetric principles and regularities. We have also limited our considerations to sums of two (model *C*, Table 3) and three (model

D, Table 10) dice, since the 14- and 18-sided dice that seem to have gained importance in the fourth century BCE can be seen as symbolic, 1-die models of the throw of two and three dice respectively.

Statistical analysis. In order to test several models against the actual data, and in order to test different numerical values of parameters within the same model, I have introduced a series of fit-tests based on standard measures between statistical samples and expected frequencies derived from probabilistic models. Although using such goodness-of-fit-tests is widespread in natural sciences, they appear to be applied here for the first time to test the adequacy of a historiographical reconstruction.

In order to quantify the distance between *i* discrete points of observed frequency (O_i) and the expected frequency (E_i) calculated on the basis of the various models, the Chi square χ^2 coefficient has been calculated for all scenarios:

$$\chi^2 = \frac{\sum_i (O_i - E_i)^2}{E_i}$$

Since the absolute values of expected frequency E_i are often below 5 or even 3, the χ^2 coefficient could only be applied qualitatively to compare the various proposed models. Grouping categories with small expected frequencies together—in our case the counts of 4s or 5s and 9s in the fourth-century BCE data—does produce a comparable χ^2 coefficient for all models compared to the manuscripts other than the *Shifa*. When grouping the 4s and 9s in the *Shifa* together, we obtain lower values for χ^2 for all models, but the coefficient still fails to reflect the obviously good fit of the theoretical model to the observed frequencies. The same holds for the logarithmic transformation of the χ^2 coefficient:

$$G = 2\sum_i O_i \bullet \ln\left(\frac{O_i}{E_i}\right)$$

For these reasons, the Bray-Curtis dissimilarity *BC*, a non-Euclidean measure often used to compare ecological samples from different sites with highly disparate counts (Bray and Curtis 1957), has also been used. It proved well adapted to the kind of distributions in our database, where small expected frequencies occurred and the majority of counts was divided between two numbers, as can be seen in particular for the *Shifa* (Table S5). Rescaled to a percentage (sample numerosity = *N*), this measure takes on values between 0% for identical counts ($O_i = E_i$ for all *i*) and 100% to quantify completely disjoint counts (i.e., exactly one value of any pair (O_i, E_i) is equal to zero):

$$BC = \frac{\sum_i |O_i - E_i|}{2N}$$

In Tables 5, 6, 7, 8, and 9 we have highlighted the best fits in dark grey with bold script, good fits in an intermediary shade of grey, and the ones that we can still

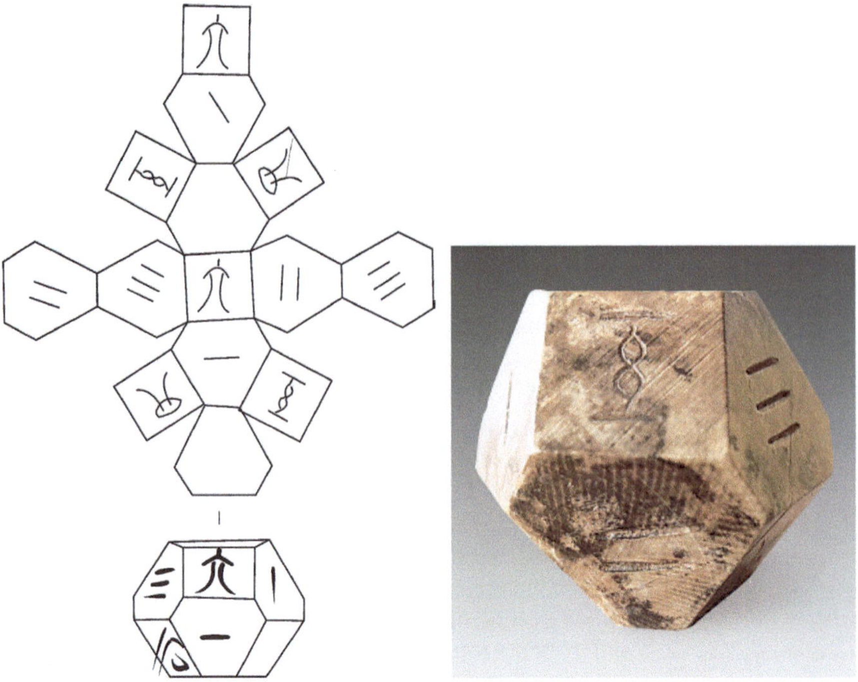

Fig. 5 An irregular 14-sided die from the Warring States, found in Shandong Province. Source: Shandong (2014, 27, fig. 58)

Fig. 6 An irregular 18-sided die from Mawangdui tomb 3. Source: Fu and Chen 1992

Table 4 Relative frequencies of numbers as recorded on artifacts from the Shang dynasty (12th cent. BCE) to fourth-century manuscripts. *N* indicates the total number of counts

	1	4	5	6	7	8	9	*N*
Shang	13.50%		6.00%	41.50%	23.50%	14.00%	1.50%	**200**
early-mid Western Zhou	24.71%		6.27%	33.73%	15.29%	20.00%		**255**
late Western-early Eastern Zhou	54.86%		3.47%	19.44%		19.44%	2.78%	**144**
Shifa	45.47%	1.02%	1.90%	46.20%	0.15%	1.61%	3.65%	**684**
4th cent. BCE (other than Shifa)	40.51%		2.25%	50.80%		5.47%	0.96%	**311**

Table 5 Relative frequencies (in %) of $N = 200$ numbers as recorded on oracle bones from the Shang dynasty compared to the theoretical probabilities calculated for two scenarios involving 54 stalks with 4 repetitions (model A_4), or combined with a special m-sided die (model B_4^m for $m = 6$, 14 and 18). The use of 59 stalks with model A_5 or B_5^m is statistically equivalent to using 53 stalks with model A_4 or B_4^m (see Appendix 3. Methods: Modeling the stalk procedure A_n)

N = 200	5	6	7	8	9	**Chi-square χ^2**	**Bray-Curtis**
Shang	**6.00**	**41.50**	**37.00**	**14.00**	**1.50**		
54 stalks (A_4)	2.86	17.78	37.10	32.22	10.05	105.4	26.86%
special die & 54 stalks (B_4^6)	0.95	39.26	45.70	10.74	3.35	61.1	10.55%
14-sided die & 54 stalks (B_4^{14})	0.41	45.40	48.16	4.60	1.44	197.5	15.05%
18-sided die & 54 stalks (B_4^{18})	0.32	46.42	48.57	3.58	1.12	271.1	40.11%
59 stalks (A_5) or 53 stalks (A_4)	5.60	23.78	37.57	26.22	6.83	46.2	18.12%
special die & 53 stalks (B_4^6)	1.87	41.26	45.86	8.74	2.28	28.6	9.63%
14-sided die & 53 stalks (B_4^{14})	0.80	46.25	48.22	3.75	0.98	130.4	15.98%
18-sided die & 53 stalks (B_4^{18})	0.62	47.09	48.62	2.91	0.76	180.5	17.21%
52 stalks (A_4)	8.45	30.02	38.06	19.98	3.48	21.4	11.48%
special die & 52 stalks (B_4^6)	2.82	43.34	46.02	6.66	1.16	24.0	10.86%
14-sided die & 52 stalks (B_4^{14})	1.21	47.15	48.29	2.85	0.50	135.7	16.94%
18-sided die & 52 stalks (B_4^{18})	0.94	47.78	48.67	2.22	0.39	193.2	17,95%

consider relatively acceptable in the context of our study in the brightest shade of grey.

Modeling the division of the stalks. In the above models, it has always been assumed that the division of the stalks is done uniformly, i.e. that the probability of dividing a bunch of say $n = 45$ stalks into 1 and 44 stalks is the same as the probability of dividing it more symmetrically into 22 and 23 stalks. Intuitively one might think that there is a tendency to divide the stalks into bunches of fairly equal size. Such non uniform division can be modeled by a binomial law, whereby no preference is assumed between right and left hand. Overall, due to the cyclic nature of the remainders, there is no significant difference between the uniform and the

Table 6 Relative frequencies (in %) of $N = 255$ numbers as recorded on artifacts from early to mid-Western Zhou dynasty compared to the theoretical probabilities calculated for scenarios involving 46, 45, or 44 (model A_3) stalks only, or combined with a special m-sided die (model B_3^m) for $m = 6$, 14 and 18. The use of 45 stalks with models A_3 or B_3^m is statistically equivalent to using 51 stalks with model A_4 or B_4^m (see Appendix 3. Methods: Modeling the stalk procedure A_n). Since models B_3^{14} and B_3^{18} generally provide very bad fits in this period, we have not included all scenarios here

$N = 255$	5	6	7	8	χ^2	Bray-Curtis
early-mid Western Zhou	**6.27**	**33.73**	**40.00**	**20.00**		
46 stalks (A_3)	**5.85**	**30.20**	**44.18**	**19.78**	**2.1**	**4.18%**
special die & 46 stalks (B_3^6)	1.95	43.40	48.06	6.59	102.9	17.73%
14-sided die & 46 stalks (B_3^{14})	0.84	47.17	49.17	2.83	370.6	22.61%
18-sided die & 46 stalks (B_3^{18})	0.65	47.80	49.35	2.20	507.0	23.43%
45 (A_3) or 51 (A_4) stalks	11.42	36.51	38.57	13.49	14.6	7. 93%
special die & 45 (B_3^6) or 51 (B_4^6) stalks	3.81	45.50	46.19	4.50	150.2	17.97%
44 stalks (A_3)	11.69	36.82	38.30	13.19	16.2	8.51%
special die & 44 stalks (B_3^6)	3.90	45.61	46.10	4.40	154.9	17,98%

Table 7 Relative frequencies of $N = 144$ numbers as recorded on artifacts from the late Western to the early Eastern Zhou dynasty compared to the theoretical probabilities calculated for all scenarios involving 54, 53, or 52 stalks only (model A_4) or combined with a special m-sided die (model B_4^m) for $m = 6$, 14 and 18. The use of 59 stalks with models A_5 or B_5^m is statistically equivalent to using 53 stalks with model A_4 or B_4^m (see Appendix 3. Methods: Modeling the stalk procedure A_n). Finally, model C (see Table 3) considering the sum of two conventional dice is tested

$N = 144$	5	6	7	8	9	χ^2	Bray-Curtis
late Western - early Eastern Zhou	**3.47**	**19.44**	**54.86**	**19.44**	**2.78**		
54 stalks (A_4)	2.86	17.78	37.10	32.22	10.05	27.5	20.04%
special die & 54 stalks (B_4^6)	0.95	39.26	45.70	10.74	3.35	37.0	20.39%
14-sided die & 54 stalks (B_4^{14})	0.41	45.40	48.16	4.60	1.44	126.6	25.95%
18-sided die & 54 stalks (B_4^{18})	0.32	46.42	48.57	3.58	1.12	173.7	26.98%
59 stalks (A_5) or 53 stalks (A_4)	5.60	23.78	37.57	26.22	6.83	19.7	17.29%
special die & 53 stalks (B_4^6)	1.87	41.26	45.86	8.74	2.28	40.2	21.82%
14-sided die & 53 stalks (B_4^{14})	0.80	46.25	48.22	3.75	0.98	138.5	26.81%
18-sided die & 53 stalks (B_4^{18})	0.62	47.09	48.62	2.91	0.76	185.2	27.64%
52 stalks (A_4)	8.45	30.02	38.06	19.98	3.48	13.7	16.80%
special die & 52 stalks (B_4^6)	2.82	43.34	46.02	6.66	1.16	58.0	23.90%
14-sided die & 52 stalks (B_4^{14})	1.21	47.15	48.29	2.85	0.50	184.7	27.70%
18-sided die & 52 stalks (B_4^{18})	0.94	47.78	48.67	2.22	0.39	248.8	28.34%
sum of 2 regular dice (C)	**2.78**	**19.44**	**55.56**	**19.44**	**2.78**	**0.3**	**0.69%**

Table 8 Relative frequencies of N = 311 numbers as recorded in three fourth century BCE manuscripts (*Tianxingguan*, *Baoshan*, and *Xincai Geling*) compared to the theoretical probabilities calculated for all scenarios involving 54, 53, or 52 stalks only (model A_4) or combined with a special m-sided die (model B_4^m) for m = 6, 14 and 18

N = 311	5	6	7	8	9	χ^2	Bray-Curtis	χ^2 5&9	G-Test
4th cent. BCE (other than Shifa)	**2.25**	**50.80**	**40.51**	**5.47**	**0.96**				
54 stalks (A_4)	2.86	17.78	37.10	32.22	10.05	286.8	36.44%	283.8	276.3
special die & 54 stalks (B_4^6)	0.95	39.26	45.70	10.74	3.35	31.2	12.84%	39.3	32.7
14-sided die & 54 stalks (B_4^{14})	0.41	45.40	48.16	4.60	1.44	32.7	8.11%	35.3	19.4
18-sided die & 54 stalks (B_4^{18})	0.32	46.42	48.57	3.58	1.12	45.2	8.20%	52.1	23.8
59 stalks (A_5) or 53 stalks (A_4)	5.60	23.78	37.57	26.22	6.83	169.2	29.97%	174.8	181.1
special die & 53 stalks (B_4^6)	1.87	41.26	45.86	8.74	2.28	15.2	9.93%	13.5	16.1
14-sided die & 53 stalks (B_4^{14})	0.80	46.25	48.22	3.75	0.98	15.9	7.72%	19.5	13.0
18-sided die & 53 stalks (B_4^{18})	0.62	47.09	48.62	2.91	0.76	25.5	8.10%	32.9	18.9
52 stalks (A_4)	8.45	30.02	38.06	19.98	3.48	97.9	23.24%	112.0	111.7
special die & 52 stalks (B_4^6)	**2.82**	**43.34**	**46.02**	**6.66**	**1.16**	**7.2**	**7.46%**	**7.5**	**7.1**
14-sided die & 52 stalks (B_4^{14})	1.21	47.15	48.29	2.85	0.50	16.4	7.78%	19.2	14.1
18-sided die & 52 stalks (B_4^{18})	0.94	47.78	48.67	2.22	0.39	28.0	8.16%	33.7	21.5

Table 9 Relative frequencies (in %) of N = 684 numbers as recorded in the *Shifa* bamboo manuscript compared to the theoretical probabilities calculated for all scenarios involving 56, 57, or 58 stalks only (model A_5) or combined with a special m-sided die (model B_5^m) for m = 6, 14, and 18. Finally, the theoretical probabilities for the sum of three conventional dice (model D, Table 10) are given and the model proposed by L. Jia in (Jia 2014), calculated with precise theoretical probabilities (Table 12) is tested

N = 684	4	5	6	7	8	9	χ^2	Bray-Curtis	χ^2 4&9	G-Test
4th cent. BCE (Shifa)	**1.02**	**1.90**	**46.20**	**45.61**	**1.61**	**3.65**				
58 stalks (A_5)	1.38	10.11	27.23	34.77	21.39	5.12	287.7	29.81%	287.7	382.0
special die & 58 stalks (B_5^6)	0.46	3.37	42.41	44.92	7.13	1.71	55.9	6.99%	55.9	65.2
14-sided die & 58 stalks (B_5^{14})	0.20	1.44	46.75	47.82	3.06	0.73	110.0	4.21%	109.9	59.5
18-sided die & 58 stalks (B_5^{18})	0.15	1.12	47.47	48.31	2.38	0.57	154.9	4.73%	154.9	71.7
57 stalks (A_5)	2.71	14.46	30.56	32.07	16.73	3.48	269.2	29.37%	264.5	365.7
special die & 57 stalks (B_5^6)	0.90	4.82	43.52	44.02	5.58	1.16	69.8	6.89%	55.6	67.6
14-sided die & 57 stalks (B_5^{14})	0.39	2.07	47.22	47.44	2.39	0.50	147.0	3.79%	113.9	64.2
18-sided die & 57 stalks (B_5^{18})	0.30	1.61	47.84	48.01	1.86	0.39	202.8	4.29%	160.3	76.7
56 stalks (A_5)	4.09	18.97	34.01	29.26	11.90	1.77	287.7	30.42%	259.9	383.5
special die & 56 stalks (B_5^6)	1.36	6.32	44.67	43.09	3.97	0.59	141.5	7.12%	58.1	92.9
14-sided die & 56 stalks (B_5^{14})	0.58	2.71	47.72	47.04	1.70	0.25	317.5	3.84%	122.9	91.3
18-sided die & 56 stalks (B_5^{18})	0.45	2.11	48.22	47.70	1.32	0.20	422.4	4.31%	172.2	104.1
sum of 3 regular dice (D)	**1.85**	**2.78**	**45.37**	**45.37**	**2.78**	**1.85**	**20.2**	**2.88%**	**7.4**	**18.6**
(Jia 2014) corrected	0.24	6.01	40.14	39.55	12.74	1.32	143.7	15.24%	140.9	173.5

Table 10 Theoretical probabilities P of the sums S of three regular six-sided dice numbered 1 to 6 in model D, compared to the relative frequencies found in the *Shifa* manuscript

S	3	4	5	6	7	8	9	10	11	12	13	14	15	16	17	18
	1/216	1/72	6/216	5/108	5/72	7/72	25/216	1/8	1/8	25/216	7/72	5/72	5/108	6/216	1/72	1/216
$P(X = S)$	0,46%	1,39%	**2,78%**	4,63%	6,94%	9,72%	11,57%	12,50%	12,50%	11,57%	9,72%	6,94%	4,63%	**2,78%**	1,39%	0,46%
$P_{compounded}$		1,85%	**2,78%**					45,37%	45,37%					**2,78%**	1,85%	
for S		3,4	5					6,8,10,12,14	7,9,11,13,15					16	17,18	
		➔	➔					➔	➔					➔	➔	
		4	**5**					**6**	**7**					**8**	**9**	
4th cent. BCE (*Shifa*)		1,02%	**1,90%**					46.20%	45,61%					**1,61%**	3,65%	

Table 11 Enumeration of all 110 theoretical possibilities to divide a bunch of $Y = 1 + s_1 + s_2 = 58$ stalks into two bunches of s_1 and s_2 stalks in the first subprocedure of A_n. After elimination of one stalk from either the left or the right bunch, the remaining $s_1 + s_2 - 1$ stalks are reduced by the total remainder $r = r_1 + r_2$, where $r_i = \begin{cases} 4 & \text{if } b_i \equiv 0 (mod\ 4) \\ \text{remainder in } b_i \equiv r_i (mod\ 4) & \text{if } b_i \not\equiv 0 (mod\ 4) \end{cases}$

initial bunches		1 stalk less (right)		rem.	1 stalk less (left)		rem.
s_1	s_2	$b_1 = s_1$	$b_2 = s_2 - 1$	r	$b_1 = s_1 - 1$	$b_2 = s_2$	r
1	56	1	55	4	0	56	8
2	55	2	54	4	1	55	4
3	54	3	53	4	2	54	4
4	53	4	52	8	3	53	4
5	52	5	51	4	4	52	8
6	51	6	50	4	5	51	4
7	50	7	49	4	6	50	4
8	49	8	48	8	7	49	4
9	48	9	47	4	8	48	8
10	47	10	46	4	9	47	4
11	46	11	45	4	10	46	4
12	45	12	44	8	11	45	4
13	44	13	43	4	12	44	8
14	43	14	42	4	13	43	4
15	42	15	41	4	14	42	4
16	41	16	40	8	15	41	4
17	40	17	39	4	16	40	8
18	39	18	38	4	17	39	4
19	38	19	37	4	18	38	4
20	37	20	36	8	19	37	4
21	36	21	35	4	20	36	8
22	35	22	34	4	21	35	4
23	34	23	33	4	22	34	4
24	33	24	32	8	23	33	4
25	32	25	31	4	24	32	8
26	31	26	30	4	25	31	4
27	30	27	29	4	26	30	4
28	29	28	28	8	27	29	4

initial bunches		1 stalk less (right)		rem.	1 stalk less (left)		rem.
s_1	s_2	$b_1 = s_1$	$b_2 = s_2 - 1$	r	$b_1 = s_1 - 1$	$b_2 = s_2$	r
29	28	29	27	4	28	28	8
30	27	30	26	4	29	27	4
31	26	31	25	4	30	26	4
32	25	32	24	8	31	25	4
33	24	33	23	4	32	24	8
34	23	34	22	4	33	23	4
35	22	35	21	4	34	22	4
36	21	36	20	8	35	21	4
37	20	37	19	4	36	20	8
38	19	38	18	4	37	19	4
39	18	39	17	4	38	18	4
40	17	40	16	8	39	17	4
41	16	41	15	4	40	16	8
42	15	42	14	4	41	15	4
43	14	43	13	4	42	14	4
44	13	44	12	8	43	13	4
45	12	45	11	4	44	12	8
46	11	46	10	4	45	11	4
47	10	47	9	4	46	10	4
48	9	48	8	8	47	9	4
49	8	49	7	4	48	8	8
50	7	50	6	4	49	7	4
51	6	51	5	4	50	6	4
52	5	52	4	8	51	5	4
53	4	53	3	4	52	4	8
54	3	54	2	4	53	3	4
55	2	55	1	4	54	2	4
56	1	56	0	8	55	1	4

Table 12 Probabilities. for possible number r_i of stalks by which the bunch is reduced in one iteration of the subprocedure in model A_n

initial number of stalks	*stalks entering subprocedure*	*deduced stalks* $1 + r_1$	*deduced stalks* $1 + r_2$	$P(r_1)$	$P(r_2)$
59	**58**	**6**	**2**	**100.00%**	**0%**
58	57	5	9	75.00%	25.00%
57	56	4	8	50.91%	49.09%
56	55	3	7	25.93%	74.07%
55	**54**	**6**	**2**	**100.00%**	**0%**
54	53	5	9	75.00%	25.00%
53	52	4	8	50.98%	49.02%
52	51	3	7	26.00%	74.00%
51	**50**	**6**	**2**	**100.00%**	**0%**
50	49	5	9	75.00%	25.00%
49	48	4	8	51.06%	48.94%
48	47	3	7	26.09%	73.91%
47	**46**	**6**	**2**	**100.00%**	**0%**
46	45	5	9	75.00%	25.00%
45	44	4	8	51.28%	48.84%
44	43	3	7	50.00%	50.00%
	40	4	8	51.28%	48.72%
	36	4	8	51.43%	48.57%
	32	4	8	51.61%	48.39%
	28	4	8	51.85%	48.15%

Table 13 Relative frequencies (in %) of $N = 255$ numbers as recorded on artifacts from early to mid-Western Zhou dynasty compared to the theoretical probabilities calculated for uniform scenarios (cf. Table 6) and non-uniform scenarios for dividing the stalks and involving 46 stalks only (model A_3), or combined with a special 6-sided die (model B_3^6)

$N = 255$	5	6	7	8	χ^2	**Bray-Curtis**
early-mid Western Zhou	**6.27**	**33.73**	**40.00**	**20.00**		
46 stalks (A_3)	**5.85**	**30.20**	**44.18**	**19.78**	**2.1**	**4.18%**
special die & 46 stalks (B_3^6)	1.95	43.40	48.06	6.59	102.9	17.73%
Non-uniform scenario: 46 stalks (A_3)	**6,25**	**31,25**	**43,75**	**18,75**	**1,5**	**3,75%**
special die & 46 stalks (B_3^6)	2,08	43,75	47,92	6,25	107,8	16,67%

non-uniform scenarios as concerns our conclusions to be drawn (Cf. Tables 6 and 13). In the case of $n = 45$ stalks, for example, instead of 75% and 25% for the probabilities to deduce 5 or 9 stalks in the uniform model, we would obtain 75.000006% and 24.999994% respectively in the non-uniform case. In the case of $n = 36$ stalks, instead of 51.43% and 48.57% for the probabilities to deduce 4 or 8 stalks in the uniform model, we would obtain 49.9998% and 50.0002% respectively in the non-uniform case.

Bibliography

Beijing daxue chutu wenxian yanjiusuo 北京大學出土文獻研究所, ed. 2014. *Beijing daxue cang Xi Han zhushu* 北京大學藏西漢竹書. Vol. 5. Shanghai: Shanghai guji chubanshe.

Bian Xiaobing 邊曉冰. 2016. "Chutu shuzi gua 'liangxi yong shu tixi' jiashuo yanjiu" 出土數字卦「兩系用數體系」假說研究. M.A. thesis, Qufu shifan daxue 曲阜師範大學.

Bray, J. Roger and J. T. Curtis. 1957. "An Ordination of the Upland Forest Communities in Southern Wisconsin," *Ecological Monographs* 27.4: 325–349.

Bréard, Andrea and Constance A. Cook. 2020. "Cracking bones and numbers: solving the enigma of numerical sequences on ancient Chinese artifacts," *Archive for History of Exact Sciences* 74: 313–343.

Cao Wei 曹瑋. 2002. "Tao pai shang de shuzi gua yanjiu" 陶拍上的數字卦研究, *Wenwu* 文物 11: 65–71.

Chen Enlin 陳恩林 and Guo Ningxin 郭寧信. 1998. "Guanyu *Zhouyi* 'Dayan zhi shu' de wenti" 關於『周易』「大衍之數」的問題, *Zhongguo zhexueshi* 中國哲學史 3: 42–47.

Cheng Hao 程浩. 2014. "Shifa zhanfa yu 'Dayan zhi shu' " 筮法占法與「大衍之數」, *Shenzhen daxue xuebao* 深圳大學學報 (Renwen shehui kexueban) 31: 62–64.

Cook, Constance A. 2006. *Death in Ancient China: The Tale of One Man's Journey.* Leiden: Brill.

Cook, Constance A. and Andrea Bréard. 2022. "Stalk and Other Divination Traditions Prior to the *Changes* Canon: Views from Newly-Discovered Texts," in *The Other* Yijing: *The* Book of Changes *in Chinese History, Politics, and Everyday Life*, ed. Tze-ki Hon. Leiden: Brill, 23–68.

Cook, Constance A. and Zhao Lu. 2017. *Stalk Divination A Newly Discovered Alternative to the I Ching*. Oxford: Oxford University Press.

Fu Juyou 傅舉有 and Chen Songchang 陳松長, ed. 1992, *Mawangdui hanmu wenwu* 馬王堆漢墓文物. Changsha: Hunan chubanshe.

Jia Lianxiang 賈連翔. 2014. "Qinghua jian Shifa yu Chu di shuzigua yanbian fangfa de tuiqiu" 清華簡『筮法』與楚地數字卦演算方法的推求, *Shenzhen daxue xuebao* 深圳大學學報 3: 57–60.

Jingzhou diqu bowuguan 荊州地區博物館. 1995. "Jiangling Wangjiatai 15 hao Qin mu" 江陵王家台15 號秦墓, *Wenwu* 文物 1: 37–41.

Kalinowski, Marc. 2009. "Diviners and Astrologers under the Eastern Zhou: Transmitted texts and Recent Archaeological Discoveries." in *Early Chinese Religion, Part One: Shang through Han (1250 BC–220 AD)*, pp. 374–385, ed. John Lagerwey and Marc Kalinowski. Leiden: Brill.

Lewis, Mark E. 2002. "Dicing and Divination in Early China," *Sino-Platonic Papers* 121: 1–22.

Li Ling 李零. 2006. *Zhongguo fangshu xukao* 中國方術續考. Beijing: Zhonghua shuju.

Li Xueqin 李學勤, ed. 2013. *Qinghua daxue cang Zhanguo zhujian* 清華大學藏戰國竹簡. Vol. 4. Shanghai: Zhongxi shuju.

Liu Bin 劉彬. 2014. "Qinghua jian *Shifa* shishu de sanzhong keneng yansuan" 清華簡『筮法』筮數的三種可能演算, *Zhouyi yanjiu* 周易研究 4: 24–28.

Ma Nan 馬楠. 2014. "Qinghua jian *Shifa* erti" 清華簡『筮法』二題, *Shenzhen daxue xuebao*深圳大學學報 (Renwen shehui kexueban) 31.1: 64–65.

Peterson, Willard J. 1982. "Making Connections: 'Commentary on the Attached Verbalizations' of the *Book of Change*," *Harvard Journal of Asiatic Studies* 42.1: 67–116.

Pischedda, Flaminia and Andrea Bréard. Forthcoming. "The Shuzi gua Corpus: A Historico-Statistical Analysis," in *The Dao Companion to the* Book of Changes, Part I, ed. Tze-ki Hon. Cham: Springer.

Röllicke, Hermann-Josef. 1999. "Von 'Winkelwegen', 'Eulen' und 'Fischziehern' - *liubo*: ein altchinesisches Brettspiel für Geister und Menschen," *Board Game Studies* 2: 24–41.
Shandong sheng wenwu kaogu yanjiusuo 山東省文物考古研究所 and Qingzhou shi bowuguan 青州市博物館. 2014. "Shandong Qingzhou Xixin Zhanguo mu fajue jianbao" 山東青州西辛戰國墓發掘簡報, *Wenwu* 文物 9: 4–32.
Shaughnessy, Edward L. 2014. *Unearthing the Changes*. New York: Columbia University Press.
Smith, Richard J. 2008. *Fathoming the Cosmos and Ordering the World*. Charlottesville, Va.; London: University of Virginia Press.
Xiong Chuanxin 熊傳新. 1979. "Tan Mawangdui san hao hanmu chutu de liubo" 談馬王堆三號西漢墓出土的陸博, *Wenwu* 文物 4: 35–39.

On the Variety of Language in Formulating the Same Procedure in *Litian* 里田 Land Measurement Problems in Excavated Qin-Han Mathematical Manuscripts

Tan Jingnan

Abstract The study of Qin-Han (221 BCE–220 CE) mathematics was once mostly limited to *The Nine Chapters on Mathematical Procedures* (*Jiuzhang suanshu* 九章算術), but all that has changed in the last three decades with the successive discovery of mathematical manuscripts in Zhangjiashan tomb 247, Shuihudi tomb 77 and in the Yuelu Academy and Peking University collections. In addition to furnishing us with new, never-before-seen materials and shedding light on the origins of those incorporated into later classics like *The Nine Chapters*, these manuscripts document a variety in the language and formulation of common procedures that stands in stark contrast with the later mathematical canon. Focusing on the case of *Litian* 里田 land measurement problems in the aforementioned sources, and a peculiar variation thereupon in the Peking University *Suanshu* 算書 that the Chūgoku kosansho kenkyūkai (2015) suggests might be a mnemonic, this paper explores how what we see in these manuscripts might be linked to oral practices and basic training and what, if that be the case, this says about the redaction of *The Nine Chapters*.

Keywords Qin-Han mathematics · Manuscripts · "*Litian*" 里田 · Mnemonics

1 Introduction

According to Liu Hui's preface of 263 CE, the *Nine Chapters on Mathematical Procedures* (*Jiuzhang suanshu* 九章算術) supposedly originated in the "Nine Reckonings" (*jiu shu* 九數) of the Duke of Zhou (r. 1042–1036 BCE). The Duke of Zhou's work was then lost during the "burning of books and burial of scholars" under the Qin empire (221–206 BCE), and the extant version comes down to us through Zhang Cang 張蒼 (253–152 BCE) and Geng Shouchang 耿壽昌 (fl. 52 BCE), who were able to reconstruct the text what from fragmentary and

Tan Jingnan (✉)
Jianghan University, Wuhan, People's Republic of China

D. P. Morgan, T. Quan (eds.), *Mathematics and Astronomy in the Ancient World*,
https://doi.org/10.1007/978-3-031-93809-2_3

incomplete manuscripts survived into the Han (206 BCE–220 CE). Whatever the actual origin of the *Nine Chapters*, it served as a model for later generations of mathematical writing in terms of problem design, algorithm construction, language, and presentation.

Thanks to the archeological discoveries of the last several decades, we too now have access to the sort of raw, early manuscripts from which Zhang and Geng reportedly worked: *Suanshushu* 筭數書 (*Writings on Calculating Numbers*) from Zhangjiashan 張家山 Tomb 247 (*ca*. 186 BCE),[1] *Shu* 數 (*Numbers*) from the Yuelu Academy collection (TAQ 212 BCE),[2] *Suanshu* 筭書 (*Writings on Calculation*) A, B, and C from the Peking University collection (TPQ 214 BCE),[3] and *Suanshu* 筭術 (*Procedures for Calculating*) from Shuihudi 睡虎地 Tomb 77 (157–141 BCE).[4] The most salient characteristic of these early manuscripts is that they are not standardized, unified works like the *Nine Chapters* but, instead, miscellanies gathering unique sets of independent, free-floating sections (Cullen 2007). In this paper, we will compare one such section appearing in multiple manuscript versions—the *Litian shu* 里田術 land measurement procedure—suggesting a correction to the transcription of the Zhangjiashan manuscript, offering a typological analysis of the different versions, and considering the role that orality may have played in the forms that this procedure takes in early manuscript sources.

2 Variants on a Problem

Litian shu land measurement problems appear in many early sources for Chinese mathematics. In the words of Peng Hao (2001: 126), the editor of the Zhangjiashan manuscript, the *litian shu* is "a method for converting the area of a field whose length is given in units of *li* 里 into *qing* 頃 and *mu* 畝," where 1 li^2 = 375 *mu*, 100 *mu* = 1 *qing*, 1 *li* ≈ 415.8 meters, and 1 *qing* ≈ 4.6 hectares. As concerns its use, the *Suanshushu* explains that, "when one marks out an allotted fief, you do it by this means" 直(置)提封，以此爲之 (slip 188), referring to the calculation of the total number of enclosed fields.[5] The exact text is as follows:

> 里田术(術)曰:里乘里，里也，廣、從(縱)各一里，即直(置)一，因而三之，有(又)三五之，即爲田三頃七十五畝。其廣、從(縱)不等者，先以里相乘，已乃因而三之，有(又)三五之，乃成。今有廣二百廿里，從(縱)三百五十里，爲田廿八萬八千七百五十頃。直(置)提封以此爲之。【算數書187–188】
>
> The procedure for a field in *li* states: if a *li* multiplies a *li*, that is a *li*; if the breadth and length are each 1 *li*; set out 1, take and 3-fold it (1 × 3 ➔ 3), then 5-fold it thrice (3 × 5 × 5 × 5

[1] Zhangjiashan ersiqi hao Han mu zhujian zhengli xiaozu (2001).

[2] Zhu and Chen (2011).

[3] Han Wei (2013).

[4] Cai and Tan (2019).

[5] According to Yan Shigu's 顏師古 (581–645) commentary in Ban Gu (1962, 28B.1640), "*tifeng* 提封 (lit. 'raise a fief') is to 'lift up' the boundary [lines around] the fief" 提封者，大舉其封疆也 (cf. Peng Hao 2001, p. 126).

➔ 375), and that gives you a field of 3 *qing* and 75 *mu* (= 375 *mu*). If the breadth and length are not equal, first let the *li* multiply each other; when that is done, then take and 3-fold it, then 5-fold it thrice, then it is complete. Now there is a breadth of 220 *li* and a length 350 *li*; it makes a field of 288,750 *qing* (= 220 × 350 × 3.75). When one marks out an allotted fief, you do it by this means.

一曰:里而乘里=，(里)也，壹三而三五之，即頃畝數也。.
One [procedure]: a *li* if multiplies a *li*, that is a *li*; 3-fold it once, then 5-fold it thrice; and that gives you the number of *qing* and *mu*.

有(又)曰:里乘里=，(里)也，X(見)里之下，即予廿五，因而三之，亦其頃【算數書189】畝數也。
Further: if a *li* multiplies a *li*, that is a *li*; **after you [see] a *li*, then [separately] prepare 25, take and 3-fold them** (1 *li* × 3 ➔ 3; 25 × 3 ➔ 75), and that is also the number of *mu* and *qing*.

曰:廣一里、從(縱)一里，爲田三頃七十五畝。【算數書190】.
[Another procedure]: a breadth of 1 *li* and a length of 1 *li* make a field of 3 *qing* 75 *mu*.[6]

In the 2000s, where was some debate about the damaged character ("X"), illustrated in Fig. 1, and the curious phrase highlighted in bold. Following Peng Hao, the editors transcribed the former as *yi* 以 ("use," "then"), giving us 以里之下，即予廿五 (Dauben: "using li, next multiply by 25"; Cullen: "below the *lǐ* then prepare 25").[7] By contrast, Ōkawa Toshitaka's 大川俊隆 team read this as two characters, *yin yi* 因以 ("then use"), giving us 因以里之下，即予廿五, which they translate "add 25 below the 1 square *li*."[8]

Graphically, it is my opinion that this is one character and that it should be read *jian* 見 ("to see"). As illustrated in Fig. 1, there is in the upper half a circle- or box-like shape crossed by two horizontal strokes for which any transcription must account. There is indeed space here for two characters (*yin yi* 因以), but comparison with *yin* 因 and *qing* 頃 from the same slip reveals that the horizontal strokes bear a

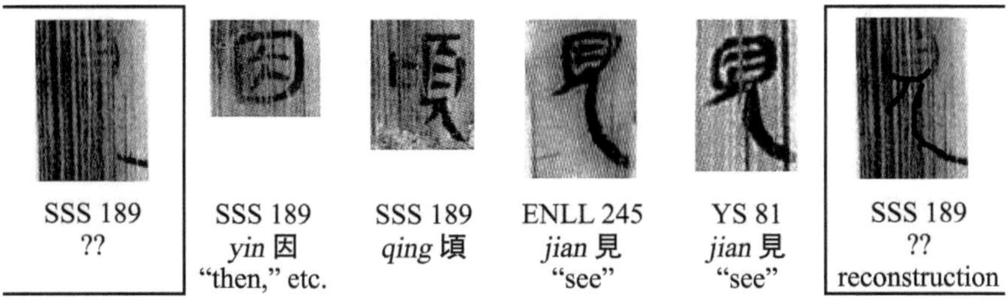

Fig. 1 Damaged character(s), *Suanshushu*, slip 189, as compared across the Zhangjiashan tomb 247 corpus. Manuscripts: SSS = *Suanshushu* 筭數書, ENLL = *Ernian lüling* 二年律令, YS = *Yinshu* 引書

[6] *Suanshushu*, slips 187–190; tr. modified from Cullen (2004, p. 110).

[7] Peng Hao (2001, p. 126), Zhangjiashan… (2001), p. 272, Dauben (2008, p. 167), Cullen (2004, p. 110).

[8] 1平方里の下に25を付け加えて, Chōkazan Kankan *Sansūsho* kenkyūkai (2006, p. 13).

closer resemblance to the component *ye* 頁 of *qing* 頃 than the slanted, disconnected strokes of a *da* 大. What its visible features most resemble within the larger Zhangjiashan tomb 247 corpus, however, is the elongated form of *jian* as it appears in the manuscripts *Ernian lüling* 二年律令 and *Yinshu* 引書, the exaggerated downstroke of which effectively occupies the space of a full second character.[9]

Indeed, this identification seems all the more likely now thanks to the parallel section in the contemporary manuscript *Suanshu* 算書 A, donated to Peking University in 2010:

> 里田述(術)曰:里乘里，一毄(也)，見一鼠(予)二，見二鼠(予)四₌，(四)者加一，因而三之，即頃畝數毄(也)。【算書04-081】
> The procedure for a field in *li* states: if a *li* multiplies a *li*, that is 1. If you see 1 rat (prepare) 2; if you see 2 rats (prepare) 4; and with 4, add 1. Take and 3-fold it, and that gives you the number of *qing* and *mu* (*Suanshu* A, s. 04-081).

> 其一述(術)曰:里乘里₌，(里)毄(也)，壹參(三)之，有(又)參(三)五之，即頃畝數毄(也)。【算書04-096】
> Another procedure states: if a *li* multiplies a *li*, that is a *li*; 3-fold it once (1 × 3 ➔ 3), then 5-fold it thrice (3 × 5 × 5 × 5 ➔ 375), and that gives you the number of *qing* and *mu* (*Suanshu* A, s. 04-096).

> 今有方一里，問田幾可(何)?曰:三頃七十五畝。方十里，田三百七十五頃。【算書04-095】
> Now there is a square 1 *li* [to a side]. Question: how much of a field does that make? 3 *qing* 75 *mu*. And for a square 10 *li* [to a side], a field of 375 *qing* (*Suanshu* A, s. 04-095).[10]

The second, alternative procedure here is identical to the first two from the *Suanshushu*, and the first uses *jian* ("to see") in a similar formulation that is somewhat more difficult to understand. Citing paleographical precedents and an explanation by Guo Shuchun, the manuscript's editor, Han Wei, suggests that *shu* 鼠 (*nhaʔ, "rat") is a phonetic loan for *yu* 予 (*laʔ, "to give" or "prepare"), and that you build to a five by multiplication and addition (1 × 2 ➔ 2; 2 × 2 ➔ 4; 4 + 1 ➔ 5), then "take, [multiply by five], and threefold it" 因而三之 to get 375.[11] Alternatively, and without adding anything in brackets, Ōkawa et al. (2015, 133) suggest that you build by place values to 125 (1 then 2 then 5), then "take and 3-fold it" for 125 × 3 ➔ 375.

Beyond the *Suanshushu* and *Suanshu* A, one finds the *litian shu* presented in a mix of similar terms in Peking University's *Suanshu* B:

> 里乘里₌，(里)也，壹三有(又)三五之，即成田畝數，爲田三頃七十五畝。積四里爲田十五頃。【算書03-047】
> If a *li* multiplies a *li*, that is a *li*; 3-fold it once (1 × 3 ➔ 3), then 5-fold it thrice (3 × 5 × 5 × 5 ➔ 375), and that gives you the number of *mu* in the complete field, which makes a field of 3 *qing* 75 *mu*. An accumulated (square) 4 *li* makes a field of 15 *qing* (s. 03-047).[12]

---

[9] Tan Jingnan (2016, pp. 66–67).

[10] Published in Han Wei (2013, p. 37).

[11] Han Wei (2013, p. 37). The phonetic reconstructions provided here are Minimal Old Chinese from Schuessler (2007).

[12] Han Wei (2013, p. 37).

And in the Yuelu Academy manuscript *Shu* 數:

> 里田述(術)曰:里乘里=，(里)也，因而參(三)之，有(又)【參】(三)五之，爲田三頃七十五畝。【數62(0947)】
> The procedure for a field in *li* states: if a *li* multiplies a *li*, that is a *li*; take and 3-fold it (1 × 3 ➔ 3), then 5-fold it thrice (3 × 5 × 5 × 5 ➔ 375), and that gives you a field of 3 *qing* and 75 *mu* (s. 62[0947]).

The same procedure, under the same title, appears in the received *Jiuzhang suanshu* 九章算術 (*Nine Chapters on Mathematical Procedures*), from the first century CE. Unlike the manuscripts, the *Jiuzhang suanshu* gives one general procedure applicable to all circumstances. What is more, rather than building up to it with threes and fives, and producing a result in *qing* and *mu*, the *Jiuzhang suanshu* deals with unit conversion via direct multiplication by 375 to obtain the number of *mu* (and *mu* alone):

> 里田術曰:廣從(縱)里數相乘得積里，以三百七十五乘之，即畝數。.
> The procedure for a field in *li* states: the number of *li* in the breadth and length multiply each other, and one gets the "accumulated *li*." Multiply it by 375; and that gives you the number of *mu*.[13]

Between the *Suanshushu*, *Suanshu* A, *Suanshu* B, *Shu*, and *Jiuzhang suanshu*, we have ten different statements of the same procedure. As illustrated in Table 1, six of these are clearly variations on the same Ur text, which we will call **Version A**:

> 里乘里，里也，【廣、縱各一里，即置一，】因而三之，又三五之，即頃畝數，爲田三頃七十五畝。.
> If a *li* multiplies a *li*, that is a *li*; [if the breadth and length are each 1 *li*; set out 1,] take and 3-fold it (1 × 3 ➔ 3), then 5-fold it thrice (3 × 5 × 5 × 5 ➔ 375), and that gives you the number of *qing* and *mu*, which makes a field of 3 *qing* and 75 *mu* (*Suanshushu*, *Suanshu*, and *Shu*, see Table 1).

Within the variations on Version A, furthermore, we can distinguish two coherent groups: the one which uses the expression "take and 3-fold it, then 5-fold it thrice" 因而三之，又三五之 and skips straight to answer, while the other uses "3-fold it once, then 5-fold it thrice" 壹三而三五之 and specifies that "that gives you the number of *qing* and *mu*" 即頃畝數也 (or "the number of *mu* in the complete field" 即成田畝數).

Procedurally, this is quite distinct from the approach that would arrive at the number 375 by first multiplying and adding ones and twos to get by place value to 125, to be multiplied by 3, which we will call **Version B**:

> 里乘里，一也，見一予二，見二予四，四者加一，因而三之，即頃畝數也。
> If a *li* multiplies a *li*, that is 1. If you see 1 [in the hundreds column], prepare 2 [in the tens column]; if you see 2 prepare 4; and with 4, add 1 [to make the units]. Take and 3-fold it (125 × 3 ➔ 375), and that gives you the number of *qing* and *mu* (*Suanshu* A).

[13] Guo Shuchun (2004), 1.10; tr. modified from Shen et al. (1999, p. 63); cf. Chemla and Guo (2004, pp. 154–155).

Table 1 *Litian shu* Version A, a text-critical comparison of six parallel versions of the *Litian shu* requiring multiplication by threes and fives to arrive at the conversion rate of 375 *mu* : 1 *li*². Abbreviations: *SSS* (*Suanshushu* 筭數書, Zhangjiashan), *SS* (*Suanshu* 算書, Peking University) manuscripts A & B, *S* (*Shu* 數, Yuelu Academy)

(1)	*SSS* 187–188	里 乘里=也	廣 從各一里即直一
	SSS 190		廣一里從 一里
	S 62(0947)	里 乘里=也	
	SSS 189	里而乘里=也	
	SS A 04-096	里 乘里=殹	
	SS B 03-047	里 乘里=也	
(2)	*SSS* 187–188	因而三之有三五之	
	SSS 190		
	S 62(0947)	因而參之有□五之	
	SSS 189	壹三 而三五之	
	SS A 04-096	壹參之有參五之	
	SS B 03-047	壹三 有三五之	
(3)	*SSS* 187–188	即	爲田三頃七十五畝
	SSS 190		爲田三頃七十五畝
	S 62(0947)		爲田三頃七十五畝
	SSS 189	即 頃畝數也	
	SS A 04-096	即 頃畝數殹	
	SS B 03-047	即成田 畝數	爲田三頃七十五畝

Or, in what appears to be a variation on this given the peculiar use of *jian* ("see") and *yu* ("prepare"), recalling to place 1 in the *qing* (i.e., hundreds) column and 25 in the *mu* column, to be multiplied by 3:

> 里乘里，里也，見里之下，即予廿五，因而三之，亦其頃畝數也。
> If a *li* multiplies a *li*, that is a *li*; after you see a *li*, then [separately] prepare 25, take and 3-fold them (1 *li* × 3 ➔ 3; 25 × 3 ➔ 75), and that is also the number of *mu* and *qing* (*Suanshushu*).

Though procedurally identical to Version A, we can also distinguish the one instance where the manuscripts instruct the user what to do with a rectangular field of any given dimensions, which we will call **Version C**:

> 其廣、縱不等者，先以里相乘，已乃因而三之，又三五之，乃成。
> If the breadth and length are not equal, first let the *li* multiply each other; when that is done, then take and 3-fold it, then 5-fold it thrice, then it is complete (*Suanshushu*).

Moving from the manuscripts to the early received tradition, lastly, we have a general procedure, in very different terms, which has us skip straight to a multiplication by 375 to get the number of *mu* (and *mu* alone), which we will call **Version D**:

> 廣縱里數相乘得積里，以三百七十五乘之，即畝數。
> The number of *li* in the breadth and length multiply each other, and one gets the "accumulated *li*." Multiply it by 375; and that gives you the number of *mu* (*Jiuzhang suanshu*).

3 Manuscripts and Mnemonics

Functionally, all of these versions and variants of the *litian shu* produce the same results with the same core operations and values (length × width × 375). This leads one to ask why all these different versions exist and why, in a manuscript like the *Suanshushu*, they are given redundantly one after the other.

First of all, Cullen (2007) reminds us that the early manuscripts are not coherent "books" such as can be directly compared to later transmitted works like the *Jiuzhang suanshu*. On the one hand, he invokes the parallel of Keegan's (1988) study of the *Huangdi neijing* 黃帝內經 in the light of early medial manuscripts, which led the later to the conclusion that…

> the elementary unit of that literature was not to be seen as the "text" in its usual meaning of an extensive piece of writing equivalent to what we would call a book, but rather what we may call a "textlet," a shorter piece of writing capable of being transmitted on its own. Different extended texts might contain overlapping but not identical collections of textlets, and one text might separate textlets contiguous in another text, while bringing together textlets separated in other collections. … By the Eastern Hàn, however, … this process had led to the formation of more than one large ossified collection of material that was no longer subject to "textlet" transmission in the old way. (Cullen 2007, p. 36)

On the other hand, trying to explain the signatures Wang 王, Yang 楊, "Checked by Wang" 王已讎, and 'Checked by Yang' 楊已讎 in the bottom margin of the Zhangjiashan manuscript, he suggests that:

> On the whole the most probable solution seems to be that Wáng and Yáng were in some sense mathematical teachers and (to a limited extent at least) authors of mathematical problems, and that material by them has been copied into the present collection. (Cullen 2007, p. 19)

In sum, multiple variant "textlets" concerning the same problem are to be expected in early mathematical manuscripts, and the *Suanshushu*, for one, may well be schoolwork rather than a treatise. Keegan's observation about "textlets," the miscellaneous nature of early manuscripts, and the codification of works under the Han is generally accepted across early Chinese studies.[14] Additionally, Morgan and Chemla (2018) have argued that an alternation of hands across the manuscript confirms Cullen's educational hypothesis and reveals one hand filling in blanks expressly left by the other, presumably from memory.[15]

As to the phrase "after you [see] a *li*, then [separately] prepare 25, take and 3-fold them" X(見)里之下，即予廿五，因而三之 (*Suanshushu*, s. 189), Cullen notes that this is a "slightly strange expression" before going on to explain how the factors

[14] See for example Shaughnessy (2006, esp. pp. 9–130), as concerns the Confucian classics, and Kalinowski (2008) and Liu Lexian (2017), as concerns daybooks (*rishu* 日書).

[15] Likewise, Yang Bo (2017, p. 63) concludes of the new Peking University Qin manuscript *Tianshu* 田書 that we are looking there at "a sort of specialized mathematical educational or reference material for the study of field surveying and taxation" 供人學習田畝、租稅計算的一種特殊算術教材或參考書.

multiply to 375, as in every other version.[16] Peng Hao 彭浩, Joseph Dauben, Karine Chemla, and Ma Biao 馬彪 offer equality little to say about the language of Version B of the *litian shu* in the Zhangjiashan M247 and Shuihudi M77 manuscripts.[17] However, Ōkawa et al. offer the following explanation in passing:

> In fact, this series of phrases tell the story of what was a mnemonic. … No one remembered the "Nine-Nines" (multiplication table) as logic, they remembered it naturally through practice and recitation, and neither is this mnemonic something to be remembered for its logic, but something to be remembered through practice and recitation.[18]

This reminds us of a second feature of early Chinese text culture, which is widely acknowledged but more or less invisible in the written sources to have survived. In the words of the famous Qing scholar Ruan Yuan 阮元 (1764–1849):

> The ancients did not have the convenience of [the modern] brush, inkstone, paper, and ink. They often cast [their writing in] metal and carved [it in] stone, which would see transmission over far and wide. Writing on bamboo slips, for its part, involved the labor of lacquer writing and scraping knives. It was not as easy as it is for the modern person to pen out a thousand words… The bamboo slips of the ancients were burdensome; much was transmitted from mouth to ear, and little was transmitted by the eye.[19]

In the words of Wei Wei (2017, pp. 133, 136):

> In the early stage of the emergence of literary style, style and text are the products of oral behavior. In early oral narration, before it was fixed by writing, content was not easy to preserve. Later, it was mostly passed down through poetry and philosophical and historical writings, not recorded independently, but we can still find traces of oral narration… The narrator not only plays the role of oral transmission, but also "the creator of secondary texts."[20]

In the last several decades, the discussion of orality in early Chinese manuscript studies has been principally focused on the *Book of Odes* (*Shijing* 詩經), the rhythm and rhymes of poetry and song seeming like an obvious place to begin.[21] However, this is not the only form of knowledge that one learns by recitation, even today, and

[16] Cullen (2004, p. 110). Cullen translates this line "below the *lǐ* then prepare 25, take and 3-fold them."

[17] Peng Hao (2001, pp. 125–128), Dauben (2008, pp. 166–167), Chemla and Ma (2011, pp. 183–185).

[18] よって、これらの事實は、これら一連の句が一種の口訣であったことを物語つている……「九々」の口訣が論理として憶える者でなく、練習や口ずさみの中で自然と憶えていくように、この口訣も論理的に憶えるものではなく、練習や口ずさみの中で記憶されていったものなのである (Ōkawa et al. 2015, p. 131).

[19] 古人無筆硯紙墨之便，往往鑄金刻石，始傳久遠。其著之簡策者，亦有漆書刀削之勞。非如今人下筆千言，言事甚易也……古人簡策繁重，以口耳相傳者多，以目相傳者少 (Ruan Yuan 1993, p. 605).

[20] 在文體的產生初期，文體與文本都是口頭行為的產物。早期的口頭敘事，在沒有被文字定型之前，內容不易保存，後來多是通過子書、史書、詩歌等作品中流傳下來，不獨立記載，但我們依然可以找到口頭敘事的痕跡。……講述者除了起到口頭傳播的作用，還是"次生文本的創造者"。

[21] Famously, see Kern (2002, 2005) and the recent critical study of Shaughnessy (2016).

Morgan (2015) reminds us that debate, face-to-face learning, and suspicion of the written word were important facets of ancient Chinese astronomy.

In his introduction, Morgan thanks scholarship on Sanskrit sources for helping us to see this—helping us see that the articulation of written and oral knowledge is normal, at least in one ancient mathematical culture, and what it looked like in practice. In sum, early Indian sources for astronomy and mathematics come down to us in condensed, rhyming verse that is difficult to understand. The student was expected to learn by first reciting and memorizing these verses, achieving theoretical comprehension only later through the oral instruction of a master, the most famous of whose prose explanations, figures, and diagrams we know about as later committed to written commentary. In addition to brevity and rhyme, one of the most striking features of a classic like Āryabhaṭa's (b. 476) *Āryabhaṭīya* is the transformation a table of astronomical constants into ten verses by breaking long numbers apart and assigning different numbers and powers of ten to different consonants and vowels, to appear in any order: where $kh = 2$, $y = 30$, $u = 10^4$, $gh = 4$, and $ṛ = 10^6$, for example, the number 4,320,000 is reduced to $(2 + 30) \times 10^4 + 4 \times 10^6$ in the two-syllable *khyughṛ*.[22] This may come down to us in writing, but we may use the comparison of different text types in India as a source of inspiration in thinking about the articulation of written and oral transmission in ancient China.

For Ōkawa et al., the most obvious example of something to be "remembered... naturally through practice and recitation" in ancient Chinese mathematics is the "Nine Nines" multiplication table. Still memorized by schoolchildren around the world today, the multiplication table is a logical prerequisite to the written works surveyed above, and it is perhaps not surprising, therefore, that it is also the most common mathematical text in the Chinese archeological record, found in tombs, administrative sites and grottos from one side of the empire to the other.[23] Hsing I-t'ien 邢義田 has argued from the presence of these tables in Han administrative sites in the far Northwest that, for the low-level officials posted there, "their literacy, numeracy, and legal-administrative knowledge were not prerequisites of these posts but, instead, things they gradually learned during their service."[24] Likewise, in arguing that it is important to distinguish the different *levels of numeracy* at which our historical subjects were operating, Yi Fusheng 衣撫生 places these first in his 3-fold scheme: "the average educated person, who would have studied the Nine Nines multiplication method and the four basic operations; mathematicians of a certain level, who researched the application of mathematics to daily life; and high-level mathematicians, who concerned themselves with astronomical calculations."[25]

---

[22] Yano (2006). On oral and written transmission in the exact sciences in other ancient civilizations, see also Proust (2015) for Mesopotamia and Keller (2015) for India in the same special volume of *Archives internationals d'histoire des sciences* as Morgan (2015).

[23] Liu Jinhua (2003), Tan Jingnan (2016, pp. 9–10).

[24] 能書、知計算和知律令的能力並不是擔任這些職務以前就必然具備，而是在擔任職務的過程裡逐漸學會的 (Hsing 2011, pp. 585).

[25] 普通的受教育者，學習九九乘法表和基本的四則運算；具有一定水準的數學家，研究數學在日常中的應用；水準最高的數學家，關注天文曆算 (Yi Fusheng 2018, p. 109).

It is worth emphasizing that the language of the Chinese “Nine Nines” table is rhythmic and repetitive, the ancient version comprising mostly four- or five-character phrases as is typical of poetry and mnemonics. Consider the following line from the wooden board J1(16)1 found at the third- to second-century BCE administrative dumpsite at Liye 里耶, and note that the stilted translation of the numbers is to best reflect their Chinese rhythm in English:

> 五=廿五，四五廿，三五十五，二五而十。
> Five “, twenty-five. Four fives, twenty. Three fives, ten-five. Two fives, then ten.

From the medieval Dunhuang manuscript S.4569:

> 五々二十五，四五二十，三五一十五，二五一十，一五如五。
> Five “, two-tens-five. Four fives, two tens. Three fives, one-ten-five. Two fives, one ten. One five, like five.

And finally, from the version taught in modern schools:

> 一五得五，二五一十，三五十五，四五二十，五五二十五。
> One five gets five. Two fives, one ten. Three fives, ten-five. Four fives, two-tens. Five fives, two-tens-five.

“Mostly” is the key word, because there are some exceptions to the pattern of parallel four- or five-character phrases, and because we cannot know how the repetition marks (=, 々) or abbreviated numbers like *nian* 廿 (“twenty,” equivalent to *er shi* 二十) were actually read. With that said, it is important to note that these tables do insert filler words to maintain parallelism and rhythm. When dealing with a one-syllable product, be it at the beginning or end, it is inevitably introduced by an *er* 而 “then,” *ru* 如 “like,” or *de* 得 “gets.” The exception to this is *shi* 十 “ten,” which can also take an unnecessary unit *yi* 一 “one” (“Two fives, one ten. Three fives, ten-five” 二五一十，三五十五).

Beyond the “Nine Nines” table, one finds similar rules of thumb formulated in four-character phrases elsewhere in the received mathematical tradition. Here, for example, is how the *Sunzi suanjing* 孫子算經 explains calculating rod notation:

> 凡算之法，先識其位。 In rod-calculation, the first thing to understand is their positions:
>
> 一縱十横， Ones vertical, tens horizontal (*gwrâŋ),
> 百立千僵。 Hundreds standing, thousands flat (*kaŋ).
> 千十相望， The thousands looking over at one another (*maŋ),
> 萬百相當。 And the myriads and hundreds corresponding (*tâŋ).[26]

In other words, one alternates between horizontal and vertical lines by decimal position on the calculating surface, with 𝍢 𝍬 𝍪 𝍢 𝍩 representing thirty-four myriads two hundred and thirty-one (340,231). The *Sunzi suanjing* chooses to express this in couplets, the exact choice and number of words determined in part by poetic structure. This is not high poetry, the rhymes are imperfect, and it stands out from

[26] *Suanjing shishu*, p. 282. The phonological reconstructions are Minimal Old Chinese, following Schuessler (2007).

the lists and technical prose before and after. Presumably, the switch to parallel four-character couplets is precisely because this is "the first thing to understand" and, thus, memorize about the subject.

That the "Nine Nines" table represented the sort of basic knowledge accessible to low-level officials is suggested by its presence at administrative sites, and in the *Shuoyuan* 說苑 (*Garden of Eloquence*) Liu Xiang 劉向 (77–6 BCE) presents someone of even lowlier status as having mastered it:

> 齊桓公設庭燎，爲使人欲造見者，期年而士不至。於是東野有以九九之術見者，
> Lord Huan of Qi set up a torch in his reception hall in order to receive men-of-service who desired to have audiences with him, but a full year passed without anyone coming. Then a rustic from the eastern wilds came, saying that he knew the art of multiplication.
>
> 桓公曰:九九何足以見乎?
> "Can skill in multiplication be enough to qualify a man for an audience?" asked Lord Huan.
>
> 鄙人對曰:臣非以九九爲足以見也，臣聞主君設庭燎以待士，期年而士不至。夫士之所以不至者，君天下之賢君也，四方之士，皆自以論而不及君，故不至也。夫九九薄能耳，而君猶禮之，況賢於九九者乎?......
> "It's not that I think that knowing multiplication qualifies me for an audience," said the man. "I came because I heard that Your Lordship had a torch set up in the reception chamber to receive men-of-service and that no one had come for a full year. The reason they have not come is that they believe that Your Lordship is the worthiest ruler in the realm. No one in the realm feels that he can compare with Your Lordship, so they have not come. Multiplication is indeed a paltry skill; but, if Your Lordship should treat even so meager an offering with courtesy, then how much greater would be the courtesy extended to those worthier than a mere arithmetician? …
>
> 桓公曰:善。乃固禮之。期月，四方之士相導而並至矣。
> "Well observed!" exclaimed Lord Huan. He then received the arithmetician with full ceremony. Within a month, worthy men from all directions came leading each other by the hand to offer their services.[27]

Manuscripts such as the *Suanshushu* are more sophisticated, and obviously imply their owners in some degree of literacy, but they appear to have preserved something of the oral language of basic mathematical education.

4 Conclusion

In summary, the *Litian shu* appears across excavated and received sources for *suan* mathematics, and in the early manuscripts it appears in three versions (A, B, C), often side by side, and as distinct from the form that it would take in the *Nine Chapters* (version D). The version one finds in the *Nine Chapters* is written in simple language and general terms, providing a single procedure to calculate an area from any "length" and "width" and converting from *li* to *mu* by multiplying by 375. The early manuscript versions, by contrast, treat the problem in more concrete

[27] Henry (2021, pp. 449–451).

terms, focusing in versions A and B on a 1 *li* by 1 *li* square, and separately, in Version C, on a rectangular plot of any given length and width. Where the manuscript versions really stand out, however, is in their prolix explanation of how to build the *a priori* conversion rate of 375 by the multiplication and addition of the factors 1, 2, 3, 4, and 5. Viewed from the *Nine Chapters*, all of this seems unnecessarily complicated; and where multiple versions appear together in a single manuscript, it seems redundant.

As to why, Ōkawa et al. offer in passing that the more strangely worded version is probably a mnemonic, which comports with Cullen (2007), Yang Bo (2017), and Morgan and Chemla's (2018) arguments that the early manuscripts may well represent raw materials used in learning: text books, notes, or, in the case of the Zhangjiashan *Suanshushu*, a written exchange. There is not much *written* about the oral component of learning and transmission in early Chinese mathematics, but we can turn to other fields and other civilizations for an idea of what it may have looked like and, in this case, we see some parallels. On the one hand, we see a decomposition of large numbers into decimal places and factors similar to how the *Āryabhaṭīya* versifies astronomical contents. On the other hand, and closer to home, we see that that the apparent four-character couplet scheme seemingly at play in the *Litian shu* resembles that of the "Nine Nines" multiplication table and a rhyming rule of thumb from the *Sunzi suanjing*.

If the variability and redundancy of the early mathematical manuscripts in their formulation of the *litian shu* stands out as compared to the later received classics, it is partially explained by their particularity—as personal copies, notebooks or even products of educational exchange—and by the "textlet"-based transmission model common to most genres of writing in the third and second centuries BCE. According to Liu Hui's preface, the current *Nine Chapters* is a product of Zhang Cang and Geng Shouchang, who "respectively trimmed and restored [the presumed Ur text] on the basis of surviving fragments of old text."[28] Early manuscripts like the *Suanshushu* give us an idea of the sort of sources upon which the *Nine Chapters* was likely working, and they give us an idea of what was "trimmed." At one level, we see the different versions of the *litian shu* distilled in the *Nine Chapters* into a single generalized procedure, the same for a rectangle as for a square. What we also see, however, is a shift in idiom reflecting a shift in genre or educational paradigm: not so much a *rationalization* of variant procedures and strange expressions, but a shift from the rational of oral learning to that of discursive writing.

Bibliography

Ban Gu 班固. 1962. *Hanshu* 漢書. 8 vols. Beijing: Zhonghua shuju.

Cai Dan 蔡丹 and Tan Jingnan 譚競男. 2019. "Shuihudi Han jian zhong de *Suanshu* jiance"睡虎地漢簡中的《算術》簡冊. *Wenwu* 文物. 2019(12): 63–70.

[28] 因舊文之遺殘，各稱删補 (Guo Shuchun 2004, p. 1).

Chemla, Karine and Guo Shuchun 郭書春. 2004. *Les neuf chapitres: le classique mathématique de la Chine ancienne et ses commentaires*. Paris: Dunod.

Chemla, Karine and Ma Biao 馬彪. 2011. "Interpreting a Newly Discovered Mathematical Document Written at the Beginning of the Han Dynasty in China (before 157 B.C.E.) and Excavated From Tomb M77 at Shuihudi (睡虎地)." *SCIAMVS* 12: 159–191.

Chōkazan Kankan *Sansūsho* kenkyūkai 張家山漢簡「算數書」研究會, ed. 2006. *Kankan* Sansūsho*: Chūgoku saiko no sūgakusho* 漢簡「算數書」:中國最古の數學書. Kyōto: Hōyū shoten.

Cullen, Christopher. 2004. *The* Suàn Shù Shū 筭數書 *"Writings on Reckoning": A Translation of a Chinese Mathematical Collection of the Second Century BC, with Explanatory Commentary*. Cambridge: Needham Research Institute.

Cullen, Christopher. 2007. "The *Suàn Shù Shū*, 'Writings on Reckoning': Rewriting the History of Early Chinese Mathematics in the Light of an Excavated Manuscript." *Historia Mathematica* 34 (1): 10–44. https://doi.org/10.1016/j.hm.2005.11.006.

Dauben, Joseph W. 2008. "算書書 *Suan Shu Shu*, a Book on Numbers and Computations: English Translation with Commentary." *Archive for History of Exact Sciences* 62: 91–178.

Guo Shuchun 郭書春, ed. 2004. *Huijiao Jiuzhang suanshu* 匯校九章算術. 2 vols. Shenyang: Liaoning jiaoyu chubanshe.

Han Wei 韓巍. 2013. "Beida Qin jian *Suan shu* tudi mianji lei suanti chushi" 北大秦簡『算書』土地面積類算題初識. *Jianbo* 簡帛 8: 29–42.

Henry, Eric P. 2021. *Garden of Eloquence, Shuoyuan*. Seattle: University of Washington Press. https://www.worldcat.org/title/1294317160.

Hsing I-t'ien 邢義田. 2011. *Zhiguo anbang: fazhi, xingzheng yu junshi* 治國安邦:法制、行政與軍事. Beijing: Zhonghua shuju.

Kalinowski, Marc. 2008. "Les livres des jours (*rishu*) des Qin et des Han: la logique éditoriale du recueil A de Shuihudi (217 avant notre ère)." *T'oung Pao* 94 (1): 1–48.

Kangshen Shen, John N Crossley, and Anthony W.-C. Lun. 1999. *The Nine Chapters on the Mathematical Art: Companion and Commentary*. Oxford; Beijing: Oxford University Press ; Science Press.

Keegan, David. 1988. "*The Huang-Ti Nei-Ching*: The Structure of the Compilation; the Significance of the Structure." Ph.D. diss., University of California, Berkeley.

Keller, Agathe. 2015. "The Mathematical Chapter of the Āryabhaṭīya : A Compilation of Oral Mnemonic Rules ?" *Archives Internationales d'Histoire Des Sciences* 65 (175). Brepols Publishers: 573–598. https://doi.org/10.1484/J.ARIHS.5.112778.

Kern, Martin. 2002. "Methodological Reflections on the Analysis of Textual Variants and the Modes of Manuscript Production in Early China." *Journal of East Asian Archaeology* 4 (1–4): 143–181.

Kern, Martin. 2005. "The *Odes* in Excavated Manuscripts." In *Text and Ritual in Early China*, edited by Martin Kern, 149–193. Seattle: University of Washington Press.

Liu Jinhua 劉金華. 2003. "Qin Han jiandu 'Jiujiu canbiao' shulun" 秦漢簡牘「九九殘表」述論. *Wenbo* 文博 2003.3: 25–28.

Liu Lexian 劉樂賢. 2017. "Daybooks: A Type of Popular Hemerological Manual of the Warring States, Qin, and Han." In *Books of Fate and Popular Culture in Early China : The Daybook Manuscripts of the Warring States, Qin, and Han*, 57–90. Leiden: Brill.

Morgan, Daniel Patrick. 2015. "What Good's a Text? Textuality, Orality, and Mathematical Astronomy in Early Imperial China." *Archives internationales d'histoire des sciences* 65 (2): 549–572.

Morgan, Daniel Patrick, and Karine Chemla. 2018. "Writing in Turns: An Analysis of Scribal Hands in the Bamboo Manuscript *Suan Shu Shu* 筭數書 (*Writings on Mathematical Procedures*) from Zhangjiashan Tomb No. 247." *Bamboo and Silk* 1: 152–190.

Ōkawa Toshitaka 大川俊隆, Tamura Makoto 田村誠, and Harikae Toshio 張替俊夫. 2015. "Pekin daigaku *Sansho* no ridenjutsu to keidenjutsu ni tsuite" 北京大學『算書』の里田術と徑田術について. *Ōsaka Sangyō daigaku ronshū: jinbun, shakai kagaku hen* 大阪産業大學論集·人文社會科學編 23: 131–144.

Peng Hao 彭浩. 2001. *Zhangjiashan Han jian* Suan shu shu *zhushi* 張家山漢簡『筭數書』註釋. Beijing: Kexue chubanshe.

Proust, Christine. 2015. "La Chanson Des Mathématiques Dans l'Edubba (Mésopotamie, Début Du Deuxième Millénaire Avant Notre Ère)." *Archives Internationales d'Histoire Des Sciences* 65 (175). Brepols Publishers: 517–547. https://doi.org/10.1484/J.ARIHS.5.112776.

Ruan Yuan 阮元. 1993. *Yanjing shiji* 揅經室集. Beijing: Zhonghua shuju.

Schuessler, Axel. 2007. *ABC Etymological Dictionary of Old Chinese*. Honolulu: University of Hawai'i Press.

Shaughnessy, Edward L. 2006. *Rewriting Early Chinese Texts*. Albany: State University of New York Press.

Shaughnessy, Edward L. 2016. "The Origin and Development of Western Sinologists' Theories of the Oral-Formulaic Nature of the *Classic of Poetry*." *Bulletin of the Jao Tsung-I Academy of Sinology* 3: 133–149.

Tan Jingnan 譚競男. 2016. "Chutu Qin Han suanshu wenxian ruogan wenti yanjiu" 出土秦漢算術文獻若干問題研究. Ph.D. diss., University of Wuhan.

Wei Wei 魏瑋. 2017. "Xianqin koukou xushi de gainian" 先秦口頭敘事的概念. *Zhonghua wenhua luntan* 中華文化論壇, no. 2012.12: 132–140.

Yang Bo 楊博. 2017. "Beida Qin jian *Tianshu* chushi" 北大秦簡『田書』初識. *Beijing daxue xuebao (zhexue shehui kexue ban)* 北京大學學報(哲學社會科學版), no. 2017.5: 63–68.

Yano Michio 矢野道雄. 2006. "Oral and Written Transmission of Exact Sciences in Sanskrit." *Journal of Indian Philosophy* 34 (1–2): 143–160.

Yi Fusheng 衣撫生. 2018. "Chutu Qin Han shuxue wenxian ruogan wenti yanjiu" 出土秦漢數學文獻若干問題研究. Ph.D. diss., Chinese Academy of Social Sciences.

Zhangjiashan ersiqi hao Han mu zhujian zhengli xiaozu 張家山二四七號漢墓整理小組. 2001. *Zhangjiashan Han mu zhujian (ersiqi hao mu)* 張家山漢墓竹簡(二四七號墓). Beijing: Wenwu chubanshe.

Zhu Hanmin 朱漢民 and Chen Songchang 陳松長, eds. 2011. *Yuelu shuyuan cang Qin jian (er)* 嶽麓書院藏秦簡(貳). Shanghai: Shanghai cishu chubanshe.

A Radical Proposition on the Origins of the Received Mathematical Classic the *Gnomon of Zhou* (*Zhoubi* 周髀)

Daniel Patrick Morgan

Abstract In 656, the *Gnomon of Zhou* (*Zhoubi* 周髀) was canonized as one of the *Ten Mathematical Classics* (*Suanjing shishu* 算經十書) that would serve as the basis of contemporary curricula, testing and, later, the study of the history of mathematics in ancient China. Itself purporting to have been transmitted from the beginning of human civilization to the Duke of Zhou 周公 (r. 1042–1036 BCE), most modern scholars follow Qian Baocong (1968) in positing that the text was redacted from earlier materials *circa* 100 BCE. In this paper, I will present linguistic evidence that supports Christopher Cullen (1996) and early Qian Baocong's (1924, 1932) position that the received *Gnomon of Zhou* came together much later, in the first or second century CE, at around the same time or later than the *Nine Chapters on Mathematical Procedures* (*Jiuzhang suanshu* 九章算術). In addition, I will present evidence connecting its early reception and transmission history to the Far North/Northwest that might explain features of this strange text and its strange position in discussions of cosmology in the Middle Period Chinese capitals further south.

Keywords Chinese mathematics · *Zhoubi* 周髀 (*Gnomon of Zhou*) · Cosmology · Text dating

D. P. Morgan (✉)
Centre national de la recherche scientifique, Paris, France
e-mail: daniel.morgan@college-de-france.fr

D. P. Morgan, T. Quan (eds.), *Mathematics and Astronomy in the Ancient World*,
https://doi.org/10.1007/978-3-031-93809-2_4

1 Introduction

The *Zhoubi* 周髀 (*Gnomon of Zhou*) is one of the earliest mathematical works to have been passed down through the Chinese written tradition. For those unfamiliar with the text, scholars typically follow the earliest commentary in dividing it into three discrete parts. Part 1 is a dialogue between the semi-legendary eleventh-century BCE figures of Shang Gao 商高 and Duke Dan of Zhou 周公旦 (r. 1042–1036 BCE) on the origin of numbers and the trysquare (*ju* 矩). Part 2 is a dialogue between the presumably later figures of Chenzi 陳子 and Rong Fang 榮方 on how to calculate the dimensions of the cosmos using gnomon shadows, the base and height (*gougu* 勾股) of right triangles, and the rule that one *cun* 寸 (≈ 2.31 cm) at the observer corresponds to a difference of 1000 *li* 里 (≈ 415.8 km) at the cosmic scale. Lastly, part 3—the body—goes on to describe the "umbrella heaven" (*gaitian* 蓋天) world model for which the text is most famous as well as the daily travel of the mean moon, a shadow table, and the lunisolar resonance periods of a "quarter-remainder" astronomical system (*sifen li* 四分曆).[1]

As to what this "umbrella heaven" looks like, the *Zhoubi* is somewhat vague, the fact of which, combined with other, conflicting versions of umbrella heaven in early sources, has led to considerable disagreement. In a nutshell, what the *Zhoubi* has to say is this: heaven and earth are like inverted pans of the same shape and size;[2] heaven is 80,000 *li* above the earth, turning around a pole centered over a 11,500-*li* radius zone called the *xuanji* 璇璣;[3] and as to account for seasonal differences in altitude, and the hour and direction of its apparent rising and setting, the sun, being affixed to heaven, shifts seasonally through seven circular tracks (the *qi heng* 七衡) of different diameters concentric with the celestial pole. As to what this configuration is meant to explain, the text focuses on geographical conditions at the edges of the Chinese world:

> 故日運行處...極西，西方日中，東方夜半。
> Therefore when the sun's rotation has brought it to a position [...] west of the pole, it is noon in the western region and midnight in the eastern region. [...].[4]

---

[1] This traditional division of the text into three chronological layers follows Qu Anjing (2002: 7–8); cf. Cullen (1996: 138–156). A "quarter-remainder astronomy" (*sifen li*) is one that uses a solar year of 365¼ days and a synodic month of 29 $^{499}/_{940}$ days.

[2] More specifically, "heaven resembles a covering rain-hat, while earth is patterned on an inverted pan" 天象蓋笠，地法覆槃 (*Zhoubi suanjing*, 2.2b; tr. Cullen 1996: 189).

[3] In the *Zhoubi*, the *xuanji* is both a point (star) 11,500 *li* from the celestial pole and the zone within the circle it describes. Depending on how one parses the phrase 極下者其地高人所居六萬里滂沲四隤而下 (*Zhoubi suanjing*, 1.19a), the terrestrial *xuanji* can be interpreted as a flat "polar circle" on a 60,000-*li* high plateau-shaped earth (Fig. 1) or a 60,000-*li* high axis mundi on a disk-shaped earth (Fig. 4, right). The term is a reference to the famous *xuanji yuheng* 璇璣玉衡 mentioned in the *Book of Documents*, on which see Cullen and Farrer (1983). As to its rationale and function in the *Zhoubi*, see Cullen (1996: 119–127).

[4] *Zhoubi suanjing*, 2.1b–2a; tr. Cullen (1996: 189).

> 北極左右，夏有不釋之氷…中衡去周七萬五千五百里。中衡左右冬有不死之草。
> Around the north pole, there is unmelting ice in summer. […] The middle track is 75,500 *li* (≈ 31,393 km) from Zhou, and near the middle track (the equator) there are plants that do not die in winter.[5]

To help visualize what this might look like, I offer Christopher Cullen's reconstruction in Fig. 1 as one plausible interpretation.[6]

In 656 CE, Li Chunfeng 李淳風 (602–670 CE) et al. canonized the *Zhoubi* as one of what we now call the *Ten Mathematical Classics* (*Suanjing shishu* 算經十書), a selection of commentated texts that would go on to serve as the basis of the contemporary mathematical curriculum and of the later historiography of early Chinese mathematics. Of these, modern historians unanimously place it *first*, prior to the famous *Jiuzhang suanshu* 九章算術 (*Nine Chapters on Mathematical Procedures*), which is curious given that the one is pseudepigraphal, the other anonymous, and that both were historically attributed to the Duke of Zhou. No modern scholar takes this attribution seriously.[7] Rather, the majority opinion over the last half-century is that the two texts were compiled from earlier sources under the Han 漢 (206 BCE–220 CE); that the *Jiuzhang suanshu* was in circulation around 100 CE; and, as per Qian Baocong's 錢寶琮 1963 critical edition of the *Ten Mathematical Classics*, that the *Zhoubi* came together two centuries earlier, around 100 BCE.[8]

In this paper, I will attempt to turn this majority opinion on its head via the critical re-examination of its premises and the introduction of new, more conclusive evidence. I begin by reminding the reader that Qian's (1963) date of *circa* 100 BCE for the *Zhoubi* relies on faulty assumptions and is vitiated by Cullen (1996) and Qian's own previous writings. Shifting the parameters of the discussion, I establish 1 BCE as the *Zhoubi* and *Jiuzhang suanshu*'s common *terminus post quem* on the basis of a lexical shift that occurs systematically across *li* 曆 astronomy and *suan* 算 mathematics between 1 BCE and 85 CE. Turning from questions of time to questions of place, I then explore what we know of the *Zhoubi*'s reception history immediately following its appearance in the written record in 178 CE, postulating that it was likely authored in the region to which its later circulation was confined, in the Far North/Northwest. This departure from previous attempts to establish the *Zhoubi*'s provenance stands to change the way that we read the text, and, in the Conclusion, I offer a brief reflection about some of the questions that it answers as well as those that it raises in their place.

---

[5] *Zhoubi suanjing*, 2.8b–9a; tr. modified from Cullen (1996: 192).

[6] For a survey of alternative interpretations, see Qu (2002: 91–143).

[7] One sees skepticism about the *Zhoubi*'s self-proclaimed antiquity as early as He Chengtian 何承天 (370–447 CE), Xu Yuan 徐爰 (395–475 CE) and Shen Yue's 沈約 (441–513 CE) histories of the field in *Song shu*, 13.679 (cf. Morgan 2017: 188–196). Note that some scholars such as Wu (1998–2004: vol. 1, 386–390) and Qu (2002: 4–8) maintain that the earliest layer—the Shang Gao dialogue—originates in oral tradition about the Duke of Zhou, as passed down from the eleventh century BCE, but the language of this dialogue resembles nothing of the early classics—the *Shijing* 詩經, *Shangshu* 尚書 and *Zhouyi* 周易—let alone the sort of Zhou-era bronze inscriptions studied in Shaughnessy (1991).

[8] See for example Jiang (1996: 43, 1997: 207), Wu (1998–2004: vol. 1, 386–390), Qu (2002: 8), and Chemla (2013a: 177; 2014).

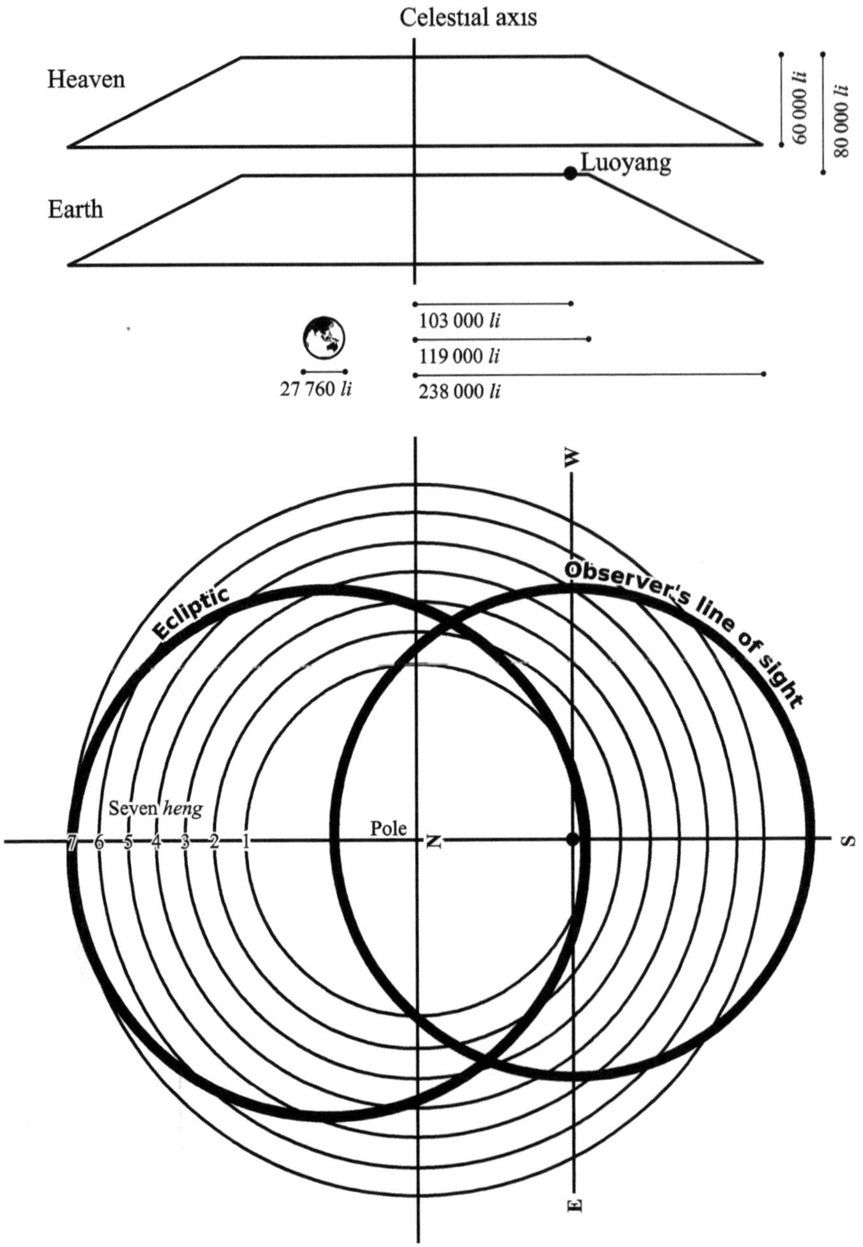

Fig. 1 *Gnomon of Zhou* "umbrella heaven" (*gaitian* 蓋天) world model, redrawn from Cullen (1996, 136, fig. 13, 222, fig. 31). Source: CC BY NP D.P. Morgan (https://hal.science/hal-04284009)

2 Prior Attempts to Date the *Zhoubi*

The one thing that no one contests about the *Zhoubi* or *Jiuzhang suanshu* is their respective *termini ante quem*. There is clear evidence that the *Jiuzhang suanshu* was in circulation by the end of the first century CE. First, Zheng Zhong 鄭眾, who died in 83 CE, lists its extant chapter titles.[9] Second, the text appears in an account of the early studies of Ma Xu 馬續 (fl. 111–141 CE), which, given the known dates of his younger brother Ma Rong 馬融 (79–166 CE), we can place in 93 CE at the very latest:

> 七歲能通『論語』，十三明『尚書』，十六治『詩』，博觀羣籍，善『九章筭術』。
> At seven *sui* of age [Ma Xu] was able to master the *Lunyu*, at thirteen he understood the *Shangshu*, and at sixteen he had perfected the *Shi*; he was broadly read in a host of books and adept at the *Jiuzhang suanshu*.[10]

The first mention of the *Zhoubi* comes a century later in Cai Yong's 蔡邕 (133–192 CE) memorial from Shuofang 朔方 in 178 CE:

> 論天體者三家，宣夜之學，絕無師法。『周髀』術數具存，考驗天狀，多所違失。惟渾天僅得其情。
> The discourse on heaven's form is comprised of three expert-lineages, but the study of "expansive night" has died out and has no master method. The procedures and numbers of the *Zhoubi* all survive, but when examined and verified against the case of heaven, there is much that misses the mark. It is only "sphere heaven" which completely grasps the true circumstances.[11]

This tells us that the *Zhoubi* was in circulation by 178 CE, so the question is how long it took for people like Cai Yong to start talking about it.

Following Qian (1963), most say three hundred years.[12] If we are going to cite Qian Baocong, however, it is important to note that his 1963 argument for dating the text to *circa* 100 BCE is taken from his "Zhoubi suanjing kao" 周髀算經攷 of 1924 and *Zhongguo suanxue shi* 中國算學史 of 1932, both of which tell a very different story. In 1924, Qian Baocong concludes that:

---

[9] As cited by Zheng Xuan 鄭玄 (127–200) in *Zhouli zhushu*, 14.212b–213a: "Minister of Agriculture Zheng [Zhong] says: … the Nine Numbers are 'Rectangular Field', 'Decorticated and Non-decorticated Grains', 'Different Shares', 'Reduced Width', 'Estimation of Works', 'Even Distribution', 'Square Arrays', 'Excess and Insufficiency', and 'Other Essentials'; and now there is double difference *xijie* [sic.], i.e. 'Base and Height'" 鄭司農云…九數:方田、粟米、差分、少廣、商功、均輸、方程、贏不足、旁要。今有重差夕桀，句股也 (cf. Chemla and Guo 2004: 51).

[10] *Hou Han shu*, 24.862. One is born at one *sui* of age and turns two at New Years, thus 16 *sui* = 14 years of age, and given that he was born before Ma Rong, that places the sixteen-*sui*-old Ma Xu sometime prior to 79 + 14 = 93 CE.

[11] *Song shu*, 23.673; *Jin shu*, 11.278; *Sui shu*, 19.505; tr. modified from Ho (1966: 49). For a translation of the full memorial, preserved only in the *Song shu*, see Morgan (2019: 152).

[12] See the studies cited in Note 8.

> The mathematics of the *Zhoubi* are already rather complete as concerns the multiplication and division of fractions and square root extraction, and compared to the procedures in the chapters "Fangtian" (Rectangular Field) and "Shaoguang" (Reduced Width) in the *Jiuzhang suanshu* it is difficult to say which is better or worse. They might well be contemporary works. (Qian 1924 [1983]: 134)[13]

Likewise, in 1932 he writes:

> I thus consider that the *Zhoubi* was written somewhere roughly between Liu Xiang's 劉向 (79–8 BCE) *Qi lüe* 七略 [of 6/1 BCE] and Wang Chong's 王充 *Lunheng* 論衡 [of 80 CE], which is to say around the turn of the [Common] Era. (Qian 1932 [1998]: 199)[14]

Why does Qian Baocong later move this date back a full century to 100 BCE? His simplified argument of 1963 cites four points:

1. That the *huntian* 渾天 (sphere heaven) cosmology (Fig. 2) that appears in the first century BCE had to have come after [and immediately replaced] umbrella heaven, [because it is better]
2. That the quarter-remainder-style astronomy in the *Zhoubi* reflects that abandoned by the government in 104 BCE [but continuously studied, written about, and ultimately re-adopted in 85 CE]
3. That for the third of the twenty-four *qi* the received text uses the old name, Qizhe 啟蟄 "Awakened Insects," rather than the new legal name, Jingzhe 驚蟄 "Startled Insects," relating to a posthumous taboo on Emperor Liu Qi's 劉啟 (d. 141 BCE) given name [as do numerous other texts from the first century CE on]
4. That the text uses the classical position of the winter solstice sun, "at the beginning of Ox" 牛初 (β Cap), rather than the new one, "five *du* before Ox" 牛前五度 [adopted by the court in 85 CE amid debate]

In conclusion, he offers that:

> From these materials and other period-specific language, we have determined that the *Zhoubi* was written *circa* 100 BCE. (Qian 1963: 4)[15]

There are a number of problems with this argument, as indicated by my use of brackets and as openly admitted to in Qian Baocong's earlier writings. Concerning the shift from Qizhe to Jingzhe in 141 BCE, for example, Qian Baocong (1924 [1983]: 134) acknowledges that the former does continue to see sporadic use into the Common Era, notably in the works of Li Chunfeng, the lead editor of the *Ten Mathematical Classics*, and that it may well be "that while editing and commentating the *Zhoubi* he changed *jing* 'startled' to *qi* 'awakened'."[16] The biggest challenge,

---

[13] 『周髀』算學于分數乘除及開方算法，已甚完備。與『九章算術』方田、少廣兩章之術，難分軒輊。或是同時代之著作。

[14] 故余以為『周髀』之撰述時代，大約在劉向『七略』以後，王充『論衡』以前。或為西曆紀元左右之書也。

[15] 我們根據這些資料和其他有時代性的文字，斷定『周髀』是公元前100年前後的作品。

[16] 李淳風造『麟德曆術』，主用啟蟄舊名。當其校注『周髀』時，改「驚」為「啟」。One notes that the ancient term Qizhe 啟蟄 also appears in Wang Chong's *Lunheng* of 80 CE, Zhang Zhouxuan's 張胄玄 modified *Daye li* 大業曆 of 608 CE and Fu Renjun's 傅仁均 *Wuyin li*

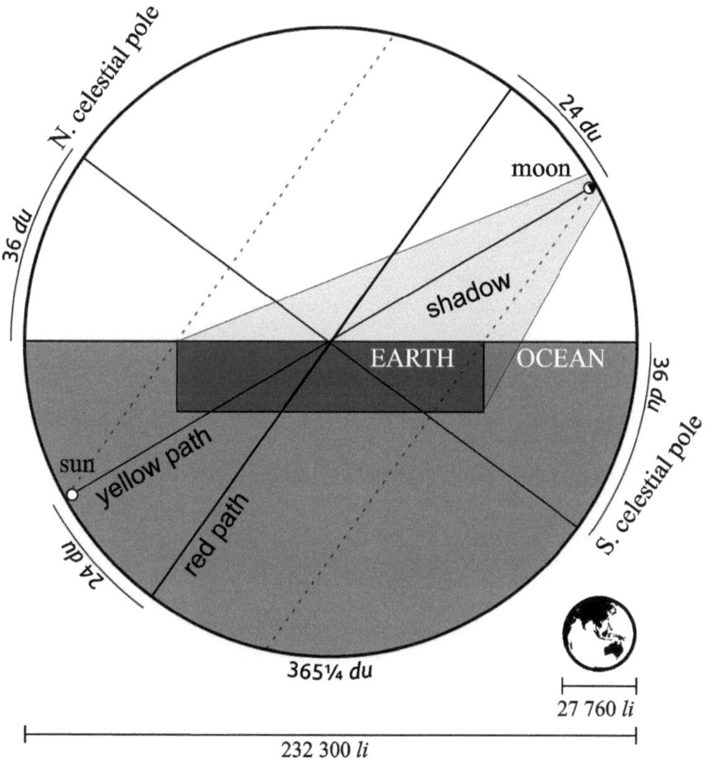

Fig. 2 Zhang Heng's 張衡 (78–139 CE) "sphere heaven" (*huntian* 渾天) world model. Dotted lines represent the daily course of the sun at the winter and summer solstices; at the equinoxes, the sun travels on the Red Path (*chidao* 赤道). The globe in the bottom right corner is to scale with Zhang Heng's dimensions of heaven. At 1000 *li* (415.8 km) in diameter, the sun is about twice the size of Sri Lanka. Source: CC BY D.P. Morgan (https://shs.hal.science/halshs-03858375)

however, is Cullen's (1996) observation that the *Zhoubi* incorporates elements of sphere heaven cosmology and instrumentation such as the measurement of polar distance (*quji du* 去極度), which goes back no earlier than Xianyu Wangren's 鮮于妄人 observation program of 78–74 BCE.[17] If, as Cullen argues, the received *Zhoubi* was written as an anti-sphere heaven polemic pseudepigraphically set in ancient times for purposes of self-legitimation, then that stands to explain its more ancient-looking technical contents: they are not so much *antique* as they are purposefully *antiquarian*.

As highlighted in Fu (1988) and Takeda (1989), there are striking parallels in contents between the *Zhoubi* and the *chenwei* 讖緯 apocrypha that begin to appear in the first century CE: the interest in cosmology, cosmic dimensions, solar tables,

戊寅曆 of 618 CE (*Lunheng jiaoshi*, 25.1062; *Daye li*, cited in *Sui shu*, 17.454 ff.; *Wuyin li*, cited in *Jiu Tang shu*, 32.1153 ff.).

[17] On Xianyu Wangren's observation program and the measurement of north polar distance by armillary sphere in 78–74 BCE, see Sun and Kistemaker (1997).

and the return to an ancient quarter-remainder astronomy rooted in eschatology and high antiquity.[18] Take for instance the cycle of 31,920 years that the *Zhoubi* adds to the typical set of quarter-remainder resonance periods:

> 陰陽之數，日月之法。十九歲為一章。四章為一蔀，七十六歲。二十蔀為一遂，遂千五百二十歲。三遂為一首，首四千五百六十歲。七首為一極，極三萬一千九百二十歲。生數皆終，萬物復始。
>
> The numbers of Yin and Yang and the factors of sun and moon [are as follows]. Nineteen years make one "rule." Four rules make one "obscuration." Twenty obscurations make one "succession," and a succession is 1520 years. Three successions make one "head," and a head is 4560 years. Seven heads make one "culmination," and a culmination is 31,920 years. [At this], the numbers of life (?) all come to an end, and the myriad creatures return to the start.[19]

Using slightly different terms, everything up to the "culmination" (*ji* 極) is covered in the Han *Sifen li* 漢四分曆 of 85 CE: the "rule" (*zhang* 章) of 19 years:235 months:6939¾ days marks the coincidence of new moon and winter solstice; the "obscuration" (*bu* 蔀) of 76 years:940 months:27,759 days marks the coincidence of new moon, winter solstice and midnight; the "era" (*ji* 紀) of 1520 years = 9253 × 60 days marks the coincidence of new moon, winter solstice, midnight and day *jiazi*$_{01}$ 甲子; and the "origin" (*yuan* 元) of 4560 = 76 × 60 years marks the coincidence of new moon, winter solstice, midnight, day *jiazi*$_{01}$, and year *jiazi*$_{01}$.[20] That the 31,920-year "culmination" marking a cosmic cycle of death and rebirth does not appear in the Han *Sifen li* is not surprising, because it is not an astronomical resonance period of the sort that typically features in *li* astronomy. Instead, where we do see it is in the *Yijing* 易經 apocryphon *Qiaozaodu* 乾鑿度:

> 孔子曰:「三萬一千九百二十歲，錄圖受命，易姓(三)〔四〕十二，(紀)〔純〕德有七」。
>
> Confucius said: "[Every] 31,920 years the Mandate is received through registers and diagrams, there are [forty]-two changes of surnames, and there are seven [pure] virtues."[21]

Citing peusdo-Confucius as its authority, the *Qianzaodu* explains that over these 31,920 years a cycle of forty-two *Yijing* hexagrams divided into Yin-Yang, "rising and ascending" (*xiaoxi* 消息) pairs that acts to determine the characteristics and ruling family of a sequence of numerologically-derived periods of, on average, 760 years, from which one can prophesy the rise and fall of dynasties.[22]

This is the sort of literature that the *Zhoubi* is mixed up in, and, in addition to its contents, there is also a striking parallel in form: both the *Zhoubi* and the apocrypha

---

[18] On the apocrypha, see Dull (1966). On the astronomical contents of the apocrypha, see also Takeda (1989), Chen (2003: 167–176; 2007: 475–483), and Wu (2007).

[19] *Zhoubi suanjing*, 2.46a–52a; tr. modified from Cullen (1996: 95).

[20] *Hou Han shu*, *zhi* 3, 3057–3058; tr. Cullen (2017: 151–153).

[21] *Zhouyi Qianzaodu*, 2.9b.

[22] Takeda (1989: 64) and Nielsen (1995: 116–146).

are blatant projections of turn-of-the-Common-Era language and preoccupations onto ancient culture heroes in the form of revelation. It was the apocrypha that would come to define the political and intellectual culture of the first and second centuries CE, and if it is these texts that the *Zhoubi* resembles more than any other extant work of *li* astronomy or *suan* mathematics,[23] then we might consider the possibility that it was a product of the same age. Still, we have yet to deal with the matter of "period-specific language."

3 Linguistic Markers

To be clear, Qian Baocong does not specify what "other period-specific language" leads him in 1963 to walk the date of the *Zhoubi* back to *circa* 100 BCE, but he does raise an excellent question: might the language of the *Zhoubi* provide us with a *terminus post quem*?

In 2015 I began work on a global analysis of the language of basic operations in all sources for *li* and *suan*, excavated and received, up to 600 CE. The way that these texts prescribe such operations as division is often variable and ambiguous, and my idea was to identify every operation in every text and individually determine what every instance of a word like *chu* 除 "eliminate" must mean in context for the sums to work out, then, from the data compiled, to search for patterns behind this variability.[24] It is in the context of this work I began to question the *Zhoubi*'s place as the earliest of the *Ten Mathematical Classics*; I summarize.

In the earliest sources we now possess, the third- and second-century BCE manuscripts *Shu* 數, **Suanshu* 筭書 and *Suanshushu* 筭數書,[25] the word *ying* 盈 (fill) is used in two senses. The first is "greater than or equal to," particularly as applied to remainders:

> 如法而一步，不盈步者，以法命之。
> [Get] one *bu* [for each time] it is up to the divisor (i.e., divide by the divisor for an integer quotient in *bu*); take that which does not **fill (*ying*)** a *bu* (i.e., the remainder) and name it by the divisor (*Shu*, slip 15).

The second sense is "excess" versus "insufficiency" (*ying buzu* 盈不足) as used in the rule of false position:

---

[23] Liu Hui's preface to the *Jiuzhang suanshu* aside (Chemla and Guo 2004: 126–127), one notes that the *Zhoubi* is the only extant work of *li* or *suan* mathematics from this period that attributes *itself* to high antiquity.

[24] I have presented this work in Morgan (2015a; 2015b) and am in the process of preparing my revised and expanded results for publication at some point.

[25] For pictures and transcriptions, see *Yuelu shuyuan cang Qin jian (er)*, Han Wei (2013), and *Zhangjiashan Han mu zhujian (ersiqi hao mu)*.

> 贏(盈)不足互乘母為實，子相從為法。
> The **excess (*ying*)** and insufficiency mount (i.e., multiply) one another's numerators to make the dividend, and the denominators go with one another (i.e., add together) to make the divisor (*Suanshushu*, slips 133–134).[26]

The next earliest mathematical text of whose date we can be certain is Liu Xin's 劉歆 (*c.*50 BCE–23 CE) *Santong li* 三統曆 of 1 BCE/5 CE,[27] which sees two additional applications of the word *ying*. One is as an auxiliary verb used in division:

> 盈章歲得一，名曰積月。
> Get one [for each time] it **fills (*ying*)** the rule years, and this is named the "accumulated months'.[28]

The other, by abbreviation, is as the operation of division itself:

> 見復餘盈其見復數，一以上見在往年
> If [the number of times] the appearance and return remainder **fills (*ying*)** the appearance and return number is one or higher, then the appearance is in the previous year.[29]

As illustrated in Table 1, however, sometime between the *Santong li* of 1 BCE/5 CE and the Han *Sifen li* 四分曆 of 85 CE there is a shift to the synonym *man* 滿 (fulfil) for all usages excepting "excess." Why? One possibility is that it relates to the posthumous taboo on Emperor Liu Ying's 劉盈 (d. 188 BCE) given name.[30] If so, the dates do not quite work out: the manuscript *Suanshushu* was found in a tomb sealed

[26] Tr. modified from Cullen (2004: 81).

[27] We know that the *Santong li* was written after 1 BCE from *Han shu*, 36.1972: "When Emperor Ai fell (in 1 BCE), Wang Mang took the reins of government. [...] The empress dowager retained [Liu] Xin as bureau head of the left and [then?] transferred him to [the posts of] colonel of the capital rampart, Xi-He, and [then] governor of the capital, putting him in charge of the Bright Hall and Circular Moat and conferring him the title of Marquis of Hongxiu. He was charged with the scholarly offices of [grand] clerk and diviner, and he studied and fixed tono-metrology and mathematical astronomy, composing the *Santong li* astronomy and *pu* chronology" 會哀帝崩，王莽持政...太后留歆為右曹太中大夫，遷中壘校尉，羲和，京兆尹，使治明堂辟雍，封紅休侯。典儒林史卜之官，考定律曆，著『三統』曆、譜. We also know that the procedure text preserved in the *Han shu* was written prior to 5 CE based on how Ban Gu 班固 (32–92 CE) identifies his source: "In the mid-Yuanshi period (1 BCE–5 CE), Wang Mang held control of the government, and wishing to glorify his reputation he summoned more than a hundred experts on bells and pitch-pipes from all under heaven and had Xi-He Liu Xin et al. take charge of memorializing [their findings] item by item. Theirs is the most detailed discussion [of the topic], thus I have expunged their more spurious words, picked out what contents are correct and written them into the present piece" 至元始中王莽秉政，欲燿名譽，徵天下通知鐘律者百(餘)餘人，使羲和劉歆等典領條奏，言之最詳。故删其偽辭，取正義，著于篇 (*Han shu*, 21A.955).

[28] *Santong li*, cited in *Han shu*, 27B.1001; cf. Cullen (2017: 91).

[29] *Santong li*, cited in *Han shu*, 27B.1002; cf. Cullen (2017: 100).

[30] According to Xun Yue 荀悅 (148–209 CE), "to avoid the character *ying* one said *man*" 諱「盈」之字曰「滿」 (*Han shu*, 2.86 comm.); cf. Wang (1997: 555–556).

Table 1 Distribution of the synonyms *ying* 盈 "fill" and *man* 滿 "fulfil." The columns are in approximate chronological order, with works of *li* 曆 astronomy marked in grey and *suan* 筭 mathematics in white. Note that texts in which these words do not appear in these capacities are omitted. The row "≥" includes remainders resulting from division ('that which does not [ful]fill" 不盈/滿, etc.), and the row "÷" includes said character as it appears anywhere in an expression that produces a quotient, be it in a primary or supporting role

Word	Use	S	SSS	STL	JZSS	ZB	SFL	QXL	JCL	SZSJ	ZBZ	YJL	DML	ZQJSJ	ZGL	XHL	ZBCS	WJSS	DYL	ZQJ.XC
滿 *man* "fulfil"	≥				10	14	20	22	21	5	10	15	8		20	19	1	4	14	1
	÷						59	43	41		2	27	33		47	48	5	5	31	
盈 *ying* "fill"	≥	15	8	24												1		2		
	÷			25				1								1				
	excess		10	1	27			38	17	6	1	23	19	6	32	32			38	3

Abbreviations in Table 1 and Fig. 3 (below): **Mathematics: S** (*Shu* 數, anon. MS, ≤ 212 BCE, Yuelu Academy); **SSS** (Suan shu shu 筭數書, anon. MS, ≤ 186 BCE, Zhangjiashan tomb 247); **JZSS** (*Jiuzhang suanshu* 九章算術, anon., 1 BCE/83 CE); **ZB** (*Zhoubi* 周髀, anon., 1 BCE/178 CE); **HDSJ** (*Haidao suanjing* 海島算經, Liu Hui 劉徽, *c.*263 CE); **SZSJ** (*Sunzi suanjing* 孫子算經, anon., 280/317 CE); **ZBZ** (*Zhoubi zhu* 周髀注, Zhao Ying 趙嬰, *c.*315 CE); **ZQJSJ** (*Zhang Qiujian suanjing* 張邱建算經, Zhang Qiujian 張邱建, 431/484 CE); **SSJYZ** (*Shushu jiyi zhu* 數述記遺注, Zhen Luan甄鸞, *c.*535/570 CE); **ZBCS** (*Zhoubi chongshu* 周髀重疏, Zhen Luan, *c.*535/570 CE); **WJSS** (*Wujing suanshu* 五經算術, Zhen Luan, *c.*535/570 CE); **WCSJ** (*Wucao suanjing* 五曹算經, Zhen Luan, *c.*535/570 CE); **ZQJ.XC** (*Zhang Qiujian suanjing xicao* 張邱建算經細草, Liu Xiaosun 劉孝孫, *c.*580 CE). **Astronomy: STL** (*Santong li* 三統曆, Liu Xin 劉歆, 1 BCE/5 CE); **SFL** (*Sifen li* 四分曆, Bian Xin 編訢 and Li Fan 李梵, 85–86 CE); **QXL** (*Qianxiang li* 乾象曆, Liu Hong 劉洪, fin. 206 CE); **JCL** (*Jingchu li* 乾象曆, Yang Wei 楊偉, 237 CE); **YJL** (*Yuanjia li* 元嘉曆, He Chengtian 何承天, 444 CE); **DML** (*Daming li* 大明曆, Zu Chongzhi 祖沖之, 465 CE); **ZGL** (*Zhengguang li* 正光曆, Li Yexing 李業興and Zhang Longxiang 張龍祥, 518/520 CE); **XHL** (*Xinghe li* 大興曆, Li Yexing 李業興, 539 CE); **DYL** (*Daye li* 大興曆, Zhang Zhouxuan 張胄玄, 597 CE)

2 years later, in 186 BCE,[31] and the *Santong li* was written a couple of years before the House of Liu was deposed and the taboo lifted in 9 CE.[32] Neither does this account for why *ying* was kept in the sense of "excess." Another possibility is that this terminological shift occurred for simple reasons of disambiguation, possibly in combination with irregular censorship.

Whatever the reason, this shift was absolute, drawing a clear dividing line between all dated sources for *li* and *suan* mathematics before and after. As to *undated sources*, it should come as no surprise that the *Jiuzhang suanshu* systematically uses *man* for "greater than or equal to," comporting with post-*Santong li* usages, but so too does the *Zhoubi*:

> 實如法得一里，不滿法者，三之，如法得百步。
> Get one *li* [for each time] the dividend is up to the divisor; that which does not **fulfil (*man*)** the divisor, triple it and get 100 *bu* [for each time] it is up to the divisor.[33]

Could it be, as Qian (1924 [1983]: 134) suggests, that this is because Li Chunfeng et al. harmonized the vocabulary of the *Ten Mathematical Classics* with seventh-century usages? This is not how one normally works with "classics": finished only 3 years earlier, in 653 CE, the *Wujing zhengyi* 五經正義 (*Correct Meaning of the Five [Confucian] Classics*) systematically ignores the taboo on the given names of Li Yuan 李淵 (r. 618–626) and Li Shimin 李世民 (r. 627–649) in the base text, commentary and subcommentary, and the same is true of the *Ten Mathematical Classics*. So why would Li Chunfeng enforce Han-era political taboos whilst ignoring those of his own day? And if Li Chunfeng harmonized the language of the mathematical canon, then why is there so much variation in the expression of basic operations like subtraction? Consider the following line from the *Zhoubi*:

> 【經】術曰:置外衡去北極樞二十三萬八千里，除璿璣萬一千五百里。
> [Base text:] The procedure: Set out the 238,000 *li* distance from the [sun's] outer track to the pivot of the north [celestial] pole and **eliminate (*chu*)** the 11,500 *li* of the *xuanji* (above).
>
> 【注】北極常近牽牛，爲樞過極萬一千五百里，此求去極，故以除之。
> [Commentary:] The north [celestial] pole is always closest to Led Ox (β Cap), [where the sun is lodged amongst the stars at winter solstice, on the track (*heng*) furthest from the north celestial pole], the [edge of the *xuanji*] pivot lying 11,500 *li* beyond the pole. Here we are looking for *polar distance*, thus we have to **eliminate (*chu*)** this.

---

[31] Note that, on slips 133–134, quoted above, the *Suanshushu* uses the orthographic variant 贏 in place of 盈 for *ying* "fill" but that, elsewhere, the manuscript mixes the two (see slips 42, 56, 58, 64, 82, 160, 165 and 184). A month calendar running up to 186 BCE found together with the *Suanshushu* at Zhangjiashan 張家山 places the sealing of the tomb in or shortly after this year (*Zhangjiashan Han mu zhujian [ersiqi hao mu]*, 1).

[32] According to *Han shu*, 21.955, 36.1972, Wang Mang charged Liu Xin with leading a reform in tono-metrology and mathematical astronomy during the Yuanshi reign (1–6 CE), in the lead-up to his usurpation of the Han throne.

[33] *Zhoubi suanjing*, 1.70b; cf. Cullen (1996: 88).

> 【重述】臣鸞曰:求牽牛星去極法，先列衡去極樞二十三萬八千里，減極去樞心一萬一千五百里。
> [Subcommentary:] Your servant [Zhen] Luan 甄鸞 (fl. 535–570 CE) says: The method to find the polar distance of the Led Ox asterism is that you first array the 238,000 *li* distance from pole to track, which you then **diminish (*jian*)** by the 11,500 *li* distance from the pole to the center of the pivot.[34]

All three layers of the text are describing the same operation of subtraction, but where both the base text and commentary use *chu* 除 (eliminate), Zhen Luan's sixth-century subcommentary uses *jian* 減 (diminish). This is not random. Looking at subtraction across sources for both *li* astronomy and *suan* mathematics in Fig. 3, we see that *chu* is gradually replaced by *jian* and *qu* 去 (remove) over the intervening centuries, so it makes sense that Zhen Luan would essentially *translate chu* as *jian*. That too is what we would expect in the base text and commentary had Li Chunfeng et al. for some reason chosen to rewrite them, but such is not the case for *chu* nor for any other antiquated operational language in the early mathematical classics.

If we accept the transition between *ying* and *man* as a reliable marker of when a mathematical text was compiled, then we must conclude that the *Jiuzhang suanshu* was finished between 1 BCE and 83 CE, and the *Zhoubi* between 1 BCE and 178 CE. In itself, this new *terminus post quem* is not particularly "radical," seeing as how it comports with the traditional dating of the former as well as Cullen (1996) and early Qian Baocong's (1924 [1983], 1932 [1998]) dating of the latter. Far more befitting the promise made in the title of this chapter is what I believe we can say about where the *Zhoubi* is from.

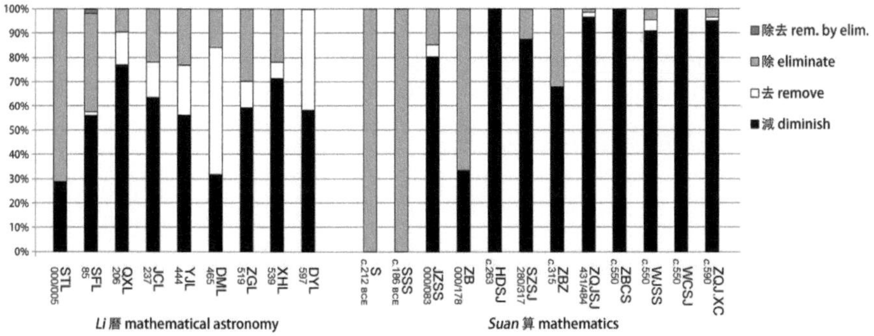

Fig. 3 Distribution of expressions for subtraction in *li* astronomy and *suan* mathematics. Abbreviations given in the caption to Table 1 (above). Note that the dates are in some cases approximate and that I have included instances of sequence subtraction in with normal subtraction

[34] *Zhoubi suanjing*, 2.16a, 2.18a.

4 Northern Transmission

One of the more curious features about the early history of umbrella heaven cosmology is that in his treatment of the topic in the *Lunheng* of 80 CE Wang Chong makes no mention of the *Zhoubi* (fin. 1 BCE/178 CE), that the *Zhoubi* likewise makes no mention of the *Lunheng*, and that Cai Yong, writing one century later, mentions only the one. As to why Cai Yong fails to mention the *Lunheng* in writing about umbrella heaven from the northern steppe in 178 CE, it turns out that it is because, in 178 CE, he did not yet have the book:

> 袁山松書曰:「充所作『論衡』，中土未有傳者，蔡邕入吳始得之，恆秘玩以為談助。…」
> The *Yuan Shan song shu* says: "[Wang] Chong's *Lunheng* had yet to see transmission in the central lands, and when Cai Yong first got a hold of it upon entering Wu 吳 (on the south-eastern seaboard) he was always secretly playing with it as a conversational aid." […]
>
> 抱朴子曰:「時人嫌蔡邕得異書，或搜求其帳中隱處，果得『論衡』，抱數卷持去。邕丁寧之曰:『唯我與爾共之，勿廣也。』」
> The *Baopuzi* says: "At the time, people envied Cai Yong's [ability] to get his hands on rare books. It was in occasionally searching the hidden corners of his home that [Wang Lang 王朗 (d. 228 CE)] indeed found the *Lunheng*, leaving with several rolls in his arms. [Cai] Yong gave him the careful instructions that 'Only I and you share this, do not spread it'."[35]

According to his biography, Cai Yong was banished to Wang Chong's hometown of Kuaiji 會稽, in Wu Commandery, only after his release from Shuofang in 179 CE.[36]

Viewed in the context of Cai Yong's life, this information is particularly revelatory. Cai Yong had spent 8 years from 170 to 178 CE working as a textual scholar in the palace archives—the Eastern Observatory (Dongguan 東觀)[37]—and we know that there were book stores in the capital since at least the day of Wang Chong.[38] If it was only in the far southeast at the age of 46+ that this avid collector of rare books finally got his hands on a copy of the *Lunheng* some hundred years after it was written, that tells us that its circulation was indeed quite regionally limited.

---

35 *Hou Han shu*, 49.1629 comm. Note that said phrase does not occur in the received *Baopuzi*.

36 *Hou Han shu*, 60B.2003.

37 *Hou Han shu*, 60B.1990 ff. On the Eastern Observatory, see Goodman (2005).

38 "[Wang Chong] was from a poor family without books, and he would often wander the market stalls of Luoyang reading the books on sale" 家貧無書，常游洛陽市肆，閱所賣書 (*Hou Han shu*, 49.1629).

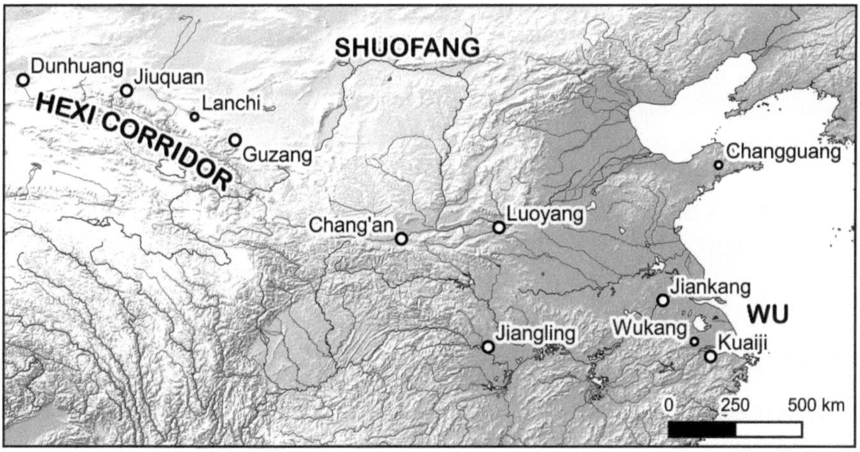

So what do we know about the transmission history of the *Zhoubi*?[39] The next evidence of its reception is the extant commentary of Zhao Shuang 趙爽 (styled Junqing 君卿). Nothing is known about Zhao Shuang, and there is even some confusion about his name, which contemporary bibliographies give as Zhao Ying 趙嬰.[40] Most scholars agree that this is an orthographic variant, posit that Zhao was active in the Eastern Han 東漢 (25–220 CE) or Three Kingdoms 三國 (220–280 CE) and go no further.[41] If we scour the written record, however, we find two men by these names active in literate circles. One is a Sui 隋 (581–618 CE) emissary by the name of Zhao Junqing who petitioned the southern court of Chen 陳 (557–589 CE) in 585 CE, as recorded in Guanding's 灌頂 (561–632 CE) *Guoqing bailu* 國清百錄.[42] The second is a district administrator active in the Hexi 河西 Corridor just prior to its independence under the Former Liang 前涼 (320–376 CE), as recorded in Cui Hong's 崔鴻 (478–525 CE) *Shiliuguo chunqiu* 十六國春秋 of 522 CE:

> 建興三年冬十月，蘭池長趙嬰上言，軍士張氷於青澗水中得一玉璽…
> In the third year of the Jianxing reign (315 CE), winter, month x, Lanchi 蘭池 [District] Chief Zhao Ying 趙嬰 memorialized the throne saying that Serviceman Zhang Bing found a jade seal in the waters of Greenbrook. […].[43]

---

[39] Between the authorship and publication of this chapter, I have developed upon the questions of transmission and regional networks discussed here in Morgan (2021).

[40] Both *Sui shu*, 34.1018, and *Jiu Tang shu*, 47.2036, record a "*Gnomon of Zhou* in 1 roll commentated by Zhao Ying" 周髀一卷趙嬰注; *Xin Tang shu*, 59.1543, records a "*Gnomon of Zhou* commentated by Zhao Ying in 1 roll" 趙嬰注周髀一卷; and *Song shi*, 207.5271, records a "Zhao Junqing's *Mathematical Classic of the Gnomon of Zhou* in 2 rolls" 趙君卿周髀算經二卷.

[41] *Chouren zhuan*, 4.48–49, Qian (1924 [1983]: 134–135), Cullen (1996: 147–148), Wu (1998–2004: vol. 3, 23), Qu (2002: 9), Chemla and Guo (2004: 59).

[42] "In the third year of the Zhide reign (585 CE)… the Young Ruler's (r. 583–589 CE) second decree was answered by a petition from the emissary Zhao Junqing" 至德三年…少主第二勅，得使人趙君卿啟 (*Guoqing bailu*, *T.* no 1934, 46:799b7).

[43] *Shiliuguo chunqiu*, 71.1b–2a.

According to the *Siku quanshu* 四庫全書 preface, the extant *Shiliuguo chunqiu* is a forgery by Tu Qiaosun 屠喬孫 (fl. 1606) and Xiang Lin 項琳 (1619–1670).[44] However, Yu Yin's 余寅 (1541–1615) *Tong xingming lu* 同姓名錄 cites the work's "Qianliang lu" 前涼録 to the same effect.[45]

The later *Jin shu* of 648 CE records the same event featuring a third orthographic variation, between Ying 嬰 and Shuang 爽:

> 蘭池長趙奭上軍士張冰得璽…
> Lanchi [District] Chief Zhao Shi 趙奭 memorialized the throne that Serviceman Zhang Bing found a seal. […].[46]

Of the two, it is the fourth-century district chief that is the most likely to be our commentator. First, the Zhao commentary twice compares the text's quarter-remainder astronomy to Liu Hong's 劉洪 (fl. 167–206 CE) *Qianxiang li* 乾象曆, which makes more sense coming from someone living in 315 CE, when the *Qianxiang li* was held as the gold standard in astronomy.[47] Second, given that Zhen Luan's *chongshu* 重述 subcommentary on Zhao's work was written under the Northern Zhou 北周 (557–581 CE), the emissary appearing in 585 CE is a bit late for his writings to have seen canonization by Zhen Luan. Third, a century later, in 437 CE, the *Zhoubi* reappears at the Northern Liang 北涼 (397–439 CE) capital at Guzang 姑臧, less than 200 km down the Hexi Corridor from Lanchi, in the context of an interstate book exchange:

> 河西人趙敺善歷算。十四年，茂虔奉表獻方物，并獻『周生子』十三卷，『時務論』十二卷，『三國總略』二十卷，『俗問』十一卷，『十三州志』十卷，『文檢』六卷，『四科傳』四卷，『燉煌實錄』十卷，『涼書』十卷，『漢皇德傳』二十五卷，『亡典』七卷，『魏駮』九卷，『謝艾集』八卷，『古今字』二卷，『（乘）〔桑〕丘先生』三卷，**『周髀』一卷**，『皇帝王歷三合紀』一卷，『趙敺傳』并『甲寅元歷』一卷，『孔子讚』一卷，合一百五十四卷。茂虔又求晉、趙起居注諸雜書數十件，太祖賜之。
>
> Zhao Fei of Hexi was good at astronomy and mathematics. In the fourteenth year [of the Liu-Song Yuanjia reign (437 CE)], [King Juqu] Maoqian 沮渠茂虔 (AKA Mujian 牧犍) [of the Northern Liang] (r. 433–439 CE) presented a memorial and a tribute of regional items. Also offered as tribute were the *Zhoushengzi* 周生子 in 13 rolls, the *Shiwu lun* 時務論 in 12 rolls, the *Sanguo zonglüe* 三國總略 in 20 rolls, the *Suwen* 俗問 in 11 rolls, the *Shisanzhou zhi* 十三州志 in 10 rolls, the *Wenjian* 文檢 in 6 rolls, the *Sike zhuan* 四科傳 in 4 rolls, the *Dunhuang shilu* 燉煌實錄 in 10 rolls, the *Liang shu* 涼書 in 10 rolls, the *Han*

[44] *Shiliuguo chunqiu*, "Tiyao" 提要,1a.

[45] *Tong Xingming lu*, 5.30a–b.

[46] *Jin shu*, 86.2227.

[47] Though Yang Wei's 楊偉 modification thereof—the *Jingchu li* 景初曆 (237 CE)—was in official service at the time, Li Chunfeng reports that "Yang Wei's planetary calculations were particularly loose and wide, thus after Emperor Yuan [of Jin] 晉元帝 (317–322 CE) crossed the left (southern) bank of the Yangtze, he had [Yang] Wei's replaced with the *Qianxiang*'s planetary models. From the Huangchu period (220–226 CE) on, [those who worked on] the reform and creation of astronomical procedures all pored over the *Qianxiang*'s Dipper parts reduction, syzygy remainder, and *yin-yang* and slow-fast [sequences] of lunar motion in search of compromise. [Liu] Hong's procedures were the exemplar of calculation for the subsequent age" 楊偉推五星尤疏闊，故元帝渡江左以後，更以乾象五星法代偉曆。自黃初已後，改作曆術，皆斟酌乾象所減斗分、朔餘、月行陰陽遲疾，以求折衷。洪術為後代推步之師表 (*Jin shu*, 17.503). On the circulation of the *Qianxiang li*, see Morgan (2015c).

> *huangde zhuan* 漢皇德傳 in 25 rolls, the *Wangdian* 亡典 in 7 rolls, the *Wei jiao* 魏駮 in 9 rolls, the *Xie Ai ji* 謝艾集 in 8 rolls, the *Gujin zi* 古今字 in 2 rolls, the *[Sang]qiu Xiansheng* 桑丘先生 in 3 rolls, **the *Zhoubi* in 1 roll**, the *Huangdi wangle sanhe ji* 皇帝王歷三合紀 in 1 roll, the *Zhao Fei zhuan* 趙𢾺傳 and [Zhao Fei's] *Jiayin yuan li* 甲寅元歷 in 1 roll and the *Kongzi zan* 孔子讚 in 1 roll, for a total of 154 rolls. [King Juqu] Maoqian also requested the court diaries of the Jin 晉 (265–420 CE) and Zhao 趙 (318–351 CE) and a various selection of several dozen other books, which Emperor Taizu [of the Liu-Song] 宋太祖 (r. 424–453 CE) granted him.[48]

In case the purpose of this book exchange is unclear, one notes that there is a preponderance of Northwestern authors and subjects among the titles sent to Jiankang 健康: Zhao Fei 趙𢾺 of Hexi (fl. 412–433 CE) was the Prefect Grand Clerk (*taishi li* 太史令) of the Northern Liang,[49] and, like Xie Ai 謝艾 (d. 353 CE), the authors of the *Han huangde zhuan*, *Zhoushengzi*, *Shisanzhou zhi*, *Dunhuang shilu*, and *Liang shu* were all natives of Dunhuang 敦煌.[50] In short, if the *Zhoubi* is on this list, it is probably because the Xiongnu 匈奴 king Juqu Maoqian thought it an appropriate "regional item."

The *Zhoubi* comes up another three times prior to 437 CE outside of the Hexi Corridor, but the circumstances in each case are somewhat odd. In 458 CE, under the Northern Wei 北魏 (386–550 CE), Yin Shao 殷紹 claims to have learned the *Zhoubi* from a Daoist mystic in Changguang 長廣 "in the age of the Yao Clan" 以姚氏之世, under the Later Qin 後秦 (384–417 CE), but the description of his "wandering studies" is somewhat fantastical and, geographically, a better fit with the reunified North of 376–383 CE under the Former Qin 前秦 (351–394).[51] In the South, citing a Prefect Grand Clerk Chen Jizhou 陳季胄, Yu Xi 虞喜 of Kuaiji's (fl. 307–346 CE) *Antian lun* 安天論 hypothesizes that the title *Zhoubi* 周髀 "might either be a person's name" 或人姓名 (i.e. Mr. Thighbone Zhou) or the description of a heaven that

---

[48] *Song shu*, 96.2416. From the Northern Liang perspective, see *Shiliuguo chunqiu*, 95.4a–b.

[49] *Sui shu*, 34.1022.

[50] Trombert (2005: 17–24). On Zhao Fei, author of the *Jiayin yuan li* 甲寅元歷, see Chen (2003: 254–255). On Hou Jin's 侯瑾 authorship of the *Han huangde zhuan*, see *Sui shu*, 33.961; for his bio, see *Hou Han shu*, 80B.2649. On Kan Yin's 闞駰 (fl. 401–444 CE) authorship of the *Shisanzhou zhi*, see *Sui shu*, 33.985; for his bio, see *Bei shi*, 34.1267. For Liu Bing's 劉昞 (fl. 420–439 CE) bio, including his authorship of the *Dunhuang shilu*, and *Liang shu*, see *Wei shu*, 52.1160–1161. For Xie Ai's bio, see *Shiliuguo chunqiu*, 75.20b–22b. On Zhousheng Lie's 周生烈 (third cent. CE) authorship of the *Zhoushengzi*, see *Sui shu*, 34.998; on his hometown and connection to Cai Yong through Wang Lang (above), see *Sanguo zhi*, 13.420. In addition, note that Yao Dan 姚澹 (fl. 317 CE), author of the *Sike zhuan*, was a Han-Zhao 漢趙 (304–318 CE) military leader who, judging from his position and surname, that was likely from among the Qiang 羌 tribes on the southern edge of the Hexi Corridor (see *Jiu Tang shu*, 46.2003; *Shiliuguo chunqiu*, 2.2b). The two known exceptions in this list are Yang Jun 楊(偉)〔俊〕 of Huojia 獲嘉 (d. 222 CE), author of the *Shiwu lun* and *Sangqiu Xiansheng*, and Zhang Ji 張揖 of Qinghe 清河 (third cent. CE), author of the *Gujin zi*. On the authorship of these works, see *Sui shu*, 32.942, 34.1006. For Yang Jun's bio, see *Sanguo zhi*, 23.663–664. On Zhang Ji's background, see *Wei shu*, 91.1963.

[51] *Wei shu*, 91.1955–1956. One notes that the Former Qin lost Changguang along with the rest of the Shandong Peninsula to Jin forces after their rout at the Battle of Fei River in 383 CE, and that it passed hands to the Later Yan 後燕 (384–407 CE) and Southern Yan 南燕 (398–410 CE) rather than the Later Qin, based in Chang'an.

"circles" (*zhou*) and an earth that is "low" (*bei* 卑), completely ignoring the *Zhoubi*'s own explanation:[52]

> 榮方曰:「周髀者，何?」陳子曰: 古時天子治周，此數望之從周，故曰「周髀。」髀者，表也。
> Rong Fang asked "What is meant by [the term] *zhou bi*?" Chenzi replied "In ancient times the Son of Heaven ruled from Zhou. This meant that these numbers were observed at Zhou, hence the term *zhou bi*. *Bi* means *biao* 表, gnomon."[53]

Lastly, in the *Wuli lun* 物理論, the hermit Yang Quan 楊泉 of Kuaiji (fl. 280 CE) offers the following description of the *Zhoubi* that corresponds *not* with the text of the *Zhoubi* but, rather, with Wang Chong of Kuaiji's *Lunheng*:

> 【楊泉】『周髀』立蓋天，言天氣循邊而行從磨石焉。
> [Yang Quan:] The *Zhoubi* established umbrella heaven [cosmology], saying that the heavenly *qi* moves along the edge (surface?) following the grindstone.[54]

> 【王充】日月附天而行，不直行也...其取喻若蟻行於磑上焉。
> [Wang Chong]: The sun and moon move affixed to (i.e. on the surface of) heaven, but they do not move in straight lines. ... By way of analogy, this is like an ant walking atop a millstone [turning in the opposite direction].[55]

Their treatment of the *Zhoubi* is so peculiar that one wonders whether these Southern authors had actually read it, or if they were simply working off of the title and what they know about umbrella heaven cosmology from other sources. Indeed, it is not until after the Sino-Xiongnu book exchange, in Shen Yue 沈約 of Wukang's 武康 (441–513 CE) *Song shu* 宋書 of 492/493 CE, that we finally get an informed description of the *Zhoubi* written in the South.[56]

The Northwest connection is considerably more solid. The *Zhoubi* would have been of little interest to the study of *li* astronomy in Zhao Ying's day except, perhaps, as a historical artefact: experts had begun to turn against the quarter-remainder *li* of the apocrypha as early as the 120s, and they set about trying to replace the failing *Sifen li* of 85 CE with something completely new as early as the 180s, the last holdout being the state of Shu 蜀 (221–263 CE); with the exception of Li Xiu 李修 and Xia Xian's 夏顯 experimental *Qiandu li* 乾度曆 of 275/280 CE, everyone had since turned to the model of Liu Hong's *Qianxiang li*.[57] Among other things, this

[52] Cited in *Taiping yulan*, 2.9b; cf. Cullen (1977: 296–297).

[53] *Zhoubi suanjing*, 1.38a; tr. modified from Cullen (1996: 179). While I would not expect Yu Xi to necessarily *agree* with this passage—he might well dismiss it as an interpolation as does Zhao Ying—I would however expect him to address the explanation that the text in question gives of its own title.

[54] Cited in *Taiping yulan*, 2.7b; cf. Cullen (1977: 272–273).

[55] *Lunheng jiaoshi*, 11.501.

[56] *Song shu*, 23.679. Another Southerner who touches upon the *Zhoubi* is He Daoyang 賀道養, who supports Chen Jizhou and Yu Xi's aforementioned claim: "the *Zhoubi* was probably authored by [a man named] Zhou Bi and does not represent the procedures of the experts of the Zhou [dynasty]" 周髀，當周髀之所造，非周家術也 (*Huntian ji* 渾天記, cited it *Taiping yulan*, 2.10a). All that we know about He Daoyang is that he was active under the Liu Song 劉宋 (420–470 CE), so it is difficult to place him vis-à-vis the book exchange of 437 CE.

[57] This sequence of events is extrapolated from the *Hou Han shu* and *Jin shu* "Lüli zhi."

meant the abandonment of the "obscuration' (*bu*), as all other *li* up to this point skipped straight from the rule (*zhang*) to some equivalent of the era (*ji*) and origin (*yuan*).[58] It therefore stands out in Li Chunfeng's later history when we read that:

> 至於西涼，亦為蔀法，事迹糾紛，未能詳記。
> In the Western Liang 西涼 they however used an obscuration factor (*bufa*), but the traces of the matter contradict one another, leaving us unable to record it in detail.[59]

It is unclear whether Li Chunfeng is referring to "the Western Liang" ruled by the Li 李 dynasty from Dunhuang and Jiuquan 酒泉, 400–421 CE, or one of the many other branches of "the Liang(s) in the West," such as the Northern Liang (397–439 CE) of Guzang. Anyways, Zhao Fei was born in Dunhuang, served in Guzang, and the *Kaiyuan zhanjing* 開元占經 (729 CE) preserves a set of astronomical constants matching "Zhao Fei of Dunhuang's *Jiayin* [origin] *li*" 敦煌趙 甲寅之曆 as adopted by the Northern Wei *circa* 452 CE.[60] Namely, between Jiang Ji's 姜岌 *Sanji jiazi yuan li* 三紀甲子元曆 of 384 CE and He Chengtian's 何承天 (370–447 CE) *Yuanjia li* 元嘉曆 of 444 CE, it records a "Zhao of the Liang's *li*" (梁)〔涼〕趙曆 using an obscuration factor (*bufa*) and a high origin (*shangyuan* 上元) of year *jiayin* 甲寅.[61] Both of these are features typical of the *Yin li* 殷曆 quarter-remainder astronomy of the Eastern Han apocrypha such as reflected in the *Zhoubi* and as explicitly connected in Zhao Ying's commentary.[62] This was apparently the tradition within which Zhao Fei was working, and it is interesting to note that the one other contribution to the outmoded quarter-remainder framework over this period, as recorded in the *Sui shu* "Jingji zhi" 經籍志, was a "Zhao the Hermit's *Sifen li* in 1 roll" 趙隱居四分曆一卷.[63]

Now if I may engage in a little speculation: if the curious "survival" (*cun* 存) of the *Zhoubi* is only first reported in the Chinese written tradition by a collector of rare books in Shuofang in 178 CE, it may well be because he discovered it on the trip north. Either way, judging from what we can reconstruct of its early reception history, it would seem that from 178 CE to 437 CE the *Zhoubi* was a "regional item" whose circulation was essentially limited to the Far North/Northwest. The same was

---

[58] See Sect. 2 above for an explanation of these cycles.

[59] *Sui shu*, 17.416.

[60] *Wei shu*, 107A.2660. Likewise, *Sui shu*, 34.1022, records a "*Hexi jiayin yuan li* in 1 roll, written by the Liang Grand Clerk Zhao Fei" 河西甲寅元曆一卷涼太史趙畋撰 and a "*Jiayin yuan li xu* in 1 roll, written by Zhao Fei" 甲寅元曆序一卷趙畋撰. In *Jiu Tang shu*, 47.2038, we also find a "*Hexi jiayin yuan li* in 1 roll, written by Li Chunfeng" 河西甲寅元曆一卷李淳風撰, where Li Chunfeng presumably goes into further detail about Zhao Fei's work.

[61] *Kaiyuan zhanjing*, 105.5b.

[62] See especially *Zhoubi suanjing*, 2.51b comm. On the *jiayin*-origin, quarter-remainder *Yin li* of the apocrypha, see Takeda (1989).

[63] *Sui shu*, 34.1022. On the transmission history of the *Zhoubi* and Zhao Ying commentary, see Cullen (1996: 157–170). Citing the appearance of Zhao Fei's works in association with the *Zhoubi* in the book exchange of 437 CE, Cullen plays with the idea that Zhao Fei might be a descendant of the commentator Zhao Ying/Shuang (p. 162).

true of the *Lunheng*, at the opposite end of the empire, and their limited regional circulation may well explain why whichever was later is completely oblivious of the other's case for umbrella heaven.

5 Conclusion

In summary, we have taken a critical look at the argument behind Qian Baocong's (1963) influential dating of the *Zhoubi*'s completion to *circa* 100 BCE, noting that it fails to account for the text's links to sphere heaven cosmology (first century BCE) and the *chenwei* apocrypha (first century CE), let alone his own previous arguments that the text was roughly contemporaneous with the *Jiuzhang suanshu* (Qian 1924 [1983], 1932 [1998]). Starting over, we affirmed the two texts' dates of first appearance as solid *termini ante quem*, and we highlighted period-specific language pointing to a common *terminus post quem*: an unexplained shift from *ying* (fill) to *man* (fulfil) across all dated sources for *li* and *suan* mathematics that occurs between 1 BCE and 85 CE. If this holds, the two texts' use of *man* places the closure of the *Jiuzhang suanshu* in 1 BCE/83 CE, and that of the *Zhoubi* in 1 BCE/178 CE. What is more, reviewing the circumstances of its first appearance and subsequent reception history up 437 CE, we have seen that the *Zhoubi* appears to have been a "regional item" of the Far North/Northwest. These precisions have a number of implications for how we read the text moving forward, and, in parting, I would like to leave you with three that I believe particularly worthy of further reflection.

First, let us consider the historical implications of dating the closure of the *Zhoubi* to *circa* 100 BCE (Qian 1963), given the constellation of texts and ideas with which it seems to be related. To review, Fu (1988) and Takeda (1989) have shown that the *Zhoubi* features extensive parallels with *chenwei* apocrypha whose titles only first appear in the Mt. Tai Stone Inscription of 56 CE, the White Tiger Hall debates of 79 CE and court discussions on mathematical astronomy in 85–102 CE.[64] As Cullen (1996) has shown, it would also seem to be unartfully appropriating elements of sphere heaven cosmology and instrumentation documented as having come together over the course of the first century BCE. Now, if Qian (1924 [1983]; 1932 [1998]), Cullen (1996), and myself are right to place its closure in the first or second century CE, that would make the *Zhoubi* a natural product of contemporary trends in classical and technical studies, unremarkable in and of itself. If we stick to *circa* 100 BCE, however, we must accept a more fantastical sequence of events: that the *Zhoubi* is the *source* of these trends, written decades if not centuries before they

[64] Namely, the *Luoshu* 洛書 apocryphon *Zhenyao du* 甄曜度, first mentioned in the Mt. Tai Stone Inscription (cited in *Hou Han shu*, *zhi* 7, 3165); the *Yijing* apocryphon *Qianzao du*, first mentioned in *Baihu tongyi shuzheng*, 9.421 (on which see Nielsen 1995); the *Shangshu* apocryphon *Kaoling yao* 考靈曜, first mentioned in Jia Kui's 賈逵 court opinion on *li* astronomy of 91 CE (*Hou Han shu*, *zhi* 2, 2027–3028; tr. Cullen 2017: 382); the *Chunqiu* apocryphon *Mingli xu* 命曆序, first mentioned in the context of the Quarter-remainder versus Triple Concordance debate of 102 CE (*Hou Han shu*, *zhi* 2, 3033; tr. Cullen 2017: 394). On the apocrypha, see Note 18.

came to prominence, and yet it was (purposefully?) omitted from Liu Xiang's *Qi lüe* catalogue of 6/1 BCE, and everything written in the various fields that it prefigured, only to mysteriously reappear two centuries later. If that were true, we would need to completely rewrite the intellectual history of the Han.

Second, if we can reliably date the closure of the *Zhoubi* to 1 BCE/178 CE, versus 1 BCE/83 CE for the *Jiuzhang suanshu*, we might want to thoroughly revisit the accepted wisdom that the *Zhoubi* necessarily came first. Admittedly, the range on their respective dates of closure leave some room for ambiguity, but their received origin stories and excavated source materials do not. Both were said to have been written by the Duke of Zhou, which is wholly implausible. What *is* plausible is Liu Hui's 劉徽 (fl. 263 CE) claim that the *Jiuzhang suanshu* was (re)compiled from scattered, earlier sources over the course of the Western Han 西漢 (206 BCE–9 CE), seeing as how we have since found such ancestral sources in tombs going back to the third century BCE.[65] We have no such sources for the *Zhoubi*, by contrast, and all that its even later commentator Zhao Ying tells us of its origins is that he deduces that parts of the text were not actually written by the Duke of Zhou.[66] Why does it matter which came first? Mostly it does not, except for the purposes of physically arranging tables and graphs like Fig. 3 above, but some scholars have built elaborate arguments upon the *Zhoubi*'s assumed chronological priority. Consider for example Chemla (2013a, 2014), who argues from the disappearance of expressions for division combining the verb *chu* (eliminate) with variations on *X er yi* X而一 (one per *y*) and *de yi* 得一 (get one [for each time]) that "the *process* of division also underwent a key change" between the *Zhoubi* and *Jiuzhang suanshu*. Obviously, this does not work if there is any ambiguity about which text came first.[67]

[65] Guo and Liu (1998: Introduction) and Chemla and Guo (2004: 43–46, 126–127). For a study of the filiation of individual problems between excavated and received sources concerning the *Jiuzhang suanshu* in particular, see Tan (2016).

[66] Namely, "Rong Fang and Chenzi were men who came after the Duke of Zhou, so this [section] does not belong to the original text of the *Zhoubi*" 榮方陳子是周公之後人，非『周髀』之本文 (*Zhoubi suanjing*, 1.20b comm.), and "This [section] does not belong to the original text of the *Zhoubi* [but is] probably the words of someone asking a teacher, because, [reading the introduction], he (the Duke of Zhou) only wanted to know about the division of the *du* and the origin of the method's procedure" 非『周髀』本文，蓋人問師之辭，其欲知度之所分、法術之所生耳 (*Zhoubi suanjing*, 2.52b comm.). Zhao Ying's deductions about "the original text" are clearly based on a false premise—that the Duke of Zhou is the text's true author—but for a recent argument as to why we should nevertheless take them seriously, see Chemla (2013a: esp. 181n17, 191; 2014: esp. 259, 267n37).

[67] As to ambiguity in chronological order, one notes that Chemla (2013b: 63) contradicts Chemla (2013a; 2014) in reaffirming her longstanding position that "[t]he oldest writing in Chinese devoted to mathematics that has been handed down through the written tradition is the book *The Nine Chapters on mathematical procedures*" (cf. Chemla and Guo 2004: xiv). In focusing solely on sources for *suan* mathematics, moreover, Chemla (2013a, 2014) fails to account for the appearance of such early language in the later Han *Sifen li*, which was composed in 85 CE and saw continued use until 263 CE: "Set out the number for the night clepsydra for the nodal *qi* of that month and multiply it by the obscuration factor; get one (*de yi*) part [for each time you can] eliminate (*chu*) it [by] two hundred, and that gives you the parts [that the sun has] travelled from midnight

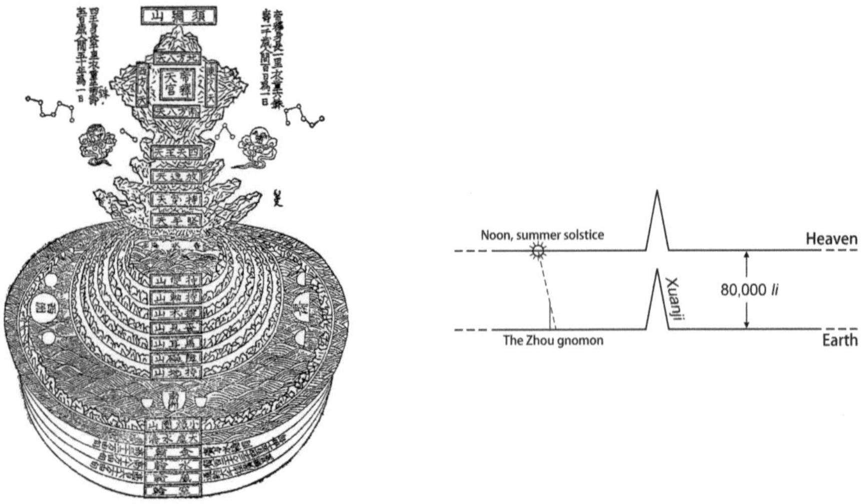

Fig. 4 **Left:** the Buddhist world model, twentieth-century woodblock illustration from the *Taishō shinshū daizōkyō* edition of *Fozu tongji*, *T.* no 2035, 31:308. **Right:** alternate reconstruction of the *Zhoubi* "umbrella heaven" world model, redrawn from Li Zhichao (1993 [2014]: 225, fig. 1). I thank Eric Huntington and Michelle McCoy for helping me select an appropriate image from Buddhist sources

Third, exploring the connection of the *Zhoubi*'s early transmission history with the Far North/Northwest likewise stands to change how we think about its contents. As noted in the opening, for example, what the text itself uses "umbrella heaven" to explain are the geographical conditions at the edges of the Chinese world, and, frankly, that makes sense if that is precisely where is it was written.[68] This naturally brings us to Jiang Xiaoyuan's (1997) famous seven-point argument that the *Zhoubi*'s peculiar cosmology bears such an uncanny resemblance to that of the Puranas (see Fig. 4) that it must have been a product of Indian influence. Qu Anjing (2002: 99–102) aptly refutes the premise of their resemblance, confirming the rather fantastical nature of this claim. To play devil's advocate, however, I would like to stress that we cannot ignore the resemblance that *medieval scholars* saw between the two: himself a Daoist, Li Chunfeng describes Emperor Wu of Liang's 梁武帝 (r. 502–549 CE) famous Buddhist-inspired world model as "completely identical to the text of the *Zhoubi*" 全同周髀之文,[69] while Guanding, the fourth patriarch of

to dawn" 置其月節氣夜漏之數，以蔀法乘之，二百除之，得一分，即夜半到明所行分也 (*Hou Han shu*, *zhi* 3, 3065; tr. modified from Cullen 2017: 177).

[68] Compare this to the flat-earth sphere heaven, which only works if the observer is at "the center of the earth" 地中 in the ancient Zhou capital city of Luoyang; see Yixing's 一行 (683–727 CE) assessment of the two theories' strengths and weaknesses as studied in Jin (1986).

[69] *Sui shu*, 19.507. On Emperor Wu of Liang's cosmology, see Yamada (1975), Jiang (2001: 212–246), Qu (2002: 102–114), and Yuan and Qu (2008).

Tiantai 天台 Buddhism, describes the *Zhoubi* in his commentary to the *Mahāparinirvāṇa-sūtra* as "the same as the Buddhist model" 同於佛法.[70] To me, the most difficult part of Jiang's argument to accept has always been the date: that a Chinese text written around 100 BCE reflects Buddhist cosmology when the first Buddhist temple in Luoyang 洛陽 was only constructed in 68 CE.[71] Well, if we accept that the *Zhoubi* was likely written *after this date* in what *was* or *would become* an important Buddhist center on the Silk Road,[72] that might make the hypothesis of Indian influence worth revisiting as well.

Acknowledgments I thank Tang Quan 唐泉 and Qu Anjing 曲安京 for graciously inviting me to the second ICHMA, where I delivered a preliminary version of this paper, and Alexis Lycas and Eric Trombert for helping with my historical geography.

Bibliography

Primary Sources

Baihu tong shuzheng 白虎通書證, Ban Gu 班固 (32–92 CE); commentary by Chen Li 陳立 (1809–1869); critical edition by Wu Zeyu 吳則虞, Beijing: Zhonghua shuju, 1994 (2 vols.).

Bei shi 北史, Li Yanshou 李延壽 (fl. 618–676 CE), 659 CE; punctuated critical edition, Beijing: Zhonghua shuju, 1974 (10 vols.).

Chouren zhuan 疇人傳, Ruan Yuan 阮元 (1764–1849) et al.; punctuated critical edition by Peng Weiguo 彭衛國 and Wang Yuanhua 王原華 in *Chouren zhuan huibian* 疇人傳彙編, Yangzhou: Guangling shushe, 2008.

*Daban niepan jing shu*大般涅槃經疏, Guanding 灌頂 (561–632 CE); edition *Taishō shinshū daizō-kyō* 大正新脩大藏經, ed. Takakusu Junjirō 高楠順次郎 and Watanabe Kaigkoku 渡邊海旭, Tōkyō: Taishō issaikyō kankō-kai, 1924–1934, *T.* no 1767.

Fozu tongji 佛祖統紀, Zhipan 志磐, 1269; edition *Taishō shinshū daizō-kyō* 大正新脩大藏經, ed. Takakusu Junjirō 高楠順次郎 and Watanabe Kaigkoku 渡邊海旭, Tōkyō: Taishō issaikyō kankō-kai, 1924–1934, *T.* no 2035.

Guoqing bailu 國清百錄, Guanding 灌頂 (561–632 CE); edition *Taishō shinshū daizō-kyō* 大正新脩大藏經, ed. Takakusu Junjirō 高楠順次郎 and Watanabe Kaigkoku 渡邊海旭, Tōkyō: Taishō issaikyō kankō-kai, 1924–1934, *T.* no 1934.

Hou Han shu 後漢書, Fan Ye 范曄 (398–445 CE) et al., commentary by Liu Zhao 劉昭 (sixth century CE) and Li Xian 李賢 (654–684 CE); punctuated critical edition, Beijing: Zhonghua shuju, 1965 (12 vols.).

Jin shu 晉書 (*Book of Jin*), Fang Xuanling 房玄齡 et al., 648 CE; punctuated critical edition, Beijing: Zhonghua shuju, 1974 (10 vols.).

Kaiyuan zhanjing 開元占經. Gautama Siddhārtha 瞿曇悉達, 729. Reprint from Wenyuange Siku quanshu 文淵閣四庫全書, 1782; edition Taibei: Taiwan shangwu yinshuguan, 1983–1986.

---

[70] *Daban niepan jing shu, T.* no 1767, 12:112a19–20.

[71] On the early transmission of Buddhism into China, see Zürcher (1959 [2007]).

[72] See Trombert (2005).

Lunheng jiaoshi 論衡校釋, Wang Chong 王充 (27–c. 100 CE), 70/80 CE; critical edition with annotations by Huang Hui 黃暉 in Xinbian zhuzi jicheng 新編諸子集成, Beijing: Zhonghua shuju, 1990 (4 vols.).
Jiu Tang shu 舊唐書, Liu Xu 劉昫 (887–946 CE), 945 CE; edition Beijing: Zhonghua shuju, 1975 (16 vols.).
Sanguo zhi 三國志, Chen Shou 陳壽 (233–97 CE); edition Beijing: Zhonghua shuju, 1959 (5 vols.).
Shiliuguo chunqiu 十六國春秋, Cui Hong 崔鴻 (478–525 CE), 522 CE; reprint from *Wenyuange Siku quanshu* 文淵閣四庫全書, 1782; edition Taibei: Taiwan shangwu yinshuguan, 1983–1986.
Song shi 宋史. Tuotuo 脱脱 (1314–1355) et al., 1346; punctuated critical edition, Beijing: Zhonghua shuju, 1977 (40 vols.).
Song shu 宋書, Shen Yue 沈約 (441–513 CE), 492/493 CE; punctuated critical edition, Beijing: Zhonghua shuju, 1974 (8 vols.).
Sui shu 隋書. Wei Zheng 魏徵 (580–643 CE), Linghu Defen 令狐德棻 (582–666 CE), et al., 656 CE; punctuated critical edition Beijing: Zhonghua shuju, 1973 (3 vols.).
Taiping yulan 太平御覽, compiled by Li Fang 李昉 (925–996) et al., 984; edition *Sibu congkan* 四部叢刊, Shanghai: Shangwu yinshuguan, 1919–1936.
Tong xingming lu 同姓名錄. Yu Yin余寅 (1541–1615). Reprint from Wenyuange Siku quanshu 文淵閣四庫全書, 1782; edition Taibei: Taiwan shangwu yinshuguan, 1983–1986.
Wei shu 魏書, Wei Shou 魏收 (507–572 CE), 554 CE; edition Beijing: Zhonghua shuju, 1974 (8 vols.).
Xin Tang shu 新唐書, Ouyang Xiu 歐陽修 (1007–1072) and Song Qi 宋祁 (998–1061), 1060; edition Beijing: Zhonghua shuju, 1975 (20 vols.).
Yuelu shuyuan cang Qin jian (er) 嶽麓書院藏秦簡(貳), ed. Zhu Hanmin 朱漢民 and Chen Songchang 陳松長 (Shanghai: Shanghai cishu chubanshe, 2011).
Zhangjiashan Han mu zhujian (ersiqi hao mu) 張家山漢墓竹簡(二四七號墓), ed. Zhangjiashan ersiqi hao Han mu zhujian zhengli xiaozu 張家山二四七號漢墓整理小組 (Beijing: Wenwu chubanshe, 2001).
Zhoubi suanjing 周髀算經, anon., 1 BCE/178 CE, commentary by Zhao Shuang 趙爽 (fl. 314 CE), subcommentary by Zhen Luan 甄鸞 (fl. 535–570 CE) and Li Chunfeng 李淳風 (602–670 CE) et al.; edition *Sibu congkan* 四部叢刊, Shanghai: Shangwu yinshuguan, 1919–1936.
Zhouli zhushu 周禮注疏, pre-Qin (< 221 BCE), commentary by Zheng Xuan 鄭玄 (127–200 CE), subcommentary by Jia Gongyan 賈公彥 (seventh century CE); edition *Shisanjing zhushu* 重刊宋本十三經注疏, 1815; reprint Taibei: Yiwen yinshuguan, 1965.
Zhouyi Qianzaodu周易乾鑿度, anon., *c.* first century CE, commentary by Zheng Xuan 鄭玄 (127–200 CE); Reprint from Wenyuange Siku quanshu 文淵閣四庫全書, 1782; edition Taibei: Taiwan shangwu yinshuguan, 1983–1986.

Secondary Sources

Chemla, Karine. 2013a. Shedding some light on a possible origin of a concept of fractions in China: division as a link between the newly discovered manuscripts and "The Gnomon of the Zhou [Dynasty]." *Sudhoffs Archiv* 97(2): 174–198.
Chemla, Karine. 2013b. Ancient writings, modern conceptions of authorship: reflections on some historical processes that shaped the oldest extant mathematical sources from ancient China. In *Writing science: medical and mathematical authorship in ancient Greece*, ed. Markus Asper, 63–82. Berlin: de Gruyter.
Chemla, Karine. 2014. Observing mathematical practices as a key to mining our sources and conducting conceptual history: division in ancient China as a case study. In *Science after the practice turn in the philosophy, history, and social studies of science*, ed. Léna Soler, Sjoerd Zwart, Michael Lynch, and Vincent Israel-Jost, 238–268. New York: Routledge.
Chemla, Karine, and Guo Shuchun 郭書春. 2004. *Les neuf chapitres: le classique mathématique de la Chine ancienne et ses commentaires*. Paris: Dunod.

Chen Meidong 陳美東. 2003. *Zhongguo kexue jishu shi: tianwenxue juan* 中國科學技術史:天文學卷. Beijing: Kexue chubanshe.

Chen Meidong 陳美東. 2007. *Zhongguo gudai tianwenxue sixiang* 中國古代天文學思想. Beijing: Zhongguo kexue jishu chubanshe.

Cullen, Christopher. 1977. Cosmographical discussions in China from early times up the T'ang Dynasty. Ph.D. diss., University of London.

Cullen, Christopher. 1996. *Astronomy and mathematics in ancient China: the* Zhou bi suan jing. Cambridge: Cambridge University Press.

Cullen, Christopher. 2004. *The* Suàn shù shū 筭數書 *"Writings on reckoning": a translation of a Chinese mathematical collection of the second century BC, with explanatory commentary*. Cambridge: Needham Research Institute.

Cullen, Christopher. 2017. *The Foundations of celestial reckoning: Three ancient Chinese astronomical systems*. New York: Routledge.

Cullen, Christopher, and Anne S. L. Farrer. 1983. On the term *hsüan chi* and the three-lobed jade discs. *Bulletin of the School of Oriental and African Studies* 46(1): 53–76.

Dull, Jack L. 1966. A historical introduction to the apocryphal (*ch'an-wei*) texts of the Han Dynasty. Ph.D. diss., University of Washington.

Fu Daiwie 傅大為. 1988. Lun *Zhoubi* yanjiu chuantong de lishi fazhan yu zhuanzhe 論『周髀』研究傳統的歷史發展與轉折. *Qinghua xuebao* 清華學報 18(1): 1–41.

Goodman, Howard L. 2005. Chinese polymaths, 100–300 AD: the Tung-kuan, Taoist dissent, and technical skills. *Asia Major* 3rd ser., 18(1): 101–174.

Guo Shuchun 郭書春, and Liu Dun 劉鈍. 1998. *Suanjing shishu* 算經十書. 2 vols. Shenyang: Liaoning jiaoyu chubanshe.

Han Wei 韓巍. 2013. Beida Qin jian *Suan shu* tudi mianji lei suanti chushi 北大秦簡『算書』土地面積類算題初識. *Jianbo* 簡帛 8: 29–42.

Ho Peng Yoke 何丙郁. 1966. *The astronomical chapters of the* Chin shu. Paris: Mouton.

Jiang Xiaoyuan 江曉原. 1996. *Zhoubi suanjing*: Zhongguo gudai weiyi de gonglihua changshi『周髀算經』—中國古代唯一的公理化嘗試. *Ziran bianzhengfa tongxun* 自然辯證法通訊 18(3): 43–48.

Jiang Xiaoyuan 江曉原. 1997. *Zhoubi suanjing* yu gudai yuwai tianxue 『周髀算經』與古代域外天學. *Ziran kexue shi yanjiu* 自然科學史研究 16(3): 207–212.

Jiang Xiaoyuan 江曉原. 2001. *Jiang Xiaoyuan zixuan ji* 江曉原自選集. Guilin: Guangxi shifan daxue chubanshe.

Jin Zumeng 金祖孟. 1986. Yixing bu shi huantianjia 一行不是渾天家. In *Zhongguo tianwenxue shi wenji* 中國天文學史文集, ed. Zhongguo tianwenxue shi wenji bianjizu 中國天文學史文集編輯組, vol. 4, 49–151. Beijing: Kexue chubanshe.

Li Zhichao 李志超. 1993 [2014]. *Tianren gu yi: Zhongguo kexueshi lungang* 天人古義:中國科學史論綱. 3d ed. Zhengzhou: Daxiang chubanshe.

Morgan, Daniel Patrick. 2021. Regional Networks in Chinese Mathematics and Astronomy, 317–618 CE." *East Asian Science, Technology, and Medicine* 53(1–2): 1–53.

Morgan, Daniel Patrick. 2019. Heavenly patterns. In *Monographs in Tang official history: Perspectives from the technical treatises of the* History of Sui *(*Sui shu*)*, ed. Daniel Patrick Morgan and Damien Chaussende, with the collaboration of Karine Chemla. Cham: Springer Nature Switzerland AG.

Morgan, Daniel Patrick. 2017. *Astral sciences in early imperial China: observation, sagehood and the individual*. Cambridge: Cambridge University Press.

Morgan, Daniel Patrick. 2015a. By process of elimination: further remarks on the operation *chú* 除 in early imperial mathematical astronomy. Paper delivered at Mathematical Practices in Relation to the Astral Sciences, Université Paris Diderot, 26–31 March 2015 (https://halshs.archives-ouvertes.fr/halshs-01333725v1).

Morgan, Daniel Patrick. 2015b. Reflections upon the presentation of parallel algorithms across the astral and mathematical sciences in first-millennium China. Paper delivered at the International Conference on History of Ancient Mathematics and Astronomy, Northwestern University, Xi'an, 23–29 August 2015 (https://halshs.archives-ouvertes.fr/halshs-01333721).

Morgan, Daniel Patrick. 2015c. What good's a text? Textuality, orality, and mathematical astronomy in early imperial China. *Archives internationales d'histoire des sciences* 65(2): 549–572.

Nielsen, Bent. 1995. The *Qian Zuo Du* 乾鑿度: A Late Han Dynasty (202 BC–AD 220). Ph.D. diss., Copenhagen University.

Qian Baocong 錢寶琮. 1963. *Suanjing shishu* 算經十書. 2 vols. Beijing: Zhonghua shuju.

Qian Baocong 錢寶琮. 1924 [1983]. Zhoubi suanjing kao 周髀算經考. In *Qian Baocong kexueshi lunwen xuanji* 錢寶琮科學史論文選集, ed. Zhongguo kexueyuan ziran kexueshi yanjiusuo 中國科學院自然科學史研究所, 119–136. Beijing: Kexue chubanshe.

Qian Baocong 錢寶琮. 1932 [1998]. *Zhongguo suanxue shi* 中國算學史. In *Li Yan, Qian Baocong kexueshi quanji* 李儼、錢寶琮科學史全集, vol. 1, ed. Du Shiran 杜石然, Guo Shuchun 郭書春, and Liu Dun 劉鈍. Shenyang: Liaoning jiaoyu chubanshe.

Qu Anjing 曲安京. 2002. Zhoubi suanjing *xinyi* 『周髀算經』新議. 2 vols. Xi'an: Shaanxi renmin chubanshe.

Shaughnessy, Edward L. 1991. *Sources of Western Zhou history: inscribed bronze vessels*. Berkeley: University of California Press.

Sun Xiaochun 孫小淳 and Jacob Kistemaker. 1997. *The Chinese sky during the Han: constellating stars and society*. Leiden: Brill.

Takeda Tokimasa 武田時昌. 1989. Isho rekihō kō: Zenkan matsu no keigaku to kagaku no kōryū 緯書曆法考:前漢末の經學と科學の交流. In *Chūgoku kodai kagakushi ron* 中國古代科學史論, ed. Yamada Keiji 山田慶兒, 55–120. Kyōto: Kyōto daigaku jinbun kagaku kenkyūjo.

Tan Jingnan 譚競男. 2016. Chutu Qin Han suanshu wenxian ruogan wenti yanjiu 出土秦漢算術文獻若干問題研究. Ph.D. diss., Wuhan University.

Trombert, Éric. 2005. Dunhuang avant les manuscrits: conservation, diffusion et confiscation du savoir dans la Chine médiévale. *Études chinoises* 24: 11–55.

Wang Yankun 王彥坤, ed. 1997. *Lidai bihuizi huidian* 歷代避諱字匯典. Zhengzhou: Zhongzhou guji chubanshe.

Wu Jiabi 武家璧. 2007. *Yiwei: Tonggua yan* zhong de guiying shuju 『易緯·通卦驗』中的晷影數據. *Zhouyi yanjiu* 周易研究 2007.3: 89–94.

Wu Wenjun 吳文俊, ed. 1998–2004. *Zhongguo shuxueshi daxi* 中國數學史大系. 10 vols. Beijing: Beijing shifan daxue chubanshe.

Yamada Keiji 山田慶兒. 1975. Ryō Bu no gaitenron 梁武の蓋天説. *Tōhō gakuhō* 東方學報 48: 99–134.

Yuan Min 遠敏, and Qu Anjing 曲安京. 2008. Liang Wudi de gaitianshuo moxing 梁武帝的蓋天説模型. *Kexue jishu yu bianzhengfa* 科學技術與辯證法 25: 85–89, 104.

Zürcher, Erik. 1959 [2007]. *The Buddhist conquest of China the spread and adaptation of Buddhism in early medieval China*. 3d ed. Leiden: Brill.

What Is "the Science of *Du* and *Shu*" 度數之學 in the *Chongzhen lishu* 崇禎曆書?

Deng Kehui

Abstract The *Chongzhen lishu* 崇禎曆書 introduced classical European astronomy, mathematics, and metrology to late-Ming China. One of the editors, Xu Guangqi 徐光啟 (1562–1633), studied "the science of *du* and *shu*" from translating Euclid's *Elements* and applied it to calendar reform. This article discusses the meaning and peculiarity of "the science of *du* and *shu*" in Xu Guangqi's thought. This article establishes three points. First, Xu criticized basic Chinese mathematical theory as unclear, since every dynasty inevitably amended the calendar once or several times, unlike the Western calendar, which was perfected by expansion and teaching by hundreds of generations. As a result, the "meaning and reason" 義理 was progressively better understood, and the "methods and numbers" 法數 became distinct. Moreover, the whole system went from the simple to the complex, which is one of the characteristic of axiomatic science. Second, the science of *du* and *shu* was introduced in the *Dace* 大測, *Celiang quanyi* 測量全義, *Cetian yueshuo* 測天約說, and *Bili guijie* 比例規解, which provided basic knowledge for calendar reform. Third, at the foundation of "the science of *du* and *shu*" was geometrical deduction and logic, its main application to astronomy being the establishment of the geometrical models of the heavenly bodies via observational data, demonstration of its model parameters, and empirical verification.

Keywords The science of *du* and *shu* (度數之學) · Euclid's *elements* · *Chongzhen lishu*崇禎曆書 · Calendar reform

1 Introduction

The late-Ming *Chongzhen lishu* 崇禎曆書 introduced a certain amount of European astronomical knowledge in Chinese, and it contained a great deal of pre-modern mathematics and observational knowledge. Xu Guangqi 徐光啟 (1562–1633) was

Deng Kehui (✉)
Donghua University, Shanghai, People's Republic of China

D. P. Morgan, T. Quan (eds.), *Mathematics and Astronomy in the Ancient World*,
https://doi.org/10.1007/978-3-031-93809-2_5

one of its chief editors and a leader of the related calendar reform. From his first exposure to ultimate mastery of this Western knowledge, Euclid's *Elements* played an important role for Xu Guangqi.

This knowledge was referred to as "the science of *du* and *shu*" (*dushu zhi xue* 度數之學), or "measure and number," a term derived from the translation of Euclid's *Elements*. Its practical application lied in calculation and proving the calendar, but the science of *du* and *shu* also included a set of logical deductions which require mathematical proof. However, how is the science of *du* and *shu* reflected in the editing of *Euclid's Elements*? How does the system operate? There is a hidden clue in the *Chongzhen lishui* concerning the science of *du* and *shu* that rarely gets attention, and elsewhere I have published a study arguing that the science of *du* and *shu* had its source in Greece and was adapted in Medieval Europe, mainly in the field of astronomy and mathematics.[1] Most scholars focus on the *Elements*' influence on the mathematics of the Ming and Qing Dynasties, but few consider its impact on Chinese astronomy.

2 The Growth of Xu Guangqi's Calendar Reform Thought

Matto Ricci (Li Madou 利瑪竇, 1552–1610) and Xu Guangqi produced a collaborative translation of the first six volumes of Euclid's *Elements* in 1604. They wrote separate prefaces and forewords to the work wherein they explain the science of *du* and *shu*. According to Ricci:

> 幾何家者，專察物之分限者也，其分者若截以為數，則顯物幾何眾也；若完以為度，則指物幾何大也。
> A geometrician is someone specialized in investigating the portions and extent of things. If a portion is cut up to make [a] *shu* (number), then it manifests how numerous the thing is; if it is whole and made into [a] *du* (measure), then it indicates how big the thing is.

And as to the work itself:

> 曰「原本」者，明幾何之所以然，凡為其說者，無不由此出也。...題論之首，先標界說，次設公論，題論所據；次乃具題，題有本解，有作法，有推論，先之所征，必後之所恃。
> It is called "the *Elements*" because it explains the reason why geometry is as it is, and because of all theories/explanations thereof, there are none that issue from this [book]. … Before the propositions (*ti* 題) and the demonstrations (*lun* 論), the definitions (*jieshuo* 界說) have been laid down. After that, the axioms (*gonglun* 公論) have been formulated on which the propositions and the demonstrations rest. Next, the propositions follow. They give a first explanation (*jie* 解), a manner of construction (*zuofa* 作法) and a proof (*tuilun* 推論). What comes later is based on the results that have been obtained before.[2]

[1] Deng Kehui (2017a).

[2] Ricci, "Jihe yuanben xu" 幾何原本序, in *Li Madou Zhongwen zhuyi ji*, 303.

No one had really heard of the *Elements* in China prior to Ricci's translation, but it would go on later to play an important role in calendar reform.[3]

Likewise, in Xu Guangqi's preface we read:

> 『幾何原本』者，度數之宗，...利先生...獨謂此書未譯，則他書俱不可得論。
> The *Jihe yuanben* is the ancestor of the science of *du* and *shu*… and Mr. Ricci… emphasized that no other [Western] book can be discussed without first translating this book.[4]

Clearly, Xu believed that the science of *du* and *shu* in the *Elements* was the starting point of all knowledge and, especially, for innovation, and this is evident in his later mathematical activities.

First, Xu Guangqi thought that a thorough understanding and mastery of the basic principles and the timeless methods of the *Elements* would permit their extension to another place and time. Therefore, he divided the *Chongzhen lishu* into "six orderly sections" 節次六目 and "five essential parts" 基本五目 in order from easy to difficult.[5]

Second, Xu proposed "ten affairs worked out by the science of *du* and *shu*" 度數旁通十事 in his memorial of Chongzhen year 2 (1629), "A Point-by-Point Discussion of the Calendar and Correcting for Precession" (Tiaoyi lifa xiuzheng suicha shu 條議曆法修正歲差疏).[6] These affairs included astronomy, geography, water conservation, observation, music, military affairs, architecture, manufacture, medicine, and making sundials.

Third, Xu presented three different calendar reform proposals to the emperor. The first involved indiscriminately imitating the Western calendar. The second was "to distinguish theories and establish methods following the [six] orderly sections and to separate tasks to manage its operation [by] the five essential parts" 依循節次，辨理立法，基本五事，分任經營. The third was that…

> 一義一法，必深言所以然之故，從流溯源，因枝達幹，不止集星曆之大成，兼能為萬務之根本。
> … for one [and every] meaning and one [and every] method one must profoundly speak to the reason why it is the way it is, following its course and tracing its origin, going from its branches to arrive at the trunk, not merely a great achievement in collecting calendars but at once capable of serving as the foundation of ten thousand duties.[7]

However, they were pressed for time, and the emperor only adopted the second proposal.

[3] *Li Madou Zhongguo zhaji*, 303.

[4] Deng Kehui (2017a).

[5] *Chongzhen lishu* 崇禎曆書, "Yuanqi yi" 緣起一, cited in *Xinfa suanshu*, 1.17b; cf. Pan Nai (1994).

[6] *Xu Guangqi ji*, vol. 2, 337–338.

[7] *Xu Guangqi ji*, vol. 2, 377.

3 The Science of *Du* and *Shu* in the *Dace* 大測, *Celiang quanyi* 測量全義, and *Cetian yueshuo* 測天約說

Jean Terrenz (Deng Yuhan 鄧玉函, 1576–1630) compiled the *Dace* 大測 (*The Great Measurement*) in 2 volumes in February 1631. In his preface to the *Dace*, he says:

> 「大測」者，測三角形法也。...三角形之理，非句股可盡，故不名句股也。...測天則圜面、曲線，非句股所能得也。故有弧、矢、弦割圜之法，弧者，曲線，弦、矢者，直線也。以弧求弧，無法可得，必以直線、曲弧相當相準，乃可得之，「相當相準」者，圍徑之法也。而圍與徑，終古無相準之率。
>
> "The Great Measurement" (*Dace*) is the method of determining triangles… The theory of triangles is not exhausted by *gougu* 句股, thus we do not name it *gougu*. It measures heaven by its round face and curved lines, of which *gougu* is incapable. Accordingly, [the *Dace*] has methods of cutting circles into bow (arc), arrow, and chord, the "bow" being a curved line, and the "arrow" and "cord" straight lines. There is no method to find an arc from an arc; one must mutually match and determine [it] using the straight lines and curved arc, and only then can it be obtained. "Mutually matching and determining" [refers to] the perimeter and diameter method, and there was no determining ratio for perimeter and diameter throughout all of [Chinese] antiquity.[8]

Later down, he explains, "It is named 'the Great Measurement' because the measurement of heaven is necessarily greater than other measurements" 測天者所必須，大於他測，故名大測.[9]

Therefore, trigonometry was introduced into China by the *Dace*, which refers to it as "eight lines cutting the circle" 割圓八線 (Fig. 1). It points out that this method was the right meaning of *du* and *shu*. The error of the *Dace* table was not one in ten

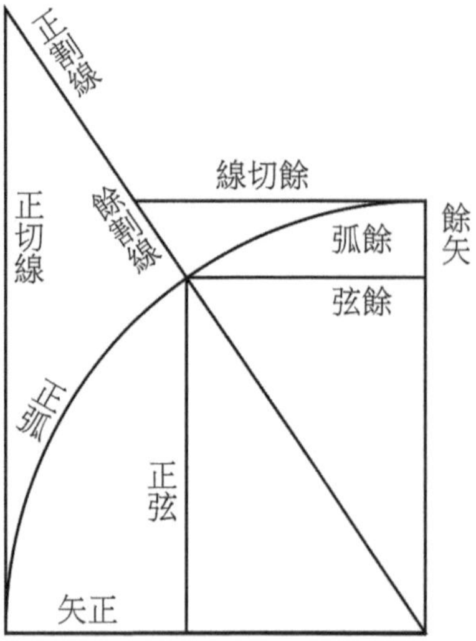

Fig. 1 "Baxian biao quantu" 八線表全圖. Redrawn from *Chongzhen lishu: Fu Xiyang xinfa lishu zengkan shizhong*, 1208

[8] *Dace*, 9.1a–b.

[9] *Dace*, 9.2b.

thousand (1/10000). According to Bai Shangshu (1963), the basis of the *Dace* was *Trigonometriae Sive De Dimensione Triangulae Libri Quinque* of 1595 by B. Pitiscus (1561–1613), but it was Terrenz who invented the Chinese term for trigonometry. The *Dace* has six sections and chapters, which explain how to geometrically cut circles with eight lines. These contents point to a deep influence by the *Elements*, and Xu Guangqi et al. recognized this, after studying the *Elements* themselves, when translating the *Dace*.

For example, one of the spherical trigonometry problems in the *Dace* involves "finding an arc from an arc" 以弧求弧. The author's knowledge of the "six purpose ratios" 六宗率 and "three important methods" 三要法 comes from Euclid's *Elements*. In the chapter "Geyuan pian" 割圓篇, Terrenz defines the basic trigonometric lines as the "eight lines cutting the circle," corresponding to the Western functions of sine, cosine, tangent, cotangent, secant, and cosecant. Since this was equivalent to the ancient Chinese procedure of dividing the circle into "bow" (arc) and "arrow" (sagitta), however, Xu Guangqi considered this a case of "using another's terms to get at what we mean" 以彼條款，就我名義.

This leads us to believe that the method of determining triangles comes from the science of *du* and *shu* in Euclid's *Elements*. The reasons for this are as follows:

(a) The section on trigonometric functions is titled "Section on Cutting Circles" (Geyuan pian 割圓篇), and its contents follow a logical layout of "proof" (*lun* 論), "corollary" (*xi* 系), and "explanation" (*jie* 解).
(b) The "method of three ratios" (*sanlü fa* 三率法) is to set two ratios as two given numbers and then work out the other. Although it solves problems of arithmetic, it in fact comes from the geometry of determining triangles.
(c) The *Dace* table is sometimes named "Triangular Table" (Sanjiaoxing biao 三角形表), and sometimes "*Du* and *Shu* Table" (Dushu biao 度數表).

What then does the *Dace* do? (1) It illuminates in principle the relationship between the science of *du* and *shu* and trigonometry. (2) It explains the methods of making tables of trigonometric functions based on the *Jihe yuanben*. (3) It provides the method to settle a problem of plane-triangles.

Jacobus Rho (Luo Yagu 羅雅谷, 1593–1638) compiled the *Celiang quanyi* 測量全義 in 10 volumes in 1631. The book discusses the determination of various geometric figures, including their area and volume, as well as the basics of plane trigonometry, spherical trigonometry, spherical astronomy, the manufacture of survey instruments, and cartography.

In the preface, Rho writes:

> 夫歷法之原有二:其一則象緯之原也，天事也，其一則推測之原也，人事也。...此書所論則推測之原也。古今言推測者又有二:其可以形察、可以度審者，謂之叀(專)術，不可以形察、不可以度審者，謂之綴術。此所論者，又綴術也。綴術之用，又有二:其一總物以為度，論其幾何大，曰量法也，其一截物以為數，論其幾何衆，曰算法也。歷象之家，兼用二法，如鳥之縛兩翼也，則無所不可之矣。
> The calendar has two origins: one is its origin in [celestial] signs and [solar] declination, which are matters of heaven; the other is its origin in calculation and measurement, which are matters of man. The subject of this book is its origin in calculation and measurement. Discussions of calculation and measurement in ancient and modern times can also [be

> divided into] two [types]: one concerns that which can be investigated by form and examined by *du*, which we call *zhuanshu* 專術; the other concerns that which cannot be investigated by form or examined by *du*, which we call *zhuishu* 綴術. The application of *zhuishu* [can also be divided into] two: one gathers things to make *du* and assess how big they are, which is called "mensuration" (*liangfa* 量法); the other divides things to make *shu* and assess how many they are, which is called "calculation" (*suanfa* 算法). [In each case], astronomers uses both methods together, like a bird depends on two wings, thus there is nothing to disapprove about [either].[10]

Above measuring the magnitude of things via "mensuration" (*liangfa* 量法) and calculating the quantity of things via "calculation" (*suanfa* 算法) was the science of *du* and *shu*.

Jean Terrenz said in the *Cetian yueshuo* 測天約說:

> 度數之學凡有七種，共相連綴。初為二本，曰數，曰度，數者，論物幾何衆，其用之，則筭法也；度者，論物幾何大，其用之，則測法量法也。……七者在西土庠士俱有專書，今翻譯未廣。僅有『幾何原本』一種，或多未見未習，然欲略舉測天之理與法，而不言此理此法，即說者無所措其辭，聽者無所施其悟矣。
> The science of *du* and *shu* has a total of seven types that are all interconnected. At the beginning it had two roots: *du* and *shu*. *Shu* are for discussing how many things there are, and its application is calculation (*suanfa*), while *du* are for discussing how big things are, and its application is measurement (*cefa* 測法), [or] mensuration (*liangfa*).... These seven [aspects] were all the subjects of specialized monographs among Western scholars, [but] they have yet to be widely translated [in Chinese]. There is only the *Jihe yuanben*, which as yet barely appears and is barely studied. However, if one wants to discuss the theory and method of measuring the sky without mention of this theory and this method, then the speaker will have nothing by which to form his words, and the listener nothing by which to realize his understanding.[11]

Clearly, the "science of *du* and *shu*" of the *Elements* was an important basis for Terrenz's work: the *Cetian yueshuo* inherited its "theory and method," which it further elaborated into a "theory and method of measuring the sky," with the understanding that, without the foundation laid in the former, the latter application to questions of astronomy would not stand.

4 The Science of *Du* and *Shu* in the Other Volumes of the *Chongzhen lishu*

Giulio Aleni (Ai Rulüe 艾儒略, 1582–1649) wrote the *Jihe yaofa* 幾何要法 in 1631, which was later compiled in the *Chongzhen lishu*. The book is about the pure science of figures or geometry apart from things. It says: "Geometry is the storehouse of *du* and *shu*" 幾何者，度與數之府也.[12]

---

[10] Rho, "Celiang quanyi xu" 測量全義叙, cited in *Xinfa suanshu*, 87.1b–2a.

[11] Terrenz, *Cetian yueshuo*測天約說, cited in *Xinfa suanshu*, 11.2a–2b; cf. Deng Kehui (2017b: 217).

[12] Aleni, *Jihe yaofa* 幾何要法, "Jihe zonglun" 幾何總論, cited in *Xinfa suanshu*, 83.1a.

The first chapter of the book is "Jieshuo" 界說, which covers primary concepts and definitions, including those of various geometrical figures. The second chapter is "Beiqi" 备器, which introduces the drawing tools *chi* 尺, *gui* 規 and *ju* 矩. The third chapter is "Yaofa" 要法, which explains how to draw figures with these three tools. This chapter is an essential portion of the book, since drawing has a rigorous procedure and includes logical proof.

The *Huntianyi shuo* 渾天儀說 was written by Johann Adam Schall von Bell (Tang Ruowang 湯若望, 1592–1638) and then included in the *Xiyang xinfa suanshu* 西洋新法算數 (*Book of New Calendar in the Western*) in 1645. In his preface, Li Tianjing 李天經 (1579–1659) writes:

> 夫晰其理者求其故，考其數者步其行，器固未可廢也，而用斯要焉。
> This book will explain the reason of things and for the reason of them,.... the armillary sphere can help the observation sky and the mathematician was able to calculate and measure the *du* 度 of it."[13]

Chousuan 筹算 is about arithmetic, but it is related to the science of *du* and *shu*. It thinks that arithmetic could play an important part in solving problems of the science of *du* and *shu*.

The *Bili guijie* 比例規解 (*Explanation of the Proportional Compass*) was written by the Italian missionary Jacobus Rho in the third year of Chongzhen. It introduces the proportional compass and illustrates its mathematical design principles and the methods of drawing sundials. In its preface, Rho writes:

> 天文曆法等學，舍度與數則授受不能措其辭，故量法、算法恒相符焉。...今系『幾何』六卷六題，推顯比例規尺一器，其用至廣，其法至妙，前諸法器不能及之。因度用數開合其尺，以規取度得算最捷。
> We cannot speak about teaching or learning such studies as astronomy and the calendar without *du* and *shu*, [because] mensuration (*liangfa*) and calculation (*suanfa*) are always mutually in accordance therewith. ... Here, following the sixth proposition of the sixth volume of the *Jihe* [*yuanben*], we expound upon the instrument of the proportional compass ruler, whose application is supremely vast, whose method is supremely subtle, and to which no previous instrument can compare. Use *shu* according to *du* to open and close its ruler; take *du* using the compass and that is the quickest way to obtain a calculation.[14]

5 The Geometrical Model in the *Chongzhen lishu*

The geometrical model in Greece was introduced into China along with Ptolemy, Copernicus, and Tycho Brahe's cosmologies in the *Chongzhen lishu*. It not only connected observation and theory, it also allowed for the examination and emendation of theory through observation.

[13] "Hunyishuo yongfa xu" 渾儀說用法序, in *Chongzhen lishu: Fu Xiyang xinfa lishu zengkan shizhong*, vol. 2, 1801.

[14] Rho, "*Bili guijie* xu" 比例規解序, in Deng Kehui (2017b: 272).

Xu Guangqi advocated Western mathematical astronomy, saying:

著述則有法有論，有度有數，講究推步，動經歲月。

We wrote [this book, which] has both method and theory, *du* and *shu*, a careful study of observation and calculation as accumulated over time.[15]

This paper will offer an example of lunar theory concerning the selection of eclipse observations and the demonstration of the equivalency of geometrical models.

5.1 "Method of Selecting Eclipses": Two Ideal Hypotheses Introduced

In Rho's *Yueli lizhi* 月離曆指, it says:

是故曆家設擇食之法，擇者，導則也。去其不齊之緣，以求其齊也。第一在日躔經度，...則擇食之第一法，宜擇兩食之日躔經度所在等，既免此緣，則餘二緣在月之本行本輪，日無與也。

It is for this reason that the calendricists set up the [two] methods of selecting [lunar] eclipses, their selection being a careful choice. [In short,] one eliminates the unaligned factors (*yuan* 緣) to find alignment (i.e., the minor effects of uneven motion). [The method of the choice of the lunar eclipses is divided into two steps]: The first thing lies in solar longitude… this is the first method for selecting an eclipse. One should select two eclipses at the same solar longitude, and once this factor has been avoided (i.e., eliminated), then the remaining two factors lie on the lunar epicycle, independent of the sun.[16]

As illustrated in Fig. 2, one begins with an alignment of "factors" (*yuan*) between two eclipses at the same longitude, with the moon at the same position in it its epicycle, i.e. in the anomalistic month (*yue zhuanfen* 月轉分), etc. If the time between the two eclipses chosen is sufficiently long to eliminate the factor of lunar and solar anomaly, it allows the calculation of their lease common multiple (*qitong zhi lü* 齊同之率).

The second method of eclipse selection involves choosing two eclipses at which the moon is the same distance to the centre of the earth, at perigee or apogee. In order to determine its distance away at apogee and perigee, one must measure by longitude. So if the longitude, time, and right and left distance of the moon are same between the two eclipses, so too will be the position of the Moon in its epicycle. This can also be measured by the diameter of the moon before the eclipse, which will be the same if the moon is at the same distance from the center of the earth at eclipse. This allows us to eliminate the second and third "factors" (*yuan*).

It is my opinion that these methods of eclipse selection derive from the *Almagest*. Ptolemy thought it impossible to select two completely identical eclipse

[15] *Xu Guangqi ji*, vol. 2, 331.

[16] Rho, *Yueli lishi*, in *Chongzhen lishu: Fu Xiyang xinfa lishu zengkan shizhong*, 135.

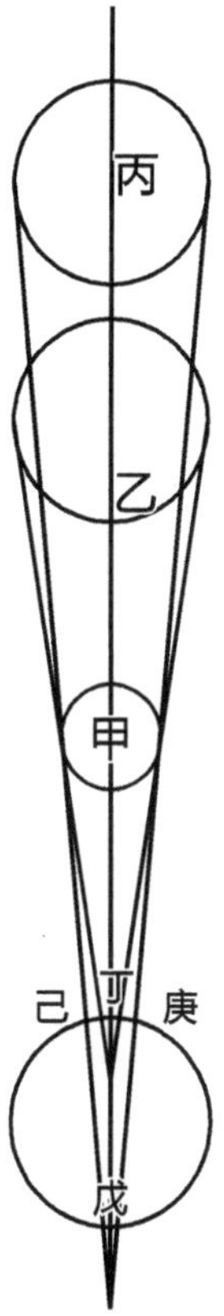

Fig. 2 “Qu qi buqi zhi yuan tu” 去其不齊之緣圖 (Diagram of the Elimination of Unaligned Elements). Redrawn from *Chongzhen lishu: Fu Xiyang xinfa lishu zengkan shizhong*, 1.135

observations, and thus decomposed the problem into “two hypotheses.”[17] These “two hypotheses” are in fact the same as the two methods of eclipse selection in Rho’s *Yueli lizhi*.

5.2 *Evidence of Equivalence on the Epicycle and Eccentric*

In what follow in Rho’s *Yueli lizhi*, we find more evidence of the Ming-era transmission of Ptolemy’s method, founded on observation and hypothesis, and Euclidian demonstration. This is especially the case as concerns the equivalence of the epicycle and eccentric, which, as Rho explains, are “different names for the same theory” 異名同理 (see Figs. 3 and 4).

Looking at it from this, every concept of geometrical modelling and corresponding explanation concerning ancient Greek lunar theory had already taken initial form in the *Yueli lizhi*. Founded on a combination of observation and hypothesis,

[17] Toomer (1984: 176–179).

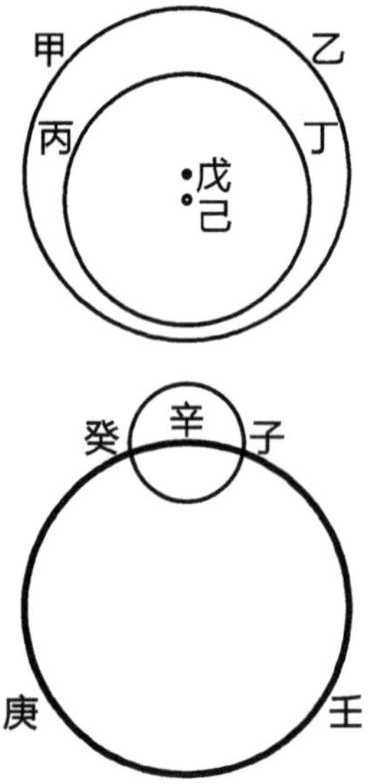

Fig. 3 "Butong xinquan moxing yu benlun moxing" 不同心圈模型與本輪模型 (Different Eccentric and Epicycle Models). Redrawn from *Chongzhen lishu: Fu Xiyang xinfa lishu zengkan shizhong*, 1.140

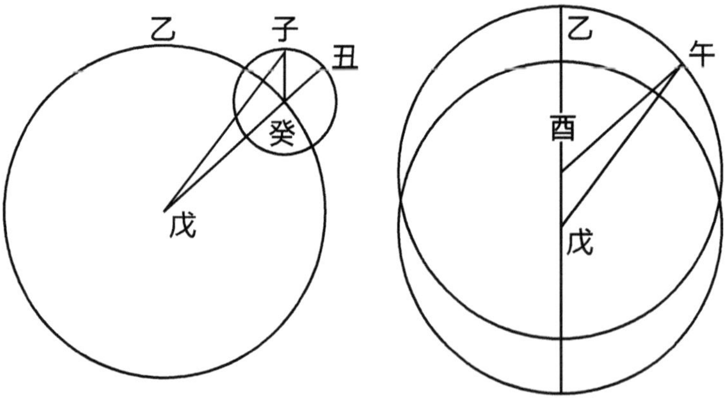

Fig. 4 Demonstration that the solar eccentric and epicycle models are "different names for the same theory" 異名同理. Redrawn from *Chongzhen lishu: Fu Xiyang xinfa lishu zengkan shizhong*, 1.141

Ptolemy's scientific method already appears therein. And the practice of proof based on Euclid's *Elements* had already begun to permeate astronomy.

6 Conclusion

"The science of *du* and *shu*" came from the translation of Euclid's *Elements*, its theory based on logical deduction and mathematical proof, and its application in practice being calculation and the proving of the calendar. Like the *Elements*, the

Jihe yaofa and certain contents of the *Celiang quanyi* treated the science of *du* and *shu* as a matter of pure geometrical forms, separate from the world. It was also applied, however, to mathematical astronomy (the calendar), which deals with observation and spherical (rather than planar) geometry. In the *Dace, Celiang quanyi, Cetian yueshuo, Bili guijie, Chousuan*, and *Huntianyi shuo*, it was also applied to trigonometry, proportion, and arithmetic and geometrical models. These works discuss the different meaning of the science of *du* and *shu*. In a more general sense, the science of *du* and *shu* played an important role in Xu Guangqi’s work on calendar reform, though it was, in the narrow sense, in the name of geometry, arithmetic, proportional arithmetic, and trigonometry.

Bibliography

Primary Sources

Chongzhen lishu: Fu Xiyang xinfa lishu zengkan shizhong 崇禎曆書:附西洋新法曆書增刊十種, ed. Pan Nai 潘鼐. Shanghai: Shanghai guji chubanshe, 2009.

Dace 大測, by Jean Terrenz (Deng Yuhan 鄧玉函, 1576–1630) and Xu Guangqi 徐光啟 (1562–1633), 1631; woodblock print edition of *Wenyuange Siku quanshu* 文淵閣四庫全書, 1782; reprinted Taibei: Taiwan shangwu yinshuguan, 1983–1986.

Li Madou Zhongguo zhaji 利瑪竇中國札記 (*De Christiana expeditione apud Sinas*), by Matteo Ricci (Li Madou 利瑪竇, 1552–1610) and Nicolas Trigault (Jin Nige 金尼閣, 1577–1628); translated by He Gaoji 何高濟, Wang Zunzhong 王遵仲, and Li Shen 李申; Beijing: Zhonghua shuju, 2010.

Li Madou Zhongwen zhuyi ji 利瑪竇中文著譯集, ed. Zhu Weijing 朱維錚; Shanghai: Fudan daxue chubanshe, 2007.

Xinfa suanshu 新法算書, compiled by Xu Guangqi 徐光啟 (1562–1633) et al., 1644; woodblock print edition of *Wenyuange Siku quanshu* 文淵閣四庫全書, 1782; reprinted Taibei: Taiwan shangwu yinshuguan, 1983–1986.

Xu Guangqi ji 徐光啟集(全二册), by Xu Guangqi 徐光啟 (1562–1633); edited by Wang Zhongmin 王重民; Shanghai: Shanghai guji chubanshe, 1984 (2 vols.).

Secondary Sources

Bai Shangshu 白尚恕. 1963. Jieshao Woguo diyi bu sanjiaoxue: “Dace” 介紹我國第一部三角學——「大測」. *Shuxue tongbao* 數學通報 1963.2: 48–51.

Deng Kehui 鄧可卉. 2017a. Mingmo de dushu zhi xue ji qi zai lifa gaige zhong de yingyong 明末的度數之學及其在曆法改革中的應用. *Ziran bianzhengfa yanjiu* 自然辯證法研究 2017.10: 101–106.

Deng Kehui 鄧可卉. 2017b. *Zhongwai tianwenxue wenxian jiaodian yu yanjiu* 中外天文學文獻交點與研究. Shanghai: Shanghai Jiaotong daxue chubanshe.

Pan Nai 潘鼐. 1994. *Chongzhen lishu* de chengshu qianhou 『崇禎曆書』的成書前後. In: *Zhongguo tianwenxue shi wenji (liu)* 中國天文學史文集(六), ed. Zhongguo tianwenxue shi wenji bianjizu 中國天文學史文集編輯組, 1–29. Beijing: Kexue chubanshe.

Toomer, Gerald J. 1984. *Ptolemy’s Almagest*. London: Gerald Duckworth & Co. Ltd

Part II
Korea

Binomial Expansions in Joseon Mathematics

Hong Sung Sa and Hong Young Hee

Abstract Extending the *zengcheng kaifangfa* 增乘開方法 for square and cube roots, Jia Xian 賈憲 (11th cent.) introduced the triangle for binomial expansions, or binomial coefficients for extractions of *n*th roots. Using *tianyuanshu* 天元術 and the *zengcheng kaifangfa* for polynomial equations in thirteenth-century China, the theory of equations became the most important development in the history of Chinese mathematics. Interpolations by difference sequences were introduced in the *Shoushili* 授時曆 (1281). Zhu Shijie 朱世傑 completed the *Suanxue qimeng* 算學啓蒙 (1299) and *Siyuan yujian* 四元玉鑑 (1303). In the latter, he showed that the triangle for binomial expansions gives rise to the fundamental structures for solving equations and the theory of finite series. Along with these, we discuss the contributions to the theory of equations and finite series in Joseon 朝鮮 (1392–1910) Korea. We deal with the contributions of Hong Jeong-ha 洪正夏 (1684–1727) in his *Gu-il jib* 九一集 (1713–1724) and then those of Lee Sang-hyeog 李尙爀 (b. 1810) in his *Igsan* 翼算 (1868).

Keywords Binomial expansions · Equations · Finite series · Joseon mathematics · Hong Jeong-ha 洪正夏 and *Gu-il jib* 九一集 · Lee Sang-hyeog 李尙爀 and *Igsan* 翼算

1 Introduction

The theory of equations is one of the most important topics in the history of East Asian mathematics, as it is in the West. The basic structure of Chinese mathematics was formalized in the *Jiuzhang suanshu* 九章算術 and the commentary of Liu Hui

Hong Sung Sa (✉)
Sogang University, Seoul, South Korea
e-mail: sshong@sogang.ac.kr

Hong Young Hee
Sookmyung Women's University, Seoul, South Korea
e-mail: yhhong@sookmyung.ac.kr

D. P. Morgan, T. Quan (eds.), *Mathematics and Astronomy in the Ancient World*,
https://doi.org/10.1007/978-3-031-93809-2_6

劉徽 (263). Indeed, the field of rational numbers was chosen as the basic field, facilitating linear equations, and then systems of linear equations were represented by matrices and solved by their reductions, known as Gaussian eliminations. Square and cube roots were introduced in the chapter "Shaoguang" 少廣, where their extractions were based on subdivisions of areas (volumes, resp.) of squares (cubes, resp.). They tried to find digits of solutions so that the processes would eventually cover those solving general quadratic and cubic equations. It is well known that most of traditional mathematics books take the format of problems and answers without any processes to reach the answers or solutions. In the following, solutions of equations mean the processes to solve equations. We note that the *Jiuzhang suanshu* contains only one general quadratic equation in the last chapter, "Gougu" 句股, but without its solution. Although Wang Xiaotong 王孝通 (7th cent.) obtained general equations up to degree 4 in his *Jigu suanjing* 緝古算經, neither did he deal with their solutions.

In the Song dynasty (960–1279), *tianyuanshu* 天元術 was introduced for representing polynomials and their mathematical structures so that mathematicians could extend the theory of equations without any restriction of degrees of polynomial equations. To solve them, they had to find a new method extending that of "Shaoguang." In this direction, the eleventh-century mathematicians Jia Xian 賈憲 and Liu Yi 劉益 made a very important contribution. Now lost, their works were transmitted by Yang Hui 楊輝. Yang Hui quoted Jia Xian's *Huangdi jiuzhang suanjing xichao* 黃帝九章算經細草 in his *Xiangjie jiuzhang suanfa* 詳解九章算法 (1261), but the first four chapters of Yang Hui's work, including "Shaoguang," were also lost. Jia's triangle for binomial expansions in the "Shaoguang" chapter of the *Xiangjie jiuzhang suanfa* was once again quoted in the *Yongle dadian* 永樂大典 (1408), which survives. Yang Hui also discussed Liu Yi's *Yigu genyuan* 議古根源 in the "Tianmu bilei chengchu jiefa" 田畝比類乘除捷法 chapter of the *Yang Hui suanfa* 楊輝算法 (1274–1275) and introduced the primitive *zengcheng kaifangfa* 增乘開方法 for quadratic equations based on geometric approaches as well as one quartic equation (Chemla 2018; Hong et al. 2011). We note that the word *zengcheng kaifangfa* is derived from that for square and cube root extraction introduced by Jia Xian, and that the procedure of *zengcheng kaifangfa* was also illustrated in Qin Jiushao's 秦九韶 (1202–1261) *Shushu jiuzhang* 數書九章 (1247) without referring to Jia Xian or Liu Yi.

Li Ye 李冶 (1192–1279) constructed *tianyuanshu* equations in his *Ceyuan haijing* 測圓海鏡 (1248, published in 1282) and *Yigu yanduan* 益古演段 (1259), but he did not include anything about their solutions other than indicating the cases *fanfa* 翻法 and *yiji* 益積 arising from the processes of *zengcheng kaifangfa*. Incidentally, Li Ye used *daoji* 倒積 and *daocong* 倒從 in the *Yigu yanduan* instead of *fanfa*. This suggests that Li Ye used the *zengcheng kaifangfa* to solve his equations and assumed the readers to know the *zengcheng kaifangfa*.

Zhu Shijie completed two books, the *Suanxue qimeng* 算學啓蒙 (1299) and *Siyuan yujian* 四元玉鑑 (1303). In the final chapter of the *Suanxue qimeng*, "Kaifang shisuo men" 開方釋鎖門, he dealt with extractions of up to quartic roots by the *zengcheng kaifangfa*, followed by constructions of equations by *tianyuanshu*. The *Suanxue qimeng* played a crucial role in the development of Joseon 朝鮮 (1392–1910) mathematics as well as in Japan. The fourth Joseon king, Sejong 世宗 (1397–1450, r. 1418–1450), wanted to improve the court's astronomy, so he decided to import Chinese works on astronomy and mathematics. Among other titles, the *Suanxue qimeng* and *Yang Hui Suanfa* were brought to Joseon, where they would become the main references for Joseon mathematicians throughout the rest of the dynasty. After the devastating invasions of the Japanese in 1592–1598 and the Later Jin 後金 in 1627 and 1636, the first effort to revive Joseon mathematics was the republication of the *Suanxue qimeng* in 1660 by Kim Si-jin 金始振 (1618–1667). At the center of this revival was the theory of equations based on *tianyuanshu* Although Zhu Shijie did not use *tianyuanshu* to construct equations for problems of finite series in the chapter "Duiji huanyuan men" 堆積還源門, finite series became one of the most important topics to relate the theory of equations along with the *gougushu* 句股術.

At the same time, one sees the introduction of the concept of difference sequences and their application to interpolation (*bujianfa* 補間法) in the Yuan 元 dynasty (1271–1368) astronomical system *Shoushili* 授時曆 (1276–1280) (Li and Du 1987). Difference sequences also played an important role in the development of the theory of series.

Zhu Shijie introduced the theory of multivariable polynomials up to four variables in his *Siyuan yujian,* where he also presumably used difference sequences to extend the theory of series. Using this and the *gougushu*, he obtained equations of much higher degrees, but his presentation in the *Siyuan yujian* was so abbreviated that it was ignored by Chinese mathematicians until the nineteenth century. We note that Mei Juecheng 梅瑴成 (1681–1763) compared *Siyuan yujian*'s *tianyuanshu* to *jiegenfang* 借根方 in his *Chisui yizhen* 赤水遺珍 (1759) and that Kim's Joseon edition of the *Suanxue qimeng* was exported to China in the nineteenth century.

Zhu Shijie began the *Siyuan yujian* with five diagrams: the "Jingu kaifang huiyao zhi tu" 今古開方會要之圖 and "Gufa qichengfang tu" 古法七乗方圖 to explain the method of solving equations; the "Siyuan zicheng yanduan zhi tu" 四元自乗演段之圖 for the multiplications in *siyuan* notation; and the "Wuhe zicheng yanduan zhi tu" 五和自乗演段之圖 and "Wujiao zicheng yanduan zhi tu" 五較自乗演段之圖 to show that the operations can be extended to the squares of five variables (Figs. 1 and 2).

As is well known, the second diagram, "Gufa qichengfang tu," is a triangle for the binomial expansions in the *Siyuan yujian* (see Figs. 3 and 4 in Sect. 2). This triangle is same as Jia Xian's, but Zhu's triangle reveals completely different structures. Indeed, Zhu Shijie obtained this triangle inductively. In its construction, he could presumably understand its combinatorial feature and, hence, describe the mathematical structures of triangular series and difference sequences. Thus, he could extend the theory of series in the *Shoushili* to arbitrary series defined by polynomials.

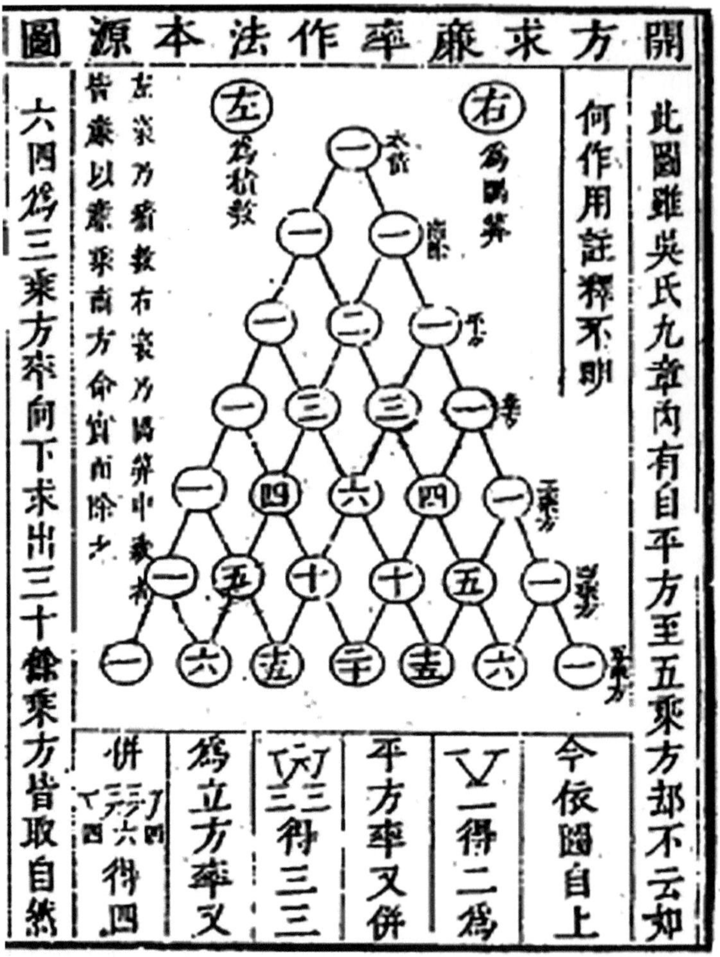

Fig. 1 Wu Jing's quote of Jia Xian's triangle

Except the books of Yang Hui, Song-Yuan mathematics were lost in the Ming 明 dynasty (1368–1643). Wu Jing 吳敬 wrote the *Jiuzhang suanfa bilei daquan* 九章算法比類大全 on the basis of Yang Hui's aforementioned works, and Wu Jing, in turn, is quoted in Cheng Dawei's 程大位 (1533–1602) *Suanfa tongzong* 算法統宗 (1592). It is interesting that in his "Shaoguang juan" 少廣卷 Wu Jing quotes Jia Xian's triangle with the title "Kaifang Zuofa benyuan" 開方作法本源, taken from Yang Hui's *Xiangjie jiuzhang suanfa*, but adds the combinatorial relations given in Zhu's triangle (see Fig. 1). What is more, Cheng Dawei also quotes Wu Jing's

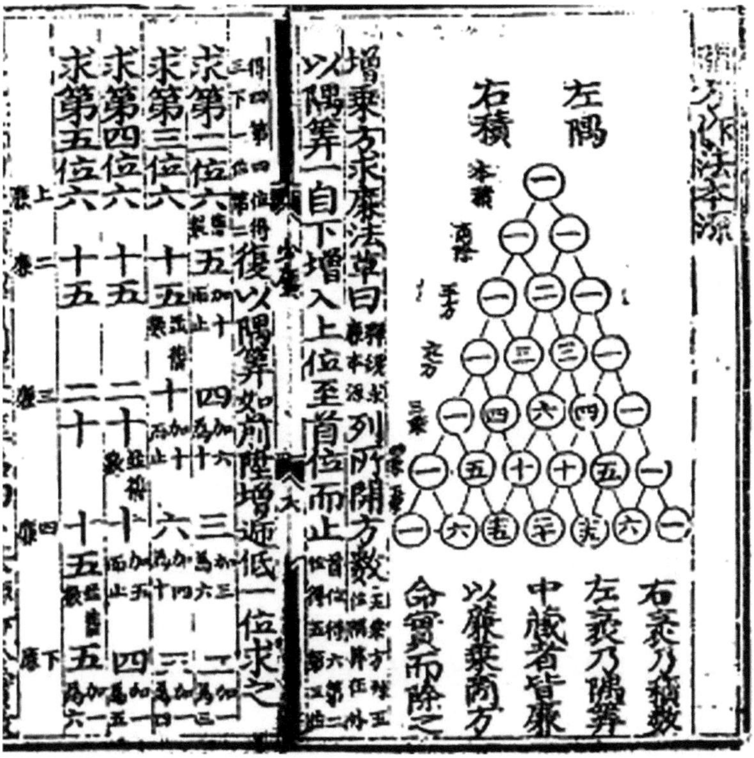

Fig. 2 Dawei's quote of Wu's triangle

quotation of the triangle with the title "Kaifang qiu lianlü zuofa benyuan tu" 開方求廉率作法本源圖 but substitutes Jia Xian's method with Zhu's, adding the remark that he did not understand Wu's quotation of Jia Xian's method and that it can be extended up to over the 30th power of the binomial expansion (see Fig. 2). This suggests that Cheng Dawei did not fully understand the *zengcheng kaifangfa*. However, thanks to Cheng Dawei's having quoted it, Zhu Shijie's triangle was used in China long before the reappearance of the *Siyuan yujian* in the nineteenth century.

The *Suanfa tongzong* was transmitted to Joseon in the seventeenth century, where Korean mathematicians would become familiar with Zhu's triangle as in China and Japan.

In the mid-nineteenth century, Luo Shilin's 羅士琳 (1784–1853) commentary *Siyuan yujian xicao* 四元玉鑑細艸 (1834) then made its way to Joseon, where it rekindled Song-Yuan mathematics (Hong et al. 2017).

Qu Anjing 曲安京 describes the historiography of mathematics in China as marked by three salient paradigm shifts: the "Li-Qian paradigm", referring to the

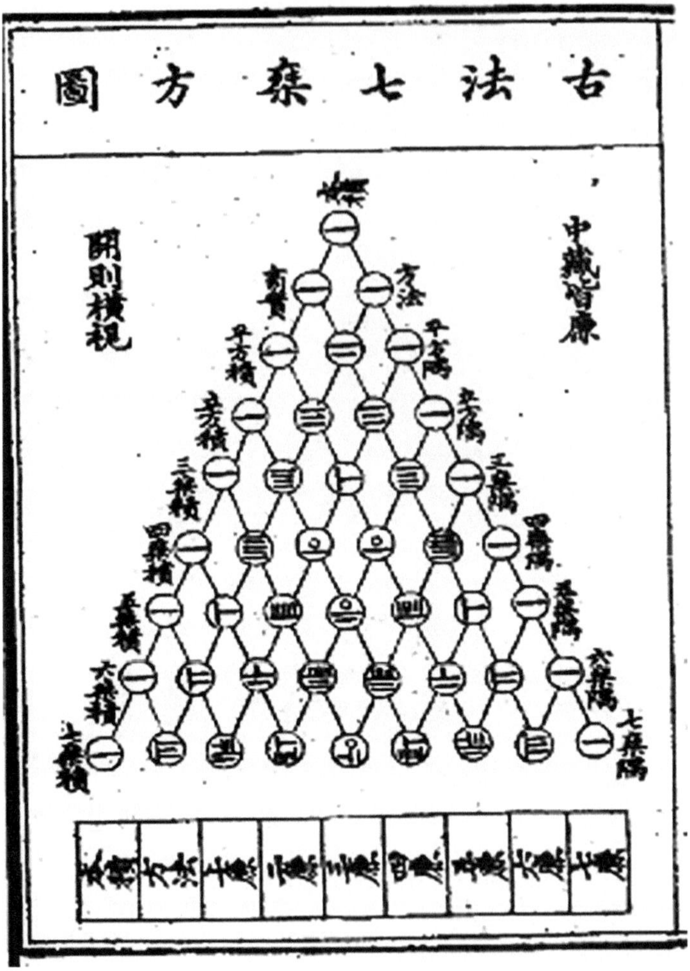

Fig. 3 Zhu's triangle

first collection and interpretation of the historical materials by Li Yan 李儼 (1892–1963) and Qian Baocong 錢寶琮 (1892–1974); the "Wu paradigm," after Wu Wentsun 吳文俊 (1919–2017), who restored historical materials on top of the Li-Qian paradigm; and finally the study of history of mathematics based on mathematical structures. We quote his paradigm shift:

有什麼 (WHAT) $K = \{$發現$\} \subset$ 如何做 (HOW) $K_1 = \{$發現,復原.$\}$
$\subset$ 爲什麼 (WHY) $K_2 = \{$發現,復原,重構$\}$ (Qu Anjing 2018)

As previously mentioned, Cheng Dawei did not understand the relation between Zhu's triangle and the theory of series. In other words, Cheng could not grasp the important mathematical structures of Zhu Shijie's triangle as relate to solving equations and the theory of series.

The aim of this paper is to discuss the mathematical structures of Zhu's triangle of binomial expansions and its influences on the development of Joseon mathematics. The paper is divided into two parts. First, we deal with the algebraic structure of the triangle in the setting of *tianyuanshu*. We note that *tianyuanshu* was completely ignored by the mathematicians of the Ming dynasty. Although the triangle contains combinatorial structures, we doubt that they were understood by East Asian mathematicians. We note that the triangle was obtained by *tianyuanshu* and *shenwai jiafa* 身外加法 and that it gives the fundamental structures of triangular series and series determined by sequence (*p(k)*) for any polynomial *p(x)* and its difference sequences. Secondly, we discuss Joseon mathematics related to Zhu's triangle. Hong Jeong-ha 洪正夏 (1684–1727) and Lee Sang-hyeog 李尙爀 (b. 1810) made the most important contributions to Joseon mathematics, particularly the theory of equations and the theory of series. We deal with the theory of equations in Hong's *Gu-il jib* 九一集 (1713–1724) and the theory of finite series in the second volume of Lee Sang-hyeog's *Igsan* 翼算 (1868), "Toetaseol" 堆垜說 (Hong 2006a). The reader may find all of the Chinese sources for this paper in Guo Shuchun (1993), individual references to which will be omitted.

2 Zhu Shijie's Triangle of Binomial Expansions

Let us begin with a brief discussion of the basic structures of the *Siyuan yujian*. Zhu's first book, *Suanxue qimeng* (1299), deals with traditional mathematics up to the theory of equations based on *tianyuanshu* somewhat in detail. His second book, *Siyuan yujian* (1303), is dedicated solely to the theory of equations. As we know, the theory of equations in the East involves constructing equations for word problems and their solutions. Word problems are given by systems of equations, and their equations are given by systems of multivariable equations. Thus, one has to have devices to represent multivariable polynomials as *tianyuanshu* for single variable polynomials, which gives rise to *eryuanshu* 二元術, *sanyuanshu*三元術, and *siyuanshu* 四元術. To solve the systems, one has to transform the systems into an equation of a single variable, namely one represented by *tianyuanshu*.

For practical problems, Zhu takes traditional word problems for geometrical areas, volumes, and *gougushu* as well as those for trading and linear equations. On top of these, Zhu deals with problems of series, using far more complicated series than any treated before the Yuan dynasty. Indeed, Zhu deals with series whose sums cannot be obtained by simple arithmetic sequences and those based on Shen Kuo's 沈括 (1031–1095) *xijishu* 隙積術 in his *Mengxi bitan* 夢溪筆談 (1086–1093). We can assume that Zhu was curious to apply and extend the interpolation theory by the difference sequences of a sequence in the *Shoushili* to the theory of series, because his results in *Siyuan yujian* cannot be completed without the theory of difference sequences and his triangle of binomial expansions. We note once and for all that Zhu did not include any explanations for constructions of equations in the main body of the *Siyuan yujian* except the four examples, “Jialing sicao” 假令四艸, all of which deal with *gougushu*.

Let us return to the triangle of binomial expansions in the *Siyuan yujian.* Zhu Shijie indicates that one should read the triangle of expansions for solving equations horizontally, as in Fig. 4. We see that each row represents the polynomial $(x + 1)^n$ in *tianyuanshu* and as given in the calculating rod notation of Song-Yuan mathematics. We can deduce that this is how Zhu Shijie would have constructed his triangle, because the *shenwai jiafa* is well established in Song-Yuan works, including in his *Suanxue qimeng*, and, thus, that it is distinct from Jia Xian’s triangle.

In the following, we will use modern polynomials in place of *tianyuanshu* notation for the convenience of the modern reader.

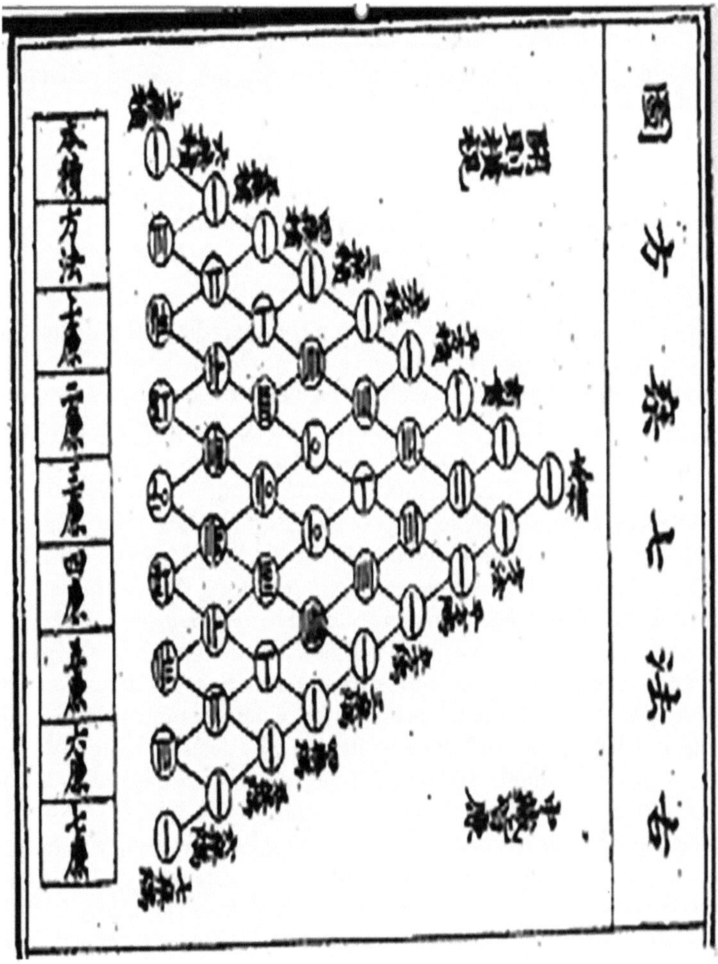

Fig. 4 Triangle in tianyuanshu

Suppose that one has $(x+1)^n = \sum_{k=0}^{n} a_k^n x^k$ for any natural number n, $0 \le k \le n$, then $(x + 1)^{n+1}$ can be obtained by *shenwai jiafa* as follows:

$$a_n^n \ a_{n-1}^n \ a_{n-2}^n \ldots a_1^n \ a_0^n$$

$$a_n^n \ a_{n-1}^n \ldots a_2^n \ a_1^n \ a_0^n \ (+$$

$$a_n^n \ a_{n-1}^n + a_n^n \ a_{n-2}^n + a_{n-1}^n \ldots a_1^n + a_2^n \ a_0^n + a_1^n \ a_0^n.$$

Thus, Zhu obtained the inductive or combinatorial relation of the triangle, namely the fundamental identities, denoted by (FI)

$$a_{k-1}^n + a_k^n = a_k^{n+1} \left(1 \le k \le n\right),$$

which are denoted by oblique connecting lines in the triangle.

In the following, the number of k-combinations of a set of n elements will be denoted by $\binom{n}{k}$, and hence the above identity is precisely

$$\binom{n}{k-1} + \binom{n}{k} = \binom{n+1}{k} \text{for all } n \text{ and } k\left(1 \le k \le n\right).$$

Using this identity and induction, one immediately arrives at the following.

Proposition 2.1 For any $m = 0, 1, 2, \ldots$ the following identities hold:

$$\sum_{k=1}^{n} \binom{k+m-1}{m} = \binom{n+m}{m+1}.$$

We now fix the following notations, which are defined inductively.

First, $S_n^{-1} = 1$ for $n = 0, 1, 2, \ldots$ and then $S_n^{m+1} = \sum_{k=1}^{n} S_k^m$.

Then using Proposition 2.1, we have

Proposition 2.2 For any $m = 0, 1, 2, \ldots$, $S_n^m = \binom{n+m}{m+1}$.

The sequence $\left(S_k^m\right)$ for a fixed $m = -1, 0, 1, 2, \ldots$ is exactly the $(m + 1)$ *th* oblique line in Zhu's triangle. These are called *jiaocaoduo*茭草垛, *sanjiaoduo* 三角垛, *saxingxingdu*撒星形垛 or *sanjiao luoyiduo* 三角落一垛 and *saxing gengluoyixingduo* 撒星更落一形垛 for the sequences $\left(S_k^m\right)$ up to $m = 4$,respectively, and their sums are given by $\sum_{k=1}^{n} S_k^m = S_n^{m+1}$. These can be inferred from the term *luoyi* 落

—, i.e. "dropping one oblique line in Zhu's triangle, one has the terms of the next series." Furthermore, we will refer to these sequences as triangular sequences.

We will now discuss the series along with the difference sequences used in the *Shoushili* and *Siyuan yujian.*

We first recall that for a sequence (a_k), the sequence $(a_{k+1} - a_k)$ is called the first order *difference sequence* of the sequence (a_k) and denoted by (a_k^1). Inductively, we shall define the pth order difference sequence of (a_k) by $(a_k^p = a_{k+1}^{p-1} - a_k^{p-1})$ for any natural number p.

The most important example is given by Zhu's triangle. Indeed, the first order difference sequence of a triangular sequence (S_k^m) is precisely $\left(S_k^{m-1}\right)(2 \le k)$ by the fundamental identity (FI). Thus, its $(m + 1)th$ difference sequence is given by the constant sequence (S_k^{-1}), and the binomial coefficients of $(x + 1)^m$ are precisely the first elements of the difference sequences of the sequence $\left(S_k^{m-1}\right)$ along its orders.

Using the induction on the degree of a polynomial $p(x)$, one has,

Proposition 2.3 For any polynomial $p(x)$ of the degree n, its nth order difference sequence is a constant sequence.

The converse of Proposition 2.3 also holds as follows.

Proposition 2.4 Suppose that the nth order difference sequence of a sequence (a_k) is a constant sequence, then for any $k = 1, 2, \ldots, a_k$ is given by

$$a_k = a_1 + \binom{k-1}{1} a_1^1 + \binom{k-1}{2} a_1^2 + \ldots + \binom{k-1}{n} a_1^n = \sum_{j=0}^{n} \binom{k-1}{j} a_1^j,$$

where a_1^j is the first element of the jth order difference sequence of the given sequence and $a_1^0 = a_1$.

One can prove this by induction and (FI) again. Collecting the above Propositions 2.3 and 2.4, one has the fundamental theorem on the sequence determined by $(p(k))$ for any polynomial $p(x)$.

Theorem 2.5 A sequence (a_k) has a constant nth order difference sequence if and only if there is a polynomial $p(x)$ of degree n with $a_k = p(k)$ for all k.

In case, the polynomial is given by

$$p(x) = a_1 + a_1^1 (x-1) + a_1^2 \frac{(x-1)(x-2)}{2!} + \ldots + a_1^n \frac{(x-1)(x-2),\ldots(x-n)}{n!},$$

where a_1^j is the first element of the jth order difference sequence of the sequence.

We recall that the above polynomial is exactly the interpolation used in the *Shoushili* for the case of the second order difference sequence being constant (Li and Du 1987).

Remark. Using triangular series, the sequence (a_k) in Theorem 2.5 can be represented by $a_k = \sum_{j=0}^{n} S_{k-j}^{j-1} \times a_1^j$ for all k.

As we mentioned, Zhu did not discuss anything about the difference sequences and their applications in his *Siyuan yujian*, but we can safely deduce that he should have used its application for the sum of the series in *Siyuan yujian* (also see Sect. 3).

Discussing the sum of series with a constant difference sequence for some order, we introduce the following notions for a sequence (a_k): $s_1 = 0, s_k = \sum_{j=1}^{k-1} a_j \left(2 \le k\right)$ and hence $s_{k+1} = \sum_{j=1}^{k} a_j$.

Then the first order difference sequence of (s_k) is exactly the given sequence (a_k) and the $(n + 1)$th order difference sequence of (s_k) is precisely the nth order difference sequence of (a_k). By Theorem 2.5, these facts immediately imply the following.

Theorem 2.6 Assuming that the nth order difference sequence of (a_k) is a constant sequence, then for any k, $\sum_{j=1}^{k} a_j = \sum_{j=1}^{n+1} \binom{k}{j} \times a_1^{j-1} = ka_1 + \sum_{j=2}^{n+1} \binom{k}{j} \times a_1^{j-1}$.

Since $\binom{k}{j} = S_{k-(j-1)}^{j-1}$, the above formula for the sums can be rewritten

$$\sum_{j=1}^{k} a_j = \sum_{j=1}^{n+1} S_{k-(j-1)}^{j-1} \times a_1^{j-1}.$$

Remark. (1) The above propositions and theorems are all based on the fundamental identity (FI) and triangular series with the mathematical induction. With the exception of the mathematical induction, all of this appears in Zhu Shijie's triangle. As mentioned above, these facts were not appreciated by Chinese mathematicians until the mid-nineteenth century, although the triangle was selected for transmission by Wu Jing and Cheng Dawei.

(2) For any polynomial $p(x)$., one can calculate the sum $\sum_{j=1}^{k} p\left(j\right)$ by the above formula. We take an example of the series $\sum_{k=1}^{n} \left(a+k-1\right)\left(b+k-1\right)$ known as the series of Shen Kuo in *Mengxi bitan*, where $p(x). = (x + a - 1)(x + b - 1)$ $(a, b; c = a + n - 1, d = b + n - 1)$. Shen compared the volume of a *chutong* 芻童 with the dimension a, b and c, d with the sum of the series. He obtained the famous formula for the sum $\sum_{k=1}^{n} \left(a+k-1\right)\left(b+k-1\right) = \frac{n}{6}\left[\left(2b+d\right)a + \left(2d+b\right)c\right] + \frac{n}{6}\left(c-a\right)$. Indeed, n is the height of the given *chutong* and the first term is its volume. We also recall

that the theory of series in China developed along areas of plane figures and volumes of solids, and Shen Kuo first showed the difference between sums of series and volumes of corresponding solids. Applying Theorem 2.5, one should be familiar with operations of polynomials, e.g. operations in *tianyuanshu*. And even if one were familiar with *tianyuanshu*, it would be difficult to arrive at the above formula, because it would require the process of factorizations, and there is hardly any trace of that in traditional East Asian mathematics. Thus, the above formula may be used for specific cases but not for general formulae like Shen's, which includes the sum $\sum_{k=1}^{n} k^2$, called *sijiao duo* 四角垛.

3 Binomial Expansions in Joseon Mathematics

Except a few records in the *Annals of the Joseon Dynasty* (*Joseon wangjo sillok* 朝鮮王朝實錄) on the study of mathematics under King Sejong, we do not have any information of the development of native Joseon mathematics until the seventeenth century. What the *Annals* do record are the importation of the *Suanxue qimeng* and *Yang Hui suanfa*, their use as examination subjects to select mathematical officers in Hojo 戶曹, called *sanhag chuijae* 算學取才, and the republication of the *Yang Hui suanfa* in 1433. We note that *sanhag chuijae* also included the *Xiangming suanfa* 詳明算法 (1373).

During the foreign invasions of the late sixteenth and early seventeenth centuries, most government documents and mathematical works were destroyed. Naturally, the Joseon government was anxious to rebuild, and it is in this context that the *Suanxue qimeng* was republished with Yang Hui's proof of Liu Hui's *Haidao suanjing* 海島算經 added to the end. The *tianyuanshu* in the final chapter, "Kaifang shisuo men" would become the principal subject of Joseon mathematics in the second half of the seventeenth century. The first extant work to come out of this was the *Sanhag wonbon* 算學原本 (1701), completed by a *yangban* 兩班 or noble mathematician Park Yul 朴繘 (1621–1668). The *Sanhag wonbon* is divided into three parts, the third of which deals with constructions of equations by *tianyuanshu*. Park includes tables for expansions of $(ax + b)^n$ for various a, b, namely integers and fractions. Judging from his explanations, he definitely knew and substituted the given cases into Zhu Shijie's triangle. Furthermore, Park's work attests to the arrival of the *Suanfa tongzong* prior to 1668.

The Qing adoption of the new *Shixianli* 時憲曆 astronomical system in 1645, garnered great attention in the Joseon government, which imported the *Xiyang xinfa lishu* 西洋新法曆書 (1645) directly from Adam Schall von Bell 湯若望 (1591–1666) in 1646 (Lee and Hong 2013). Compiled by Li Zhizao 李之藻 (1565–1630), the *Tianxue chuhan* 天學初函 (1629) was also brought into Joseon in the last decade of the seventeenth century (Hong et al. 2018). As is well known, the *Tianxue chuhan* contains the *Jihe yuanben* 幾何原本 (1607), the translation of the first six chapters of Euclid's *Elements*. The attempts of mathematicians at the National Observatory to learn the Western mathematics in the *Xiyang xinfa lishu* were unsuccessful, because they were based on the *Elements*. And since the Hojo dealt with finance and taxation, its mathematical officers did not pay any attention to Western mathematics. Thus, the first introduction of Western mathematics in seventeenth-century Joseon had little impact on contemporary practices.

Let us first discuss the theory of equations in the Joseon dynasty. The most important contribution to the theory was established by a *jungin* 中人 officer of the Hojo, Hong Jeong-ha 洪正夏 (1684–1727), in his *Gu-il jib* 九一集 (1713–1724).[1]

Starting with his grandfather, a total of 98 members in nine generations of Hong Jeong-ha's family passed the examination for mathematical officers in the Hojo. Hong Jeong-ha passed the exam in 1706, served in the Hojo as a teacher, *hundo jeong*-9 *pum* 訓導正九品, and was promoted to professor of mathematics, *ju-gyosu jong*-6 *pum* 籌敎授從六品, in 1718. The latter is the highest position for *jungin* officers in the Hojo.

The *Gu-il jib* comprises eight volumes and an appendix. The body was completed before 1713, and the appendix in 1724. Hong's focus in the *Gu-il jib* is the theory of equations, namely the construction and solution of equations. The first three volumes deal with traditional mathematics, roughly equivalent to the chapters up to "Ying buzu" 盈不足 in the *Jiuzhang suanshu*.

Hong's theory of equations begins in Volume 4 under the sections "Bangjeong jeongbumun" 方程正負門, "Gucheok haeunmun" 毬隻解隱門, "Bubyeong toetamun" 缶甁堆垛門, and "Changdon jeogsogmun" 倉囤積粟門. The first section deals with systems of linear equations, and the others treat polynomial equations of higher degrees. Hong Jeong-ha first introduces the construction of equations by *tianyuanshu* in the section "Gucheok haeunmun." He took word problems of volumes and series for the equations in the remaining two sections. In Volume 4, Hong Jeong-ha integrates systems of linear equations into the theory of equations. Furthermore, he introduces the *zengcheng kaifangfa* by indicating *fanfa* and *fanjifa* 飜積法. We recall that all *fanfa* in the *Suanxue qimeng* signify the change of signs of non-constant coefficients in the process of the *zengcheng kaifangfa*, contrary to the other mathematical works in the Song dynasty. Applying the *zengcheng kaifangfa*, Hong Jeong-ha finds many examples of the usual *fanfa*, namely the change of the sign of the constant term in the *zengcheng kaifangfa*. Thus, he introduces the term *beonjeogbeob*, or *fanjifa*.

In Volume 5, "Gugo hoeunmun" 句股互隱門, Hong Jeong-ha then takes up the theory of right triangles to reveal the mathematical structures of the theory of equations. In its 78 problems, the author discusses the traditional *gougushu*, then construction of equations by *tianyuanshu* and their solution by *zengcheng kaifangfa* (Hong et al. 2007). In Problem 64, for example, he constructs the equation by *tianyuanshu* and solves it by *zengcheng kaifangfa*, finding the case *yiji*, which he

[1] Hong Jeong-ha (1724).

calls *igjeogbeob* or *yijifa* 益積法. Furthermore, his description of *zenccheng kaifangfa* is exactly same as the modern one. Although *tianyuanshu* and *zenccheng kaifangfa* were used in Volumes 4 and 5, their problems were originated from the chapters "Fangcheng" 方程, "Gougu" 句股, and "Shanggong" 商功 in the *Jiuzhang suanfa*.

In the remaining three Volumes, titled "Gaebang gagsulmun" 開方各術門, he establishes his theory of equations, in particular their solutions based on Zhu Shijie's triangle of binomial expansions. As mentioned above, Zhu's triangle was transmitted to Joseon by the triangle in the Volume 6, "Shaoguang" 少廣, of the *Suanfa Tongzong*. Hong Jeong-ha investigates the mathematical structure of *zenccheng kaifangfa* in 166 problems of "Gaebang gagsulmun".

Let us provide a short sketch of the structure of the *zengcheng kaifangfa* for modern readers.

For a polynomial equation $p(x) = 0$, where $p(x)=\sum_{k=0}^{n} a_k x^k$, assume that one has the largest digit, say α of the solution of the equation, called *chushang* 初商, and let $x = y + \alpha$. Then the next digit of the solution, called *cishang* 次商, is the largest digit of the equation $\sum_{k=0}^{n} a_k (y+\alpha)^k = 0$. Let $\sum_{k=0}^{n} a_k (y+\alpha)^k = \sum_{k=0}^{n} b_k y^k$. Expanding the left hand, one can have the coefficients b_k $(0 \le k \le n)$.

Since $y = x - \alpha$, the equation $\sum_{k=0}^{n} b_k y^k = 0$ for the *cishang* is

$$\sum_{k=0}^{n} b_k (x-\alpha)^k = 0,$$

where $\sum_{k=0}^{n} b_k (x-\alpha)^k$ is another representation of $p(x)$

In Western mathematics, this method is called the Ruffini-Horner method, introduced in 1804 by Paolo Ruffini (1765–1822) and then in 1819 by William George Horner (1786–1837). Ruffini and Horner used the synthetic divisions of a polynomial by $x - \alpha$. Although the divisions of polynomials were first introduced in the *Shuli jingyun* (1713), they include only the cases without remainders.

Hong Jeong-ha is the first mathematician in East Asia to reveal the mathematical structure of the *zengcheng kaifangfa*. He begins with the method of expansions of $a_k(x + \alpha)^k$, called *myeong seungbangsig* 明乘方式, and then Zhu's triangles in his introduction of the *Gu-il* jib.

Let us turn first to the triangles in the *Gu-il jib* (Figs. 5 and 6).

Clearly, Hong has extended the triangle in the *Suanfa tongzong* by the same title. Adding the line "this is the diagram of the horizontal version" 此卽橫者圖, Hong indicates that the vertical *tianyuanshu* representation of polynomials is rotated. We also recall that Zhu Shijie offers a similar remark, and that Cheng Dawei did not mention anything about *tianayuanshu* in his book. Although Hong Jeong-ha did not have *Siyuan yujian*, the paragraph indicates his understanding of *tianyuanshu*.

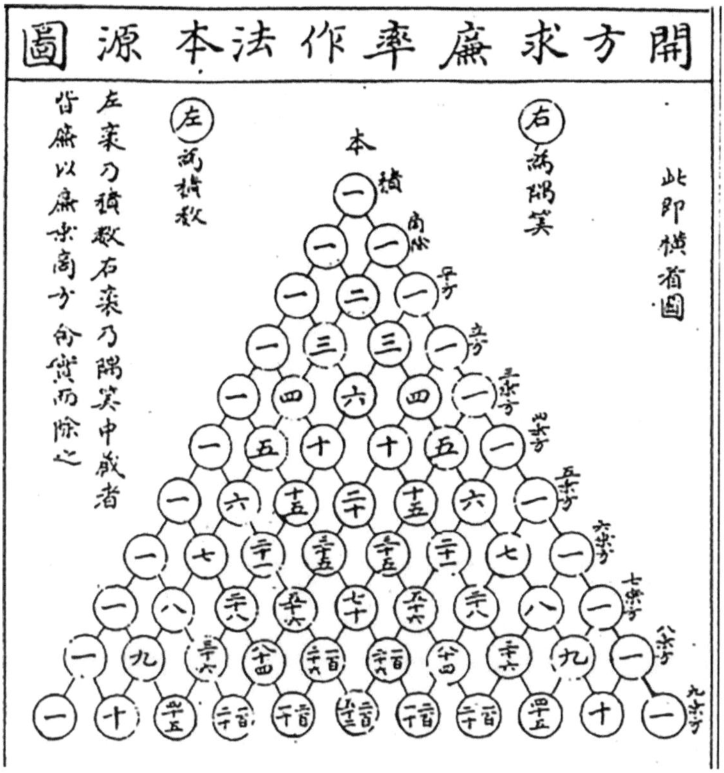

Fig. 5 Hong Jeong-ha's triangle

Figure 6 is another extension of Zhu's triangle, to $(x - 1)^n$, indicating the sign of $(-1)^k$ is positive (negative) for even (odd, resp.) k for applying the original version to the binomial expansions $(x - 1)^n$.

Let us deal with Hong's process of expansion $(x + \alpha)(x + \beta)(x + \gamma)(x + \delta)$ applying consecutive *shenwai jiafa* as follows.[2] This is treated in Problem 34 of Volume 8.

Hong uses the formula $(x + \alpha)(x + \beta) = x^2 + (\alpha + \beta)x + \alpha\beta$ by *shenwai jiafa* and then for the product of three terms, he uses a combinatorial method. The product of four terms is described for $(x + 4)(x + 8)(x + 12)(x + 16)$ in Problem 34.

Extending the case of two terms, he invented the following algorithm, where we use arbitrary letters α, β, γ and δ in place of the given numbers 4, 8, 12, and 16 in Fig. 7.

[2] See Hong et al. (2014) for the details of multiplications of linear terms as seen here.

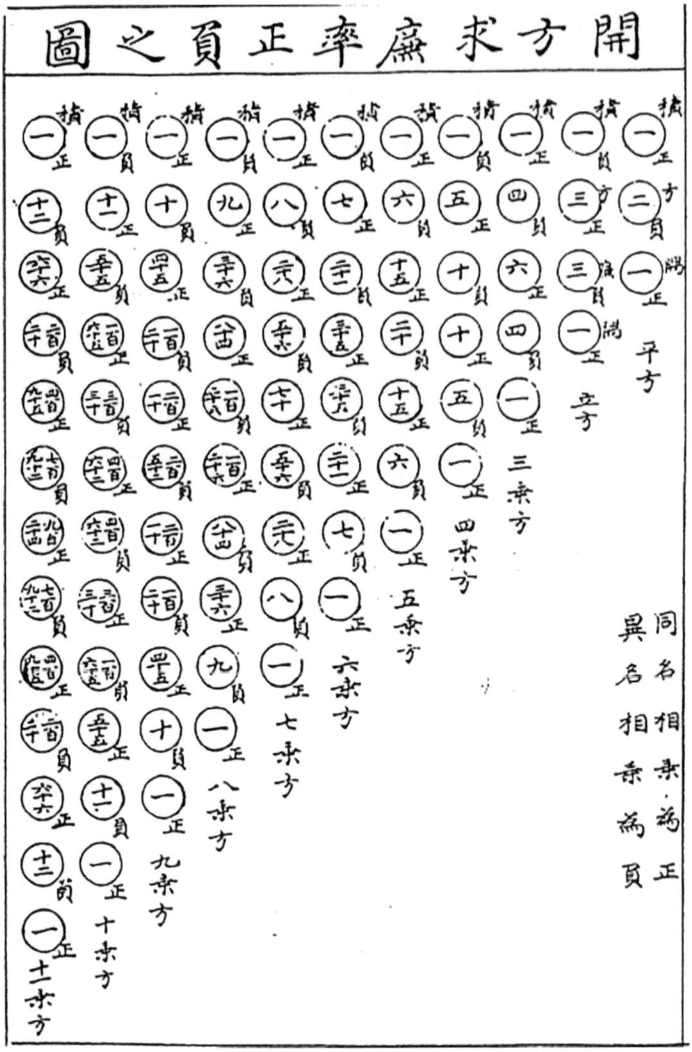

Fig. 6 Expansion of $(x - 1)^n$

1	0	0	0	0	$\alpha, \beta, \gamma, \delta$
	α	$\alpha\beta$	$\alpha\beta\gamma$	$\alpha\beta\gamma\delta$	
1	α	$\alpha\beta$	$\alpha\beta\gamma$	$\alpha\beta\gamma\delta$	
	β	$(\alpha+\beta)\gamma$	$(\alpha\beta+(\alpha+\beta)\gamma)\delta$		
1	$\alpha+\beta$	$\alpha\beta+(\alpha+\beta)\gamma$	$\alpha\beta\gamma+(\alpha\beta+(\alpha+\beta)\gamma)\delta$		
	γ	$(\alpha+\beta+\gamma)\delta$			
1	$\alpha+\beta+\gamma$	$\alpha\beta+(\alpha+\beta)\gamma+(\alpha+\beta+\gamma)\delta$			
	δ				
1	$\alpha+\beta+\gamma+\delta$				

Fig. 7 Hong Jeong-ha's synthetic expansion

In Problem 34, Hong describes Fig. 7 vertically, i.e., in *tianyuanshu*. He then puts the numbers α (4) to δ (16), four zeros (0), and the first 1 in the order of 甲 (α), 乙 (β), 丙 (γ), 丁 (δ), 戊 (0), 己 (0), 庚 (0), 辛 (0), and 壬 (1), as usual. He then evaluates the next lines as the *zengchengfa* 增乘法 but using α, β, γ and δ, in turn, instead of the single α in the *zengcheng kaifangfa*. Furthermore, for the case of α = β = γ = δ, the above algorithm gives rise to the binomial expansion of $(x + \alpha)^4$. The method can be easily extended to the arbitrary power of $(x + \alpha)^n$, and it explains Jia Xian's processes to get his triangle (Fig. 1 above). Jia Xian's triangle gives rise to the equation $(y + \alpha)^n = a$ for *cishang* y of the extractions of nth roots, i.e., the equation $x^n = a$. We note that the equation for *cishang* is a general equation.

In order to solve general polynomial equations, Hong Jeong-ha introduces the method for the expansion of $a(x + \alpha)^n$ for any a and α, and Figure 8 illustrates how

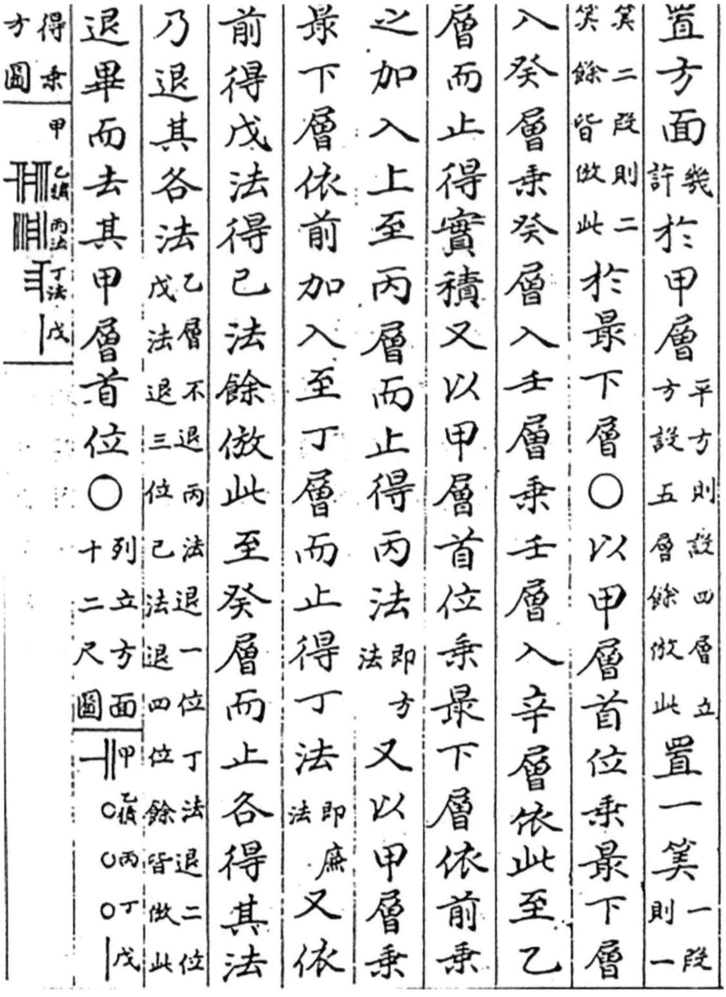
置方面許幾於甲層平方則設四層立方設五層餘倣此置一算一段則一算二段則二算餘皆倣此於最下層〇以甲層首位乘最下層入癸層乘癸層入壬層乘壬層入辛層依此至乙層而上得實積又以甲層首位乘最下層依前乘之加入上至丙層而上得丙法即方法又以甲層乘最下層依前加入至丁層而上得丁法即廉法又依前得戊法得己法餘倣此至癸層而上各得其法乃退其各法乙層不退丙法退一位丁法退二位戊法退三位己法退四位餘皆倣此退畢而去其甲層首位〇

Fig. 8 *Myeong seungbangsig* 明乘方式

to expand $a(x+\alpha)^n$. Indeed, Hong also uses *tianyuanshu* representation in said figure, putting α at the top (幾許於甲層), n 0's, and, finally, a at the lowest term (最下層). Thus, Fig. 8 is simply obtained by replacing the first 1 in Fig. 7 by a. Applying the processes to expand $(x+\alpha)^n$ to the above, we have the expansion $a(x+\alpha)^n$.

We include the beginning diagram 5 0 0 0 7 for the case of $n = 3$, $a = 5$, $\alpha = 7$ in Fig. 8, from which one can easily expand $5(x+7)^3$.

Let us now return to the mathematical structure of the *zengcheng kaifangfa*. For *cishang*, one has to calculate $\sum_{k=0}^{n} a_k (y+\alpha)^k$. Using Fig. 8, we now have the method to calculate each term $a_k(y+\alpha)^k$ of $\sum_{k=0}^{n} a_k (y+\alpha)^k$ and their sum is the desired expansion. Adding the *tianyuanshu* representation of each term $a_k(y+\alpha)^k$ in Fig. 8 and placing α on the top, the horizontal first line is

$$a_n\, a_{n-1}\, a_{n-2} \ldots a_2\, a_1\, a_0\, \alpha,$$

and the remaining processes for each term are exactly the same. Thus, applying the *zengchengfa* to the given line, one can have the expansion of $\sum_{k=0}^{n} a_k (y+\alpha)^k$.

In short, Hong Jeong-ha shows how to get the equation for *cishang* by the *zengcheng kaifangfa*.

Hong Jeong-ha always represents equations by $p(x) = 0$, even $x^n - \alpha = 0$ for the extraction of the *nth* root of α and applies the *zengcheng kaifangfa*. We recall that Zhu Shijie used the *zengcheng kaifangfa* in the form of $x^n = \alpha$ in the *Suanxue qimeng* as in the *Yang Hui suanfa* (Chemla 2018; Hong et al. 2011, 2016). Furthermore, one has to move the calculating rods of the highest power of x in the *zengcheng kaifangfa* and then shift the calculating rods given by the *zengchengfa*. However, Hong Jeong-ha uses *chushang* α in his *zengcheng kaifangfa* as in modern mathematics.

Though he had the triangle of the binomial expansion of $(x-1)^n$, Hong Jeong-ha did not deal with negative solutions of polynomial equations. Other than this, Hong Jeong-ha completes the numerical theory of polynomial equations, such as can be used today.

Let us now discuss the theory of finite series established by Lee Sang-hyeog in the nineteenth century. His theory is strongly influenced by Zhu Shijie's *Siyuan yujian*.

Lee Sang-hyeog likewise belonged to one of the Joseon's most preeminent clans of *jungin* mathematicians. We note that there were strong alliances between these clans, many sealed by marriage. Such are the ties that link Lee Sang-hyeog to Hong Jeong-ha going back for generations: Lee's great-great-great grandfather was married to Hong's cousin; Lee's great-great grandfather was Hong's brother-in-law; and Lee's father was the son-in-law of Hong Jeong-ha's nephew. Not surprisingly,

Hong's *Gu-il jib* was read by generation of Lee's family. When the *Gu-il jib* was republished by Hong Jeong-ha's great-great-great grandson, Hong Yeong-seog 洪永錫 (b. 1814) in 1868, Nam Byeong-gil 南秉吉 (1820–1869) wrote a preface at Lee's invitation, and Lee, for his part, added the postscript.

Lee Sang-hyeog first passed the state examination in astronomy (*eumyang-gwa*, 陰陽科) in 1831, then *sanhag chuijae* 算學取才 in 1832. Since the *Shuli jingyun* 數理精蘊 (1723) was one of the most important subjects for the state examination, Lee's mathematical training was aimed at the study of the astronomy as influenced by Western mathematics. We note that the *Shoushili* and its Ming-dynasty version, *Datongli* 大統曆, were used alongside the *Shixianli* in the nineteenth century. Thus, Lee would have been very familiar with interpolation given by difference sequences. He also wrote the *Chagunbang mong-gu* 借根方蒙求 (1854) and the *Sansul gwan-gyeon* 算術管見 (1855). The former deals with the construction of equations by *jiegenfang* for problems in the *Shuli jingyun*, and the latter with Western mathematics, including series of the trigonometric functions of Pierre Jartoux (Du Demei 杜德美, 1668–1720) quoted in the *Chishui yizhen*. After Luo Shilin's *Siyuan yujian xicao* arrived in Joseon, Lee Sang-hyeog found the merits of *tianyuanshu* over *jiegenfang* and studied *tianyuanshu* and *Gu-il jib*. The Yonsei University Library possesses a handwritten copy of a part of Luo's *Xicao*, entitled *Sawon og-gam* 四元玉鑑. The anonymous transcriber copied the following sections:

- "Jialing sicao" 假令四艸 (Vol. 1)
- "Liangyi hezhe" 兩儀合轍 (Vol. 3)
- "Zuoyou fengyuan" 左右逢元 (Vol. 3)
- "Sancai biantong" 三才變通 (Vol. 3)
- "Sixiang caoyuan" 四象朝元 (Vol. 3)
- "Huowen getuan" 或問歌彖 (Vol. 2)
- "Shanggong xiuzhu" 商功修築 (Vol. 1)
- "Guoduo diecang" 果垛疊藏 (Vol. 3)

The copyist adds the basic formulae for the following series:

- *sijiao luoyixing* 四角落一形
- *sanjio lanfengxing* 三角嵐峯形
- *sijiao lanfengxing* 四角嵐峯形
- *sanjiao saxing gengluoyixing* 三角撒星更落一形
- *yuanzhui duo* 圓錐垛
- *jiaocao luoyixingduo* 茭艸落一形垛
- *jiaocao saxing* 茭艸撒星
- *jiaocao lanfeng* 茭艸嵐峯
- *jiaocao saxing gengluoyi* 茭艸撒星更落一
- *jiaocao lanfeng gengluoyi* 茭艸嵐峯更落一

The transcriber quotes parts of Luo Shilin's appendices, the *Siyuan shili* 四元釋例, *Kaifang shili* 開方釋例, and *Tianyuan siyuan geli* 天元四元各例 of Yi Zhihan 易

之澣. Finally, the transcriber quotes the entirety of the section "Ruxiang zhaoshu" 如像招數.[3]

One can easily discern that the transcriber was principally concerned with *eryuan shu* 二元術 to *siyuan shu* 四元術. We note that among the chosen sections, the section "Shanggong xiuzhu" is the only section dealing with *tianyuanshu*. His selection also indicates that the copyist was familiar with *tianyuanshu* or *jiegenfang*. Furthermore, we can see that the copyist was particularly drawn by the theory of finite series based on Zhu Shijie's triangle. In the spare pages of his copy, he also includes short statements on astronomy.

On top of the study of *Gu-il jib*, Lee was fortunate to have an exceptional supporter, Nam Byeong-gil. Nam served ministers 判書 of various departments and the chief officer (*jejo* 提調) of the National Observatory, and he imported collections of mathematics books from China. Although there was a big difference between the two in terms of social status, Nam and Lee became mathematical collaborators and coauthored many works including the *Sanhag jeongui* 算學正義 (1867). Lee Sang-hyeog published his final work, the *Igsan* 翼算 (1868), in two volumes. The first volume, "Jeongburon" 正負論, deals with the theory of equations, and the second volume, "Toetaseol" 堆垛說, the theory of finite series.

From the presentations of Lee's "Toetaseol," we can safely conclude that the copyist is none other than the author himself. In the volume "Toetaseol," Lee begins with the history of finite series in Chinese mathematics up to Luo Shilin's commentary of the *Siyuan yujian*, and this is revelatory of Lee's mathematical structural approach to the theory.

Lee starts with a discussion of Shen Kuo's *xijishu* as described in Sect. 2 and new terminologies for various series in the *Siyuan yujian*. Lee also quotes the interpolation given in the *Shoushili* as *pingliding sancha* 平立定三差 sums of sequences with constant second order difference sequences, and then he extends it to the sequences with constant fourth order difference sequences. Finally, Lee introduces a new method, called *bunjeogbeob* 分積法, to evaluate a partial sum $\sum_{k=m}^{n} a_k$ of a series $\sum_{k=1}^{n} a_k$. The partial sum is called *jeoljeog* 截積, which is derived from the areas or volumes of truncated figures, i.e. the Chinese *jieji* (see Fig. 9).

Lee Sang-hyeog presents his theory of finite series along with their mathematical structures, which is not the case for the *Siyuan yujian*.

He begins with *jiaocao ji* 茭草積, S_n^1 and then S_n^m $(2 \leq \text{m} \leq 5)$. He uses the terms *luoyi* and *saxing* 撒星 of S_n^m to represent S_n^{m+1} and S_n^{m+2}, e.g. S_n^3, which is called *sanjiao luoyi* or *jiacao saxing*. He also adds S_n^5 to show that Zhu's triangle can define S_n^m for any *m*.

[3] Lee Sang-hyeog 李尙爀, *Sawon og-gam* 四元玉鑑 (manuscript of the Yonsei University Library).

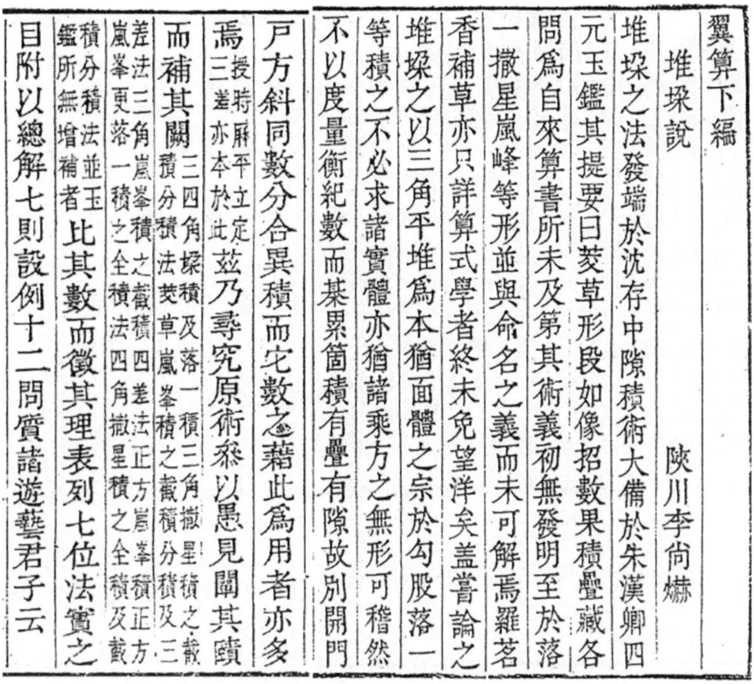
翼算下編
堆垛說　陝川李尚爀
堆垛之法發端於沈存中隙積術大備於朱漢卿四元玉鑑其提要曰茭草形段如像招數果積疊藏各問爲自來算書所未及第其術義初無發明至於落一撒星嵐峰等形並與命名之義而未可解焉羅茗香補草亦只詳算式學者終未免望洋矣蓋嘗論之堆垛之以三角平堆爲本猶面體之宗於勾股落一等積之不必求諸實體亦猶諸乘方之無形可稽然不以度量衡紀數而棊累箇積有疊有隙故別開門戶方斜同數分合異積而宅數之藉此爲用者亦多焉授時曆平立定三差亦本於此茲乃尋究原術參以愚見闡其賾而補其闕三四角垛積及落一積三角撒星積之截積分積法茭草嵐峯積之截積分積及三差法三角嵐峯積之截積四差法正方嵐峯積正方嵐峯更落一積之全積法四角撒星積之全積及截積分積法並玉鑑所無增補者比其數而徵其理表列七位法實之目附以總解七則設例十二問質諸遊藝君子云

Fig. 9 Preface to "Toetaseol"

First, Lee gives the formulae for these series as in the *Siyuan yujian*, and then using difference sequences he presents the formulae of sums. Moreover, he obtains the recursive relations of the series, namely

$$S_n^{m+2} = \sum_{k=1}^{n} (n-k+1) S_k^m.$$

For the partial sum of $\sum_{k=1}^{p} a_k$, we introduce the notion $\sum_{k=m}^{m+n-1} a_k$, $(p = m + n - 1)$ to represent the number (n) of terms of the partial sum. In the modern mathematics, the partial sum is evaluated and then simplified the result by

$$\sum_{k=m}^{m+n-1} a_k = \sum_{k=1}^{m+n-1} a_k - \sum_{k=1}^{m-1} a_k$$

For this problem, Lee Sang-hyeog invented *bunjeog-beob*, literally the "method of subdivisions" as per the following illustration of *jiaocaoji* in Fig. 10.

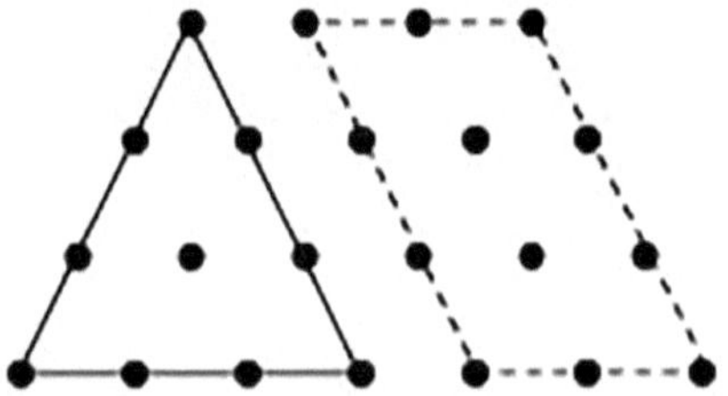

Fig. 10 "Method of subdivisions" (*bunjeog-beob*)

Indeed, Fig. 10 shows that the truncated part is divided into *jiaocaoji* S_4^1 and trivial sums $4 \times S_{4-1}^0$. For any *m, n*, Lee Sang-hyeog has

$$\sum_{k=m}^{m+n-1} S_k^0 = S_n^1 + S_{m-1}^0 S_n^0.$$

Since the triangular series were defined inductively, Lee Sang-hyeog obtains the following by *bunjeog-beob*

$$\sum_{k=m}^{m+n-1} S_k^r = S_n^{r+1} + S_{m-1}^0 S_n^r + S_{m-1}^1 S_n^{r-1} + S_{m-1}^2 S_n^{r-2} + \ldots + S_{m-1}^r S_n^0.$$

The next topic is about *sijiao duo* 四角垛. Lee defines the series of squares along with the triangular series, namely *sijiao duo*, *sijiao luoyi ji* 四角落一積, and *sijiao saxing ji* 四角撒星積 as follows:

$$T_n^1 = \sum_{k=1}^{n} k^2$$

and

$$T_n^j = \sum_{k=1}^{n} T_k^{j-1}$$

and then Lee obtains

$$T_n^1 = \frac{n(n+1)(2n+1)}{6}, T_n^2 = S_n^2 \times \frac{n+1}{2}, T_n^3 = S_n^3 \times \frac{2n+3}{5}$$

Applying *bunjeog-beob*, or difference sequences, Lee evaluates the partial sums of the above series.

Zhu Shijie introduced the terminology *lanfeng xing*嵐峯形 for a series $\sum_{k=1}^{n} a_k$ by the series $\sum_{k=1}^{n} ka_k$, and Lee extends these to have *jiacao lanfengji* 茭草嵐峯積, *sanjiao lanfengji* 三角嵐峯積, *zhengfang lanfengji* 正方嵐峯積, *zhengfang lanfeng geng luoyiji* 正方嵐峯更落一積, and *sijiao lanfengji* 四角嵐峯積. They are:

$$\sum_{k=1}^{n} kS_k^1, \sum_{k=1}^{n} kS_k^2, \sum_{k=1}^{n} k \times k^2 = \sum_{k=1}^{n} k^3, \sum_{k=1}^{n}\left(\sum_{j=1}^{k} j^3\right), \sum_{k=1}^{n} kT_k^1$$

Lee evaluates the above *lanfeng xing* series using the same procedures as those for triangular series and their partial sums. In the following, the *jiacao lanfengji* and *sanjiao lanfengji* will be denoted by L_n^1, and L_n^2, and *sijiao lanfengji* by L_n^3.

Finally Lee quotes the *Siyuan yujian*'s *yuanzhuiduo* 圓錐垛 and then reveals the mathematical structures, or relations of the formulae discussed in the body. We offer here a few interesting identities:

$$S_{n-1}^{r+1} = S_n^r \times \frac{n-1}{r+2}, S_n^{r+2} = \sum_{k=1}^{n}(n-k+1)S_k^r;$$

$$T_n^1 = S_n^2 + S_{n-1}^2, T_n^r = S_n^r \times \frac{2n+r}{r+2} = 2S_n^{r+1} - S_n^r, T_n^{r+2} = \sum_{k=1}^{n}(n-k+1)T_k^r;$$

$$L_n^1 = S_n^3 + 2S_{n-1}^3, L_n^2 = S_n^4 + 3S_{n-1}^4, L_n^3 = 2L_n^2 - L_n^1.$$

For the remaining identities on partial sums and others, we refer to Hong (2006b).

Using the theory of finite series, Lee Sang-hyeog solved twelve problems relating to series, ten of his own making and two problems selected from the section "Ruxiang zhaoshu." The first eleven problems deal with constructions of equations by *tianyuanshu*, and the last one by *eryuanshu*. Where Zhu Shijie uses finite series to construct equations, he neither includes their definitions nor arranges them systematically. Indeed, he deals with finite series in Section 1 ("Ruyi hunhe" 如意混和), Section 7 ("Jiaocao xingduan" 茭艸形段), Section 8 ("Jianji jiaocan" 箭積交參), and Section 9 ("Ruxiang zhaoshu" 如像招數) of Volume 2, as well as Section 1 ("Guoduo diecang" 果垛疊藏) of Volume 3 in the *Siyuan yujian*. As discussed above, Lee Sang-hyeog completes a commentary to the part of finite series in the *Siyuan yujian* according to the mathematical structure.

4 Conclusions

As developed over the tenth to fourteenth centuries, and as crystalized in Zhu Shijie's *Suanxue qimeng* (1299) and *Siyuan yujian* (1303), the Song-Jin-Yuan theory of equations is widely considered one of the most important contributions to the history of Chinese mathematics. Unfortunately, Zhu's works were completely ignored by Ming mathematicians, and though the *Siyuan yujian* was quoted by Mei Juecheng, Mei did not fully appreciate it in the mid-eighteenth century. The *Suanxue qimeng* was subsequently lost, only to return in the mid-nineteenth century via the Joseon edition of 1660. Without the *Suanxue qimeng*, early nineteenth-century mathematicians tended to focus on the *Siyuan yujian*, their efforts resulting in such commentaries as those of Luo Shilin and Shen Qinpei 沈欽裴.[4] As Luo's commentary was brought to Joseon, however, Joseon mathematicians had access to a complete set of Zhu Shijie's works, of which the *Suanxue qimeng* became an essential reference under the Joseon dynasty.

Explaining the construction of equations by *tianyuanshu*, *eryuanshu*, *sanyuanshu*, and *siyuanshu*, Zhu Shijie adopted the theory of finite series. He showed that the basic structures for the theory of equations and that of finite series were strongly tied with Zhu's triangle for binomial expansions, which was brought into Joseon in the seventeenth century.

Based on *tianyuanshu* for the constructions of equations, and Zhu's triangle for solving equations, Hong Jeong-ha (1684–1727) established his own theory of equations in the *Gu-il jib*. Hong used the mathematical structures of the triangle for binomial expansions given by Jia Xian's *zengchengfa* and Zhu's inductive ones by *shenwai jiafa*. Indeed, Hong showed Jia Xian's *zengcheng kaifangfa* for extractions of nth roots can be extended to solve any polynomial equation $p(x) = 0$. The method is now known as the Ruffini-Horner method with synthetic divisions, but Hong Jeong-ha revealed its true structure through the triangle of binomial expansions.

Of equal importance to the theory of finite series was another Joseon mathematician, Lee Sang-hyeog (b. 1810). Unlike Hong Jeong-ha, Lee had unprecedented access to sources for Eastern and Western mathematics, particularly the collection of Chinese imported by his collaborator, Nam Byeong-gil (1820–1869), containing almost everything from the *Jiuzhang suanshu* to the latest nineteenth-century publications. Interested in the theory of finite series in the *Siyuan yujian*, Lee first transcribed a part of the *Siyuan yujian*. He noticed that its presentation of the theory was not systematic and detected that the mathematical structures of finite series were also based on Zhu Shujie's triangle.

Lee Sang-hyeog offered a systematic treatment of the mathematical structures of finite series, extending the results in the *Siyuan yujian*. To this end, he used the triangle and difference sequences first appearing in the *Shoushili*. He also introduced a *bunjeog-beob*, or method of subdivisions, to describe the structures of the partial sums of a series along with the triangular series.

In all, we conclude that Hong Jeong-ha and Lee Sang-hyeog obtained their structural result through a perspective rooted in mathematical structures.

Acknowledgements We would like to thank Qu Anjing for his continuing support through discussions and related references. We wish to gratefully acknowledge Daniel Patrick Morgan, whose exceptionally careful reading and comments improved the quality of this paper.

[4] *Siyuan yujian* 四元玉鑑, Zhu Shijie 朱世傑, comm. Shen Qinpei 沈欽裴 (*Xuxiu Siku quanshu* 續修四庫全書, *Zibu* 子部, vol. 1042, 439–760, Shanghai Guji chubanshe, 2002).

Bibliography

Chemla K. 2018. The Proof is in the Diagram: Liu Yi and the Graphical Writing of Algebraic Equations in Eleventh-Century China, *Endeavour* 42.2–3: 60–77.

Guo Shuchun 郭書春 (ed.). 1993. *Zhongguo kexue jishu dianji tonghui, shuxue juan* 中國科學技術典籍通彙·數學卷. Zhengzhou: Henan Jiaoyu chubanshe.

Hong Jeong-ha 洪正夏. 1724. *Gu-il jib* 九一集. Seoul National University Library 서울大學校圖書館.

Hong Sung Sa 洪性士 (tr. & com.). 2006a. *Igsan* 翼算. Gyousa 教友社.

Hong Sung Sa 洪性士. 2006b. Theory of finite series in Chosun Mathematics [朝鮮 算學의 堆垜術], *The Korean Journal for History of Mathematics* 19.2: 1–24.

Hong Sung Sa 洪性士, Hong Young Hee 洪英喜, and Kim Chang Il 金昌一. 2007. *Gougushu* in the 18th century Chosun [18세기 朝鮮의 句股術], *The Korean Journal for History of Mathematics* 20.4: 1–22.

Hong Sung Sa 洪性士, Hong Young Hee 洪英喜, and Kim Chang Il 金昌一. 2016. Zeros of Polynomials in East Asian Mathematics, *Journal for History of Mathematics* 29.6: 317–324.

Hong Sung Sa 洪性士, Hong Young Hee 洪英喜, and Kim Chang Il 金昌一. 2018. Jo Tae-gu's *Juseo Gwan-gyeon* and *Jihe Yuanben*, *Journal for History of Mathematics* 31.2: 55–72.

Hong Sung Sa 洪性士, Hong Young Hee 洪英喜, and Kim Young Wook 金英郁. 2011. Liu Yi and Hong Jung Ha's *Kaifangshu* [劉益과 洪正夏의 開方術], *The Korean Journal for History of Mathematics* 24.1: 1–13.

Hong Sung Sa 洪性士, Hong Young Hee 洪英喜, and Kim Young Wook 金英郁. 2014. Hong JeongHa's *tianyuanshu* and *Zengcheng kaifangfa*, *Journal for History of Mathematics* 27.3: 155–164.

Hong Sung Sa 洪性士, Hong Young Hee 洪英喜 and Lee Seung On 李承溫. 2017. *Siyuan yujian* in the Joseon Mathematics, *Journal for History of Mathematics* 30.4: 203–219.

Lee Chang Ku 李昌九 and Hong Sung Sa 洪性士. 2013. Chinese Mathematics in Chosun, *Journal for History of Mathematics* 26.1: 1–9.

Lee Sang-hyeog 李尙爀. *Sawon og-gam* [四元玉鑑]. Yonsei University Library [延世大學校 圖書館].

Li Yan 李儼 and Du Shiran 杜石然. 1987. *Chinese Mathematics: A Concise History*. Translated by Crossley J. N., Lun A. W.-C. Oxford: Clarendon Press.

Qu Anjing 曲安京. 2018. *Jinxiandai shuxueshi yanjiu de yitiao lujing* 近現代 數學史研究的一條路徑 (A new Approach to the History of Modern Mathematics). *Kexue jishu zhexue yanjiu* 科學技術哲學研究 (Studies in Philosophy of Science and Technology) 35.6: 67–85.

The Mathematics of a Prime Minister of Joseon, Choi Seog-jeong

Kim Young Wook

Abstract In this paper, we examine what mathematics was like in seventeenth-century Joseon through the eyes of a Confucian government official. Choi Seog-jeong was a typical official, busy working in the government and researching the Confucius classics, but he was different from regular mathematicians in that he tried to mingle Confucian philosophy into mathematics. Having come into contact with Western mathematics, he also posed the very basic question as to whether there are similarities between East and West. His attempt did not succeed from our perspective, but he was successful in finding new combinatorial examples which he did not dream would be important from other viewpoints than his own.

Keywords Choi Seog-jeong · *Gusu-ryag* · Joseon mathematics · Magic squares 縱橫圖 · Battle formations 陣法 · Four symbols 四象 · Eight trigrams 八卦 · *Hetu* 河圖 · *Luoshu* 洛書 · Cayley tables · Orthogonal Latin squares

1 Introduction

This paper is the outcome of a simple question concerning how mathematics was used in military tactics in the Joseon dynasty (1392–1897). The most plausible application of mathematics in military tactics seems to be the use of numerical

The author thanks Professor Hong Sung Sa for his help studying Choi Seog-jeong and his mathematics since 2013. Special thanks go to the Organizing Committee Chair, Professor Qu Anjing, for a wonderful conference. Thanks are also due to all the organizing committee members and the participants of *The Second International Conference on History of Mathematics and Astronomy* at Xi'an.

Kim Young Wook (✉)
Korea University, Seoul, South Korea

D. P. Morgan, T. Quan (eds.), *Mathematics and Astronomy in the Ancient World*,
https://doi.org/10.1007/978-3-031-93809-2_7

squares of battle formations (*jinbeob* 陣法). Our idea to find relations between these two subjects in Korean history did not prove fruitful, but it eventually led us to a new perspective on the mathematics of Choi Seog-jeong 崔錫鼎 (1646–1715). Over the course of this work, we came across recent studies on his mathematics, which we will summarize here, introducing old and recent findings on the subject.

In this paper, we will look at some of the contents of Choi's mathematical work *Gusu-ryag* 九數略, paying special attention to the numerical diagrams. All the Chinese sources in this paper are found in the *Zhongguo kexue jishu dianji tonghui: Shuxue juan* 中國科學技術典籍通彙·數學卷 (Guo 1993) and the *Siku quanshu* 四庫全書, and all the Korean sources are found in *Hanguk gwahakgisulsa jaryo dae-gye: suhak pyeon* 韓國科學技術史資料大系·數學篇 (Kim 1985).

2 Choi Seog-jeong

2.1 Concise Biography of Choi Seog-jeong

Choi Seog-jeong was a Confucian scholar, politician, and administrator of the Joseon. His other names include Yeosi 汝時, Yeowha 汝和, his *noms de plume* Jonwa 存窩 and Myeonggok 明谷, his posthumous title Munjeong 文貞. Choi was born into a famous family and was a young prodigy. As was usual then, he was adopted to a son of his great uncle, Prime Minister Choi Myeong-gil 崔鳴吉 (1586–1647).

Choi Seog-jeong himself served under the reign of Kings Hyeonjong 顯宗 (r. 1660–1674) and Sukjong 肅宗 (r. 1674–1720). Starting as a history official, then transferred to military administration, he likewise made his way up to the position of Prime Minister in 1701 (Hong 2013). He was talented and served in various positions in the government, but people looked at him with jealousy, and his political opposition tried to control him. He was also reproached for his grandfather's political decisions during the Qing invasion of Joseon in 1636–1637. This caused him to struggle through politics and to resign his positions again and again. However, due to his talents in literary writing and logical reasoning, he was loved by the king and was rehired each time, holding the position of Minister ten times over 1697–1711. In his day, politics were divided between two parties, and political disputes continued in the government. Choi Seog-jeong was a typical official who tried to calm down these disputes and to find ways to benefit the common people. His viewpoint was that of a career official, different from the majority who were more connected to their family politics. He was considered to be firm inside but made an external show of calmness all the time.

2.2 *Choi Seog-jeong as a Scholar*

Choi Seog-jeong was a Confucian scholar and tried to organize the Classics into a flawless theory. His scholarly background lay in both the Classical studies (*gyeong-hak* 經學) and science (*gwahak* 科學). He studied and corrected the old *Record of Rites* (*Liji* 禮記) and wrote a *Corrections and Commentaries to the Record of Rites* (*Yegi yupyeon* 禮記類篇). This is his work of a lifetime. He also wrote several writings like poems, postscripts, epitaphs, and bibliographies, which are collected in his collection of writings (*Myeonggok jip* 明谷集). In addition, he wrote a book on mathematics, the *Gusu-ryag*, which he is believed to have finished in his later years. However, this book contains various topics and includes several findings of his own, which would have required long years of study. Thus, the contents of the *Gusu-ryag* must have been written over the course of his life. Choi also tried to collect history systematically and to organize historical facts away from the official history in what is called "private history" (*yaseung* 野乘).

In spite of the busy administrative life, Choi studied science and mathematics throughout his life. His time was after two big wars, and general life was difficult. Thus, there were not many who could devote themselves to science or mathematics. We do not know how Choi got acquainted with mathematics. Our first guess was that he became interested in the mathematical theory of battle formations during his time in the military. Since he was a prodigy and had enough chance to read and see advanced civilization from China, Choi could have conceived that science is in the core of people's life. Probably, his chance to visit Peking in 1686 and 1697 was one such opportunity for him. In these visits, he was the vice-envoy and the envoy, respectively. Being interested in mathematics and science, he must have taken advantage of the chance to bring back books from China in the fields. We know that he studied the *Tianxue chuhan* 天學初函, and this is one of the books that he could have brought back with him at that time.

Choi's background was in Confucius philosophy, and this worked as a limitation in his study of mathematics. However, his attitude of rationality and fairness in his office is likely to be from his study of mathematics and science, and this could have resulted in his steps toward justice-oriented politics.

3 The Mathematics of Choi Seog-jeong

3.1 *Study of Mathematics*

Eastern Mathematics Choi Seog-jeong must have been interested in mathematics in his early years. When Korean nobles taught their children, mathematics was one of the subjects on the typical curriculum. However, Choi studied beyond the usual curriculum and must have studied from various sources. In the *Gusu-ryag*, Choi mentions several mathematicians of China and Korea, such as Yang Hui 楊輝 (Ch.;

*c.*1238–1298) and Gyeong Seon-jing 慶善徵 (Kr.; 1616–1690). He also mentions scholars in the field of Neo-Confucianism (*daoxue* 道學) such as Zhu Xi 朱熹 (1130–1200), Cai Yuanding 蔡元定 (1135–1198), and his son Cai Shen 蔡沈 (1167–1230), from China, and Seo Gyeong-deog 徐敬德 (1489–1546), etc., from Korea. This clearly shows that Choi Seog-jeong's view of mathematics does not distinguish between mathematics and the theory of the *I Ching* 易經, or *Book of Changes*. Choi, as well as many other Joseon scholars of the time, followed the "Images and Numbers" (*xiangshu* 象數) school of the *I Ching*, which emphasizes symbolic and numerological parallels between the natural world and the hexagrams.

Western Mathematics As previously mentioned, Choi had been to China twice, and this must have been his chance to come into contact with mathematics and science from the West. His position as a high official also allowed him to access the national libraries in the Hojo 戶曹 and Gwansanggam 觀象監, where he could have read all the then recent books imported from China. This made him one of the earliest scholars in Joseon to have contact with Western mathematics and science. In the bibliography of his *Gusu-ryag*, Choi includes the *Tianxue chuhan* 天學初函 (1629)[1] and *Chousuan* 籌算.[2] This is the first time any work of Western mathematics was ever mentioned in a Korean book on mathematics (Hong 2006).

Tianxue chuhan is a collection of books written by various authors and is divided into two parts. Part 1, "Libian" 理編, is a collection of theory of Catholic doctrine, and Part 2, "Qibian" 器編, is a collection of mathematical and scientific books. According to Hong Young Hee, this is a huge collection and it is thought that Choi Seog-jeong could not have read all the way through it. Also, in his bibliography, Choi Seog-jeong mentions the authors of *Tianxue chuhan* as Matteo Ricci and Li Zhizao, while in fact only two books in the collection, the *Tongwen suanzhi* 同文算指 and *Yuanrong jiaoyi* 圜容較義, were actually written by these two authors. This suggests that Choi Seog-jeong read only these two books (Hong 2006, 2013).

Importance of Choi Seog-jeong's Mathematical Works As we have seen, Choi Seog-jeong was very enthusiastic towards both the research and the diffusion of mathematics in Joseon scholarly society. He lived at a time after two devastating wars, when even professional mathematicians from middle-class mathematical families were ill-prepared due to many difficulties.[3]

Historically Choi Seog-jeong's work is important because he and his book played an important role in reviving lost mathematics after two great wars, one against Japan (1592–1598), and another against the Qing (1626, 1636–1637). Other than

[1] This book is written by Li Zhizao 李之藻 (1565–1630) and contains *Tongwen suanzhi* 同文算指 (1614), a translation by Matteo Ricci and Li Zhizao of the mathematical work *Epitome Arithmeticae Practicae* (1583) by Christopher Clavius.

[2] This book is a part of the *Xiyang xinfa lishu* 西洋新法曆書.

[3] We see this from the works of Gyeong Seon-jing 慶善徵 (1616–1690), studied in Jin (2009) and Jin and Kim (2009).

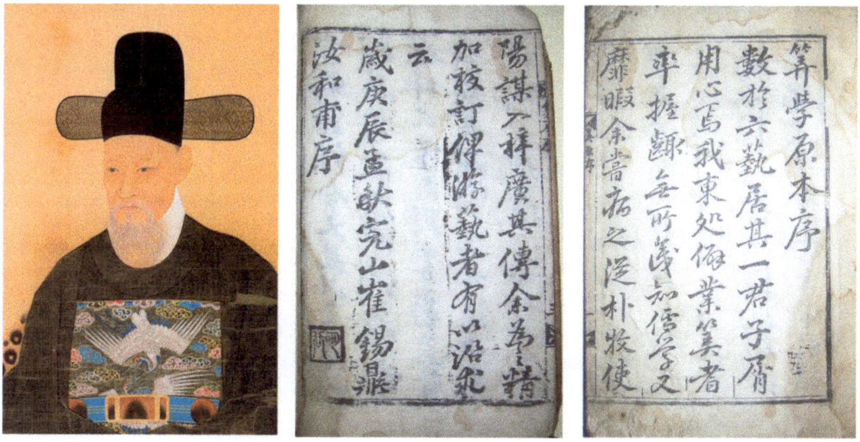

Fig. 1 Portrait of Choi Seog-jeong (left) and his calligraphy, preface to *Sanhag wonbon*, first and last page (from right to middle). Image source: Kim (1985)

writing his book, Choi Seog-jeong also encouraged Park Du-se 朴斗世 (1650–1733) to publish his father's book *Sanhak wonbon* 算學原本 (1700) and wrote the preface thereto (Fig. 1).[4]

Choi also studied Western mathematics through Chinese translations. This was probably the first time Korean scholars ever made contact with Western mathematics. He studied the *Tongwen suanzhi* and *Suanfa tongzong* 算法統宗 (1593) and summarized them in his own book. In so doing, he introduced the mathematics of the *Tongwen suanzhi* to the Joseon. He also studied Yang Hui and Zhu Shijie's thirteenth-century Chinese works. His broad studies enabled him to collect tools as useful as Napier's rods calculation or abacus calculation in his *Gusu-ryag* (Hong 2013, 2006).

Compared to other, contemporary mathematical books of the Joseon, the *Gusu-ryag* is not mathematically mature. Considering that Choi Seog-jeong was a busy official, and not a professional mathematician, this is not surprising. Still, we will see how he was trying to interpret mathematics based on his philosophical beliefs, and how he succeeded and failed mathematically will reveal that Choi Seog-jeong's mathematics is worth examining, and it exposes where his creativity comes into his mathematics.

3.2 *Choi Seog-jeong's View of Mathematics*

Due to his background, Choi Seog-jeong was influenced by the perspective of the *I Ching* and followed the philosophy of Shao Yong 邵雍 (1011–1077), all his theories splitting in binary ways (or in mathematical duality) (Jung 2009). On the other

[4] Park Yul 朴繘 (1621–1668) was a mayor of a small city and was an excellent mathematician. His book was published by his son more than 30 years after his death. This book was known to scholars and was mentioned in books like the *Gusu-ryag*.

hand, Choi's mathematical ideas came from the philosophy of Cai Shen, which posits that numbers and their properties are based on the two basic diagrams, namely, the *Hetu* 河圖 and *Luoshu* 洛書, and that life follows laws in the form of variations on these two diagrams (Cho 2017). Choi must have had the idea that mathematics itself is also governed by these laws. In fact, in his *Gusu-ryag*, he includes a nine-by-nine matrix diagram of Cai Shen as a basic one. He also categorizes the traditional Nine Chapters (*jiuzhang* 九章) and mathematical techniques according to the Four Symbols (*sixiang* 四象) and the Eight Trigrams (*bagua* 八卦). Furthermore, in the last chapter of the book, he includes as many as 50 variations of two basic diagrams, 20 of which are from the "Xugu zhaiqi suanfa" 續古摘奇算法 section of Yang Hui's *Yang Hui suanfa* 楊輝算法 (1275), while the rest are creations of Choi himself. These characteristics reveal Choi's unique application of philosophy to mathematics. In this way, he tried to interpret mathematics according to this perspective, but he also tried to structure his philosophy according to his mathematics.

Choi Seog-jeong's new view of mathematics, based on Shao Yong and Cai Shen, seems to have won considerable support from Joseon scholars chiefly because it provides a new theoretical (mathematical) foundation for the Teachings of Master Zhu [Xi] (*Zhuzi xue* 朱子學), and thus found many followers. Among them was Bae Sang-Yeol 裵相說 (1759–1789), who wrote a book on writing and mathematics that includes many such theories and diagrams.[5] Bae also wrote many articles on mathematics and astronomy that show a strong influence from Choi Seog-jeong.

This trend gave the general Joseon public theoretical grounds for the belief that Zhu Xi's theory is scientific, and it was not very helpful in the development into modern mathematics. However, in some degree, we find Choi Seog-jeong's theory to be mathematically and pedagogically important.

In the following pages, we will look at some of the diagrams in the *Gusu-ryag* and explain the mathematical structures found therein. We cannot interpret them using the theories of Shao Yong and/or Cai Shen, but we will report recent research findings that they include modern combinatorial structures. From these findings, we will try to understand Choi Seog-jeong's work as one of creative activity in mathematics, which is important in modern mathematics pedagogy.

4 *Gusu-ryag*

The title *Gusu-ryag* means "Concise Edition of the *Nine Chapters*." In this book, Choi Seog-jeong tries to categorize mathematical computation methods according to the four symbols and the eight trigrams. The *Gusu-ryag* consists of six chapters and one appendix in two volumes. The first five short chapters and part of Chapter 6 are bound in the first volume ("Geon" 乾). The rest is bound in the

[5] Title of the book is *Seogye swaerog* 書計鎖錄 (1786).

second volume ("Gon" 坤). Each volume is divided into two parts, so the whole book consists of four parts (*gap* 甲, *eul* 乙, *byeong* 丙, *jeong* 丁, or A, B, C, D). Here, we will examine the contents of the first three parts of the *Gusu-ryag*. First, we will start from the beginning, where definitions and terminologies are introduced and where relations with the Four Symbols are explained before continuing to the multiplication tables, which shows Choi's idea of connections between East and West. We will not go into the mathematical problems, which comprise most of the second and the third part of the book. Next, we will find another contrast between East and West in the fourth part, where two calculation tools are explained, namely Napier's rods and the Abacus, and we will leave the rest of the book—the diagrams—to the later sections.

4.1 Contents of the Gusu-ryag

The first chapter of the *Gusu-ryag* is titled "Origin of Numbers/Mathematics" (Suwon 數原);[6] the second "Names" (Sumyeong 數名), the third "Positions/Digits" (Su-ui 數位), the fourth "Symbols" (Susang 數象), the fifth "Tools" (Sugi 數器), the sixth "Methods" (Subeob 數法). In Chapter 1, Choi Seog-jeong explains the two basic diagrams *Hetu* and *Loushu*. In Chapter 2, he explains the naming methods of numbers and weights and measures. In Chapter 3, Choi Seog-jeong shows how to construct numbers using digits, here first small numbers being explained using tenths (*bun* 分), hundredths (*li* 釐), and so on. The chapters "Symbols" and "Tools" explain how to use counting rods as mathematical symbols (*sang* 象) and computational tools (*gi* 器). Chapter 6 is the main part, and it is all about how to do mathematics. This chapter consists of five sections, and here Choi Seog-jeong classifies numbers and computational methods into four or eight. In the last part ("Jeongpyeon" 丁編), he explains paper calculation (*munsan* 文算), abacus calculation (*jusan* 珠算), Napier's rods (*jusan* 籌算), and variations of *Heto* and *Loushu* (*Ha Nak byeonsu* 河洛變數).

4.2 Categorization of Calculations

In the first part ("Gabpyeon" 甲編) at the beginning of Chapter 6, Choi first explains four basic number facts, namely, addition and subtraction, multiplication and division, the multiplication table, and the multiplication mnemonic (*gugu gugyeol* 九九口訣).

---

[6]The character 數 means "number" but is also used to mean mathematics. Here we just use "numbers."

In Section 3, he then goes into basic calculation methods, where he mentions eight basic methods. Section 3.1 says that multiplication and division are based on addition and subtraction, and Section 3.2 is about methods for calculating fractions.

Section 5 of Chapter 6 begins categorization using the Four Symbols. This is the beginning of the second part ("Eulpyeon" 乙編), and Choi Seog-jeong develops the concepts underlying computations by binary division. For example, he explains multiplication and division in four steps: first, multiplication as repeated addition, then division as repeated subtraction, and, lastly, multiplication and then division. In this process, he uses four numbers or four ratios from the *Tongwen suanzhi*. He must have believed that the method in the *Tongwen suanzhi* corresponds to the Four Symbols and thus borrowed it to explain the philosophical structures underlying mathematics (see Fig. 2). He also established the correlations that repeated addition is "Greater Yang" (*taiyang* 太陽), repeated subtraction is "Greater Yin" (*taiyin* 太陰), regular multiplication is "Lesser Yang" (*shaoyang* 少陽), and regular division is "Lesser Yin" (*shaoyin* 少陰). This completes the Four Symbols correspondences (see Fig. 2).

The next subsection explains repeated multiplication and repeated division, which are concepts developed from previous ones. The third and the fourth subsections are multiplication followed by division, or division followed by multiplication, both appearing in equal ratio problems such as $a: b = c: d$. These are again

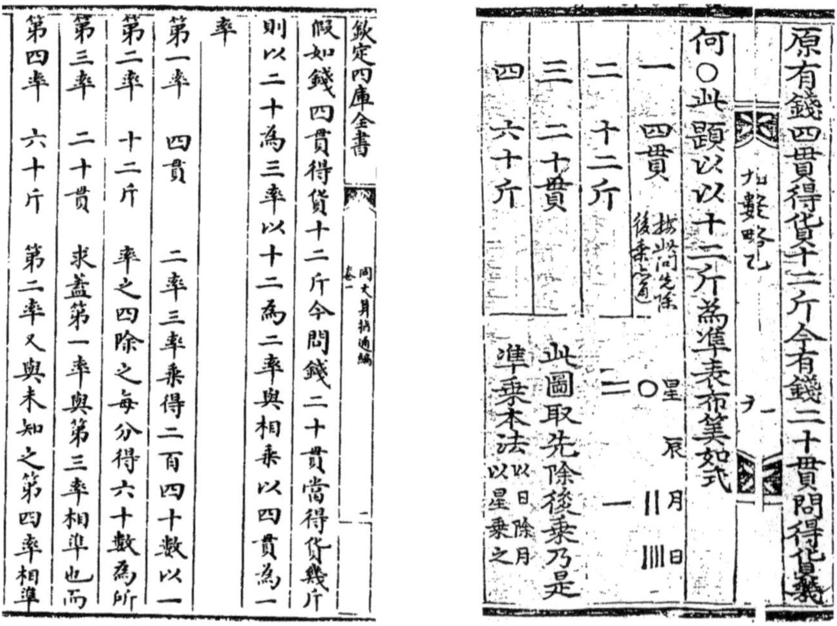
欽定四庫全書 同文算指通編 卷一
假如錢四貫得貨十二斤今問錢二十貫當得貨幾斤
則以二十為三率以十二為二率與相乘以四貫為一
率
第一率 四貫 二率三率乘得二百四十數以一
第二率 十二斤 率之四除之每分得六十數為所
第三率 二十貫 求益第一率與第三率相準也而
第四率 六十斤 第二率又與未知之第四率相準

原有錢四貫得貨十二斤今有錢二十貫問得貨幾
加減略乙
何○此題以以十二斤為準表布算如式
一 四貫
二 十二斤
三 二十貫
四 六十斤
此圖取先除後乘乃是準乘本法

Fig. 2 The same problem from *Tongwen suanzhi*, "Tongbian" 通編, Volume 1 (**left**) and from *Gusu-ryag*, Part 2 ("Eulpyeon" 乙編) (**right**). In the *Gusu-ryag*, we see the labels "sun", "moon," "stars," and "planets" 日月星辰 on the right of the counting rod symbols at the bottom. Image source: Guo (1993) (left) and Kim (1985) (right)

Table 1 Correspondences of the individual chapters of the *Nine Chapters on Mathematical Procedures* (*Jiuzhang suanshu* 九章算術) with the Four Symbols (*sixiang* 四象) according to the *Gusu-ryag* (Jung 2009)

Four Symbols			*Nine Chapters*		
Greater Yang 太陽	sun 日	1	"Fangtian" 方田		
Greater Yin 太陰	moon 月	2	"Sumi" 粟米	"Shaoguang" 少廣	
Lesser Yang 少陽	stars 星	3	"Shanggong" 商功	"Cuifen" 衰分	"Ying buzu" 盈不足
Lesser Yin 少陰	planets 辰	4	"Junshu" 均輸	"Gougu" 勾股	"Fangcheng" 方程

correlated with the Four Symbols. Then comes variations of the eight above, making a total of sixteen. This way, he considers the division of concepts into two, then again into four, so that the process goes through the 8 trigrams toward the 64 hexagrams.

Among the first four variations are problems of the forms in the chapters "Fangtian" 方田, "Shaoguang" 少廣, "Cuifen" 衰分, and "Gougu" 勾股 from the *Nine Chapters on Mathematical Procedures* (*Jiuzhang suanshu* 九章算術). In this way, Choi explains all nine problem types in the *Nine Chapters* as embodying one of these 16 methods (see Table 1).

5 Diagrams

Most of the Appendix of the *Gusu-ryag* is devoted to diagrams of numbers called magic squares and magic circles. The first appearance of such diagrams in a mathematics book is in Yang Hui's *Yang Hui suanfa*, where one finds about 20 of them. Choi Seog-jeong constructed many more and essentially completed Yang Hui's list to 46 diagrams. In fact, he lists:

- Magic squares sized 3 × 3 to 10 × 10;
- Those with names which include "Yong-odo" 用五圖, "Yong-yugdo" 用六圖, or "Saodo" 四五圖, "Chil-odo" 七五圖, "Pal-odo" 八五圖, "Guodo" 九五圖.
- Those with names which include "Sagudo" 四九圖, "Ogudo" 五九圖, "Yuggudo" 六九圖, "Chilgudo" 七九圖, "Palgudo" 八九圖, "Gugudo" 九九圖;
- Those with names which include "Yong-odo" 用五圖, "Yong-yugdo" 用六圖, "Yongchildo" 用七圖, "Yongpaldo" 用八圖, "Yonggudo" 用九圖 (other kinds than above)
- And four special tables named "Gugu mosu byeon-gung yangdo" 九九母數變宮陽圖, "Gugu mosu byeon-gung eumdo" 九九母數變宮陰圖, "Gugu jasu byeon-gung yangdo" 九九子數變宮陽圖, and "Gugu jasu byeon-gung eumdo" 九九子數變宮陰圖.

For many of the diagrams he presents the corresponding dual diagrams ("Eumdo" 陰圖), whose meaning and relations we do not currently understand.

5.1 Variations on Basic Diagrams

Like Yang Hui, Choi Seog-jeong started with the two basic diagrams of the *Hetu* and *Luoshu*. He then presents two 4 × 4 magic squares, and so on. From the names of the diagrams, we infer that some of them are based on these two basic diagrams (Fig. 3).

Magic squares were known to have had been used as sacred objects to dispel evil spirits around the times of the Song (960–1279).[7] We will look at one or two of Choi Seog-jeong's diagrams that turn out to be important in later sections.

The fact that Choi Seog-jeong's magic squares are related to the philosophy of Shao Yong and Cai Shen is obvious from the names he gave to them. First of all, each of the magic squares comes in pairs of Yin and Yang. This is the basis of Shao and his followers' writings. Secondly, some of the magic squares were named "Yanshu tu" 衍數圖 and "Yishu tu" 易數圖, and they used terms like "Yin and Yang" 陰陽, "permutation" (*cuozong* 錯綜), "double trigram" (*chonggua* 重卦), and "variation of position" (*biangong* 變宮), which all come from the Shao's theory.

On the other hand, many of the variant diagrams other than the square ones were named using the terms *Hetu* and *Luoshu* to clarify that they are variations on the basic ones. Therefore, the basis of Choi's diagrams lies in the *Hetu*, *Luoshu*, Shao and Cai's theory, and Western tables in the *Tongwen suanzhi*. Furthermore, the tools he used to generate them are based on the ideas of oriental philosophy such as duality (Yin and Yang), "variations of positions" (*biangong*), and "permutations" (*cuozong*). These are similar to basic modern combinatorial tools, and they allowed Choi Seog-jeong to generate many meaningful variations.

In the following sections we will explain a few diagrams from the *Gusu-ryag* that have caught the attention of the author as well as of other modern mathematicians.

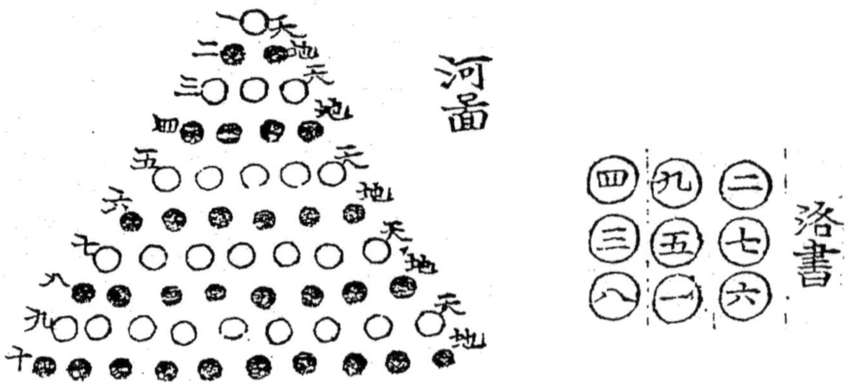

Fig. 3 Choi Seog-jeong's version of the *Hetu* (**left**) and *Luoshu* (**right**). Image source: Kim (1985)

[7] Li and Du (1987).

5.2 Battle Formation Diagrams

Among the diagrams in the *Gusu-ryag* we find the one named "Jung-gwae yong-paldo" 重卦用八圖, which corresponds to the "Bazhen tu" 八陣圖 from the *Yang Hui suanfa*. This diagram is famous thanks to the legend that it was made by Zhuge Liang 諸葛亮 (181–234), the military advisor of the Shu-Han 蜀漢 (221–263). This leads one to suspect that some of the diagrams were military oriented. There is further background for this claim: namely, the seventeenth century was a time of war and reconstruction for the Joseon. As such, there must have been more discussions of defense among military officials. Also, in his early years in office, Choi Seog-jeong served in the Ministry of War (Byeongjo 兵曹).

One way to measure the amount of discussion of defense plans after the wars is to look at the number of times the subject comes up in court records. To this end, we counted the occurrences of the term *jinbeob* ("battle formation") in *The Annals of Joseon* (*Joseon wangjo sillok* 朝鮮王朝實錄). Contrary to our expectations, the discussions were not concentrated after the wars but were in fact evenly distributed over the 600 years covered therein (see Fig. 4). It thus seems that the Joseon were constantly threatened by neighboring countries, and that there were worries of invasion from both the north and south. *Jinbeob* were prepared mainly for defense against small, fast, and dispersed invasion forces from the north.

Therefore, the wars were not especially the reason triggering discussions of *jinbeob,* but there were plenty of reasons for Choi Seog-jeong to consider the topic in his book.

For this paper, we discussed with military researchers about the meaning of oriental battle formations and their use in war.[8] The mathematical explanation that first comes to mind is that the restrictions requiring that the fact that the sums of numbers in partitions are made equal in the "Bazhen tu," or that the sums of each row and column are equal in magic squares, looks useful if an army confronts a strong enemy force. Their tactic is to number friendly soldiers by order of combat skills and then arrange them at the positions of the numbers of the "Bazhen tu." As such, the "Bazhen tu" as a defense formation can be used to confront one strong enemy soldier with eight friendly forces in a circle of the formation, and this way the power of our forces are evenly distributed in the sense of average combat skills in eights. We agreed on this idea, and this seems to explain why magic squares were considered sacred or to have mysterious power. In *fengshui* 風水, people use various symbolic methods to block the flow of bad forces and open and let flow the good forces. The magic squares or diagrams like the "Bazhen tu" offer even power of defense from all Eight Directions in a perfect block. And it has always been a mystery that such arrangements of numbers exist. Choi Seog-jeong comments at the bottom how to allocate eight troops of soldiers or multiply by three or sixteen to get a total of 66,560 soldiers. This is titled the "Warrior Battle Formation Diagram" (Muhyeop jindo 武俠陣圖; Fig. 5).

[8] Y.-k. Roh, personal communication, June 9, 2017.

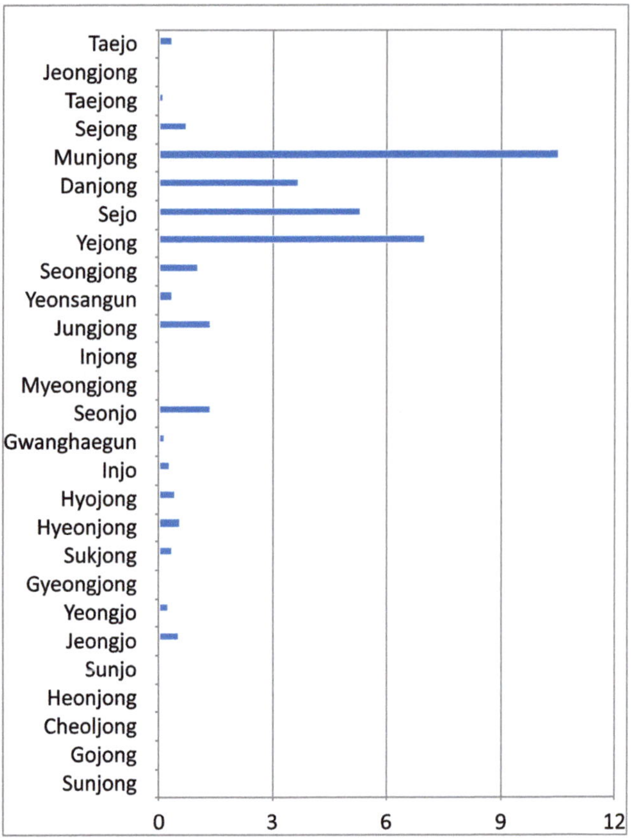

Fig. 4 The number of discussions of "battle formations" (*jinbeob* 陳法) appearing in *The Annals of Joseon* (average count per year for each king). This shows more occurrences in the early years after the authorship of the book *Jinbeob* 陳法 (1451) by King Munjong 文宗 (r. 1450–1452) and King Sejo 世祖 (r. 1455–1468)

One more thing to be noted is that in one copy (Fig. 5, right), one finds added to this the name "Bazhen tu" as well as bearings, in Eight Trigrams notation, and lines connecting all the number pairs amounting to the common sum of 65. This tells us that there are scholars oriented toward philosophical interpretations who followed Choi's line of thinking. Also, we would like to mention that the "Bazhen tu" has many more circular symmetries, as is explained in Lee (2010).

5.3 *Diagrams from the* Hetu *and* Luoshu

The *Gusu-ryag* features a couple of diagrams formed using hexagonal arrays of numbers. A smaller one is the "Jisu yongryukdo" 地數用六圖, and a larger one is the "Nakseo ryukgudo" 洛書六九圖 or "Jisu gumundo" 地數龜文圖. The latter

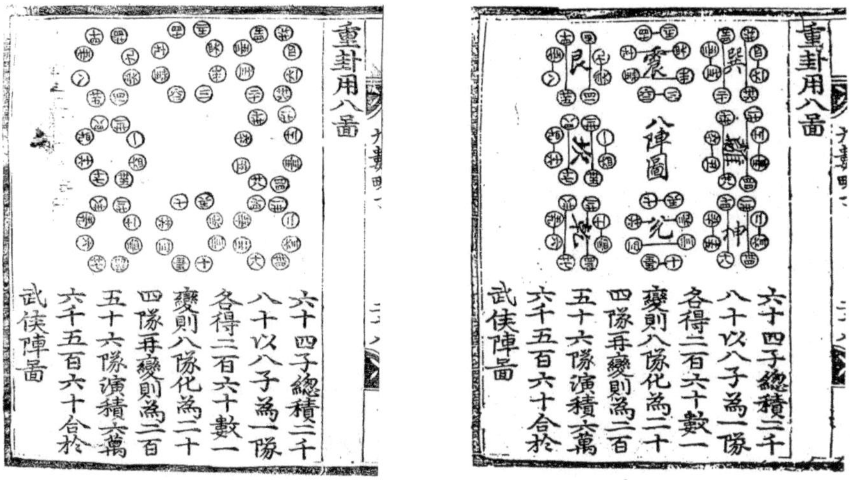

Fig. 5 The "Bazhen tu" 八陣圖is a legendary battle formation and is now lost. The history of this diagram coming to be called "Bazhen tu" is unknown. Image source: Kim (1985)

has been studied by several authors, which have showed how to construct them or how many of them exist. This one can also be thought of as a military battle formation. This kind of diagram is unique in the world. Many solutions are introduced by various authors (Kwon et al. 2014; Lee 2010; Park 2018) (Fig. 6).

5.4 Circular Diagrams

There are several circular diagrams introduced in the *Gusu-ryag* wherein diagonal sums or circular sums are the same. Thus, this is a kind of magic circle so to speak. Some of them are from Yang Hui's book, but Choi Seog-jeong completed the list changing sizes. The simplest one, of Choi's own creation, is the circular diagram made of the nine numbers 1 to 9. This one has circular sums together with the center- and diagonal sums, which all equal 25. However, we find more symmetries if we look at this in a three-dimensional format, and this seems more natural (Fig. 7).

It is easily seen that all six faces have vertices-sums of 20, and with the center 5, they are 25. Therefore, this is a kind of magic cube. If one looks for all such arrangements with 5 at the center, there are essentially 3 distinct diagrams, but if one tries with other number as centers there are even more. We think circle diagrams with more numbers can also be arranged in three dimensions with symmetries.

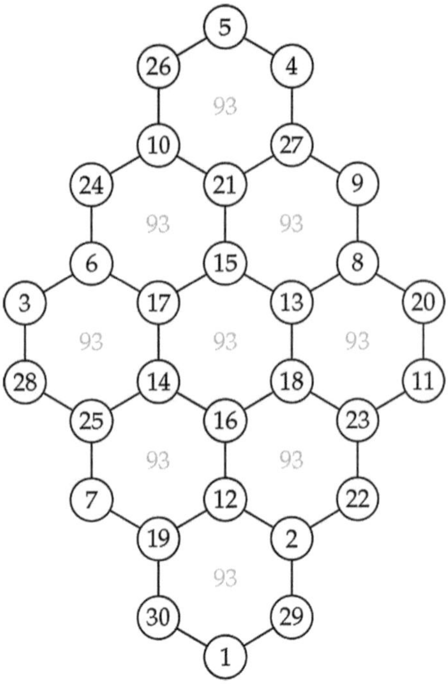

Fig. 6 The Tortoise Diagram ("Jisu guimundo" 地數龜文圖). Here, sums of six equal 93. There are others with different sums. Coding by Park Pu-Sung

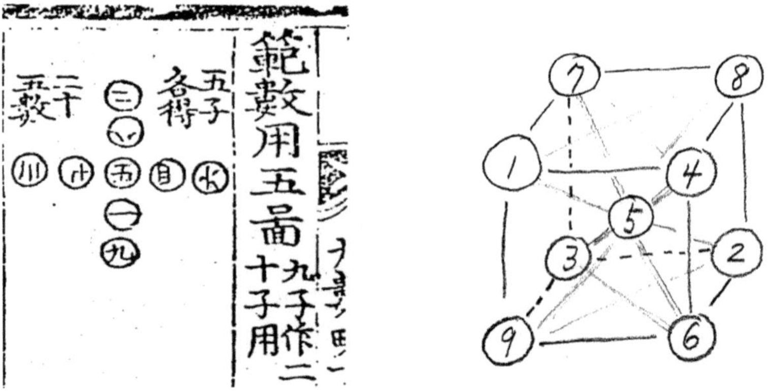

Fig. 7 Circular diagram of Choi Seog-jeong's creation. Is seen to have more symmetries if drawn in three dimensions. Image source: Kim (1985)

6 Last Diagrams: Mathematical Heritage Lost

There are four square tables of size 9 × 9 and a couple of double square tables of size 10 × 10 that are Choi Seog-jeong's creation. The four 9 × 9 ones are variations on or permutations of the "Gugu mosudo" 九九母數圖 and "Gugu jasudo" 九九子數

圖. In this section, we will briefly describe the construction of one or two, and then we will see how the "10 × 10" ones match modern combinatorial and/or group theoretic tables.

6.1 The 9 × 9 Tables

Choi Seog-jeong claims to have found a way to change the original 9 × 9 tables of Cai and the *Tongwen suanzhi*. There are two ways, Yin and Yang, each from both of the tables, making four in total. Choi uses the term "change of positions" (*byeon'gung* 變宮) in the names of them, so those from the "Gugu mosudo" are the "Gugu mosu byeon'gung yangdo" 九九母數變宮陽圖 and "Gugu mosu byeon'gung eumdo" 九九母數變宮陰圖, and those from the "Gugu jasudo" are the "Gugu jasu byeon'gung yangdo" 九九子數變宮陽圖 and "Gugu jasu byeon'gung eumdo" 九九子數變宮陰圖 (Fig. 8).

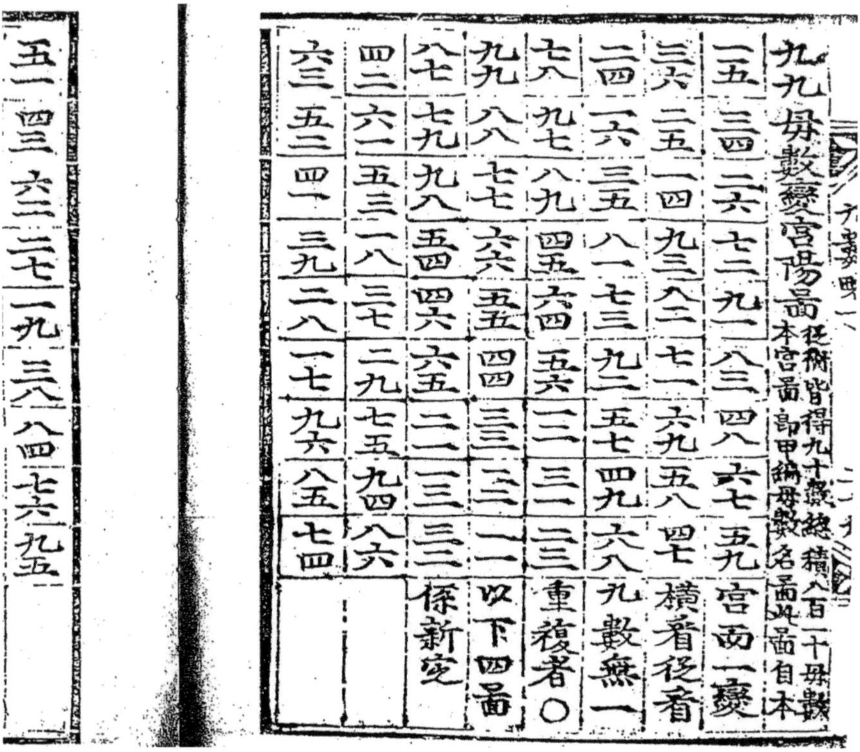

Fig. 8 This nine-by-nine table of Choi Seog-jeong's own creation is a pair of orthogonal Latin squares that precedes Euler by more than 67 years. This is presumed to be a permutation of Cai's basic nine-by-nine table (Fig. 12 below). Image source: Kim (1985)

What is interesting is that this kind of table is being studied in combinatorics nowadays. The first and the second tables are called orthogonal Latin squares or Euler squares. There is a rich theory of such tables, and it is becoming widely used in various fields including cryptology. Orthogonal Latin squares also can be used to generate a magic square. An orthogonal Latin square of size n, by definition, is two matrices overlapped together to form a matrix of number pairs such that rows and columns of each matrix are all permutations of the set of numbers $\{1, 2, \ldots, n\}$ and such that no same pair appears twice. A three by three example is as follows.

1,2	2,3	3,2
2,2	3,1	1,3
3,3	1,2	2,1

The first of Choi Seog-jeong's four diagrams is a pair of 9 × 9 orthogonal Latin squares. The second has some errors, but we can correct the errors to form a semi-magic square. We do not know why this kind of permutation intrigued him, but we know from its name that it is based on Shao Yong and Cai Shen's philosophy somehow. Choi Seog-jeong's square is internationally well known to mathematics societies, because this work is the world's first discovery of orthogonal Latin squares, and also it precedes Euler's next example by 67 years (Hahn and Oh 1993; Song 2013; Lih 2010; Kim and Khang 2010).

There is no explanation in the *Gusu-ryag* of how this was discovered, but a well-known algorithm by Allan Adler can be applied, and beginning with a 3 × 3 orthogonal Latin square, a 9 × 9 square is generated (Adler and Li 1977). We think that this method applies to at least three of these squares.

6.2 A 10 × 10 Table

Among Choi Seog-jeong's square tables there are five 10 × 10 tables. Among them, we are interested in the "Baekja jasu eumyang chakjongdo" 百子子數陰陽錯綜圖 and "Baekja saengsong jeonsudo" 百子生成純數圖. The former is a Latin square, and the latter is made of a pair of Latin squares. However, the latter is not an orthogonal Latin square, because for each entry pair the same pair appears twice (Fig. 9).

Recently, these squares have been studied and were revealed to have structures of the multiplication tables of a finite Abelian group. Such tables are called Cayley tables, and they give Latin squares (Carter 2009). To illustrate what a multiplication table is, we give simple examples of a group of order 6, i.e. with 6 elements.

One example of an order 6 group is the modulo 6 addition group. That is, if we add two numbers from {0, 1, 2, 3, 4, 5}, divide the sum by 6, and write only the remainder. Such a group is denoted by Z/6 or Z_6. The multiplication (or addition) table is like the first in Fig. 10.

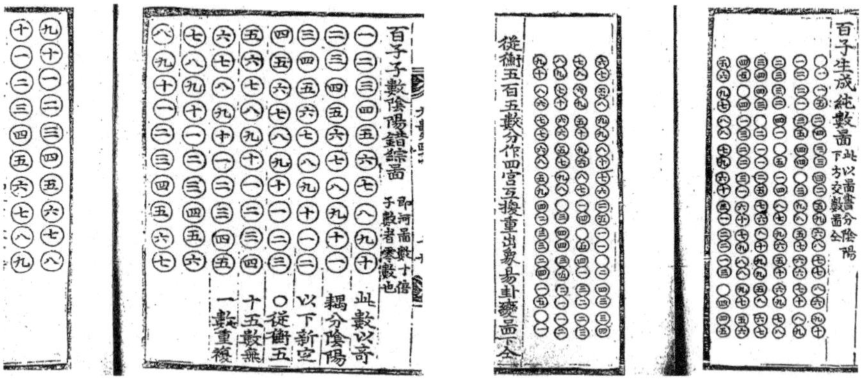

Fig. 9 The "Baekja jasu eumyang chakjongdo" 百子子數陰陽錯綜圖 (left) and "Baekja saeng-seong sunsudo" 百子生成純數圖 (right). Image source: Kim (1985)

+	0	1	2	3	4	5
0	0	1	2	3	4	5
1	1	2	3	4	5	0
2	2	3	4	5	0	1
3	3	4	5	0	1	2
4	4	5	0	1	2	3
5	5	0	1	2	3	4

+	0	1	2	3	4	5
0	0	1	2	3	4	5
1	1	2	0	5	3	4
2	2	0	1	4	5	3
3	3	5	4	1	0	2
4	4	3	5	0	2	1
5	5	4	3	2	1	0

Fig. 10 Latin Square by *Z*/6 (left) and *Z*/2 ⊕ *Z*/3 (right). Right table assignment of symbols {0, 1, …, 5} to *Z*/2 ⊕ *Z*/3: 0 = (0,0), 1 = (0,1), 2 = (0,2), 3 = (1,0), 4 = (1,1), 5 = (1,2). Addition rules in the right table: $(a,b) + (c,d) = (a + c, b + d)$ *mod* (2,3). Calculation example: 3 + 5 = (1,0) + (1,2) = (2,2) ≡ (0,2) = 2

But there are other ways to arrange this table. We can consider Z/6 as Z/2 ⊕ Z/3 and its elements as pairs (a, b) where a = 0, 1 and b = 0, 1, 2 and they add by the rules.

(a, b) + (c, d) = (a + c, b + d) mod (2, 3).

That is, a + c is divided by 2, b + d is divided by 3, and each leaves the remainder.

Now, if we name them, for example as 0 = (0, 0), 1 = (0, 1), 2 = (0, 2), 3 = (1, 0), 4 = (1, 1), 5 = (1, 2), then the addition table is like the second in Fig. 10. We will call this table the Cayley table of Z/2 ⊕ Z/3, even if it is the same group as Z/6. This representation is related to the famous Chinese remainder theorem. For example, suppose that one has an unknown number m and the remainder of $m \div 3$ is a, and $m \div 7$ is b. Then m is the element of Z/21 corresponding to the element $(a, b) \in$ Z/3 ⊕ Z/7.

In this terminology of group theory, Choi Seog-jeong's left table and the odd numbered rows of the right table in Fig. 9 are Cayley tables of Z/10 and Z/2 ⊕ Z/5, respectively.

Again, we have no idea how Choi came up with this table. The first one is simple enough that anyone could think of it, but the second one needs some work. We are sure that Choi was not led to these tables in thinking about group multiplication tables but rather some combinatorial manipulations based on oriental philosophy. In fact, he mentions the term "permutations" (*chagjong* 錯綜) in their titles and "inversions" or "transpositions" (*hohwan* 互換) in his comments.

7 Choi Seog-jeong's Unification of East and West

If we look at the *Gusu-ryag*, we immediately notice that Choi Seog-jeong borrowed and mixed some ideas from East and West. The most prominent examples have more to do with form than their contents. This is not surprising, because, until the time of Choi Seog-jeong, Joseon mathematicians including Choi had understood only a tiny part of Western mathematics and did not have a clue about the logical mathematics of the West. Below, we will note a few facts from his book that provide the basis for this claim (Fig. 11).

7.1 Choi's Basic Diagrams

In Chapter 6 of the *Gusu-ryag*, right after Choi Seog-jeong explains multiplication and division, he inserts four tables in the form of 9 × 9 matrices in the section titled "The Four 9 × 9 Tables" (Gugudo sa 九九圖四). The four diagrams are titled "Gugu mosu myeongdo" 九九母數名圖, "Gugu mosu sangdo" 九九母數象圖, "Gugu jasu myeongdo" 九九子數名圖, and "Gugu jasu sangdo" 九九子數象圖. The first two are the same table, but one is written in numbers and the other in counting rod symbols, and likewise for the second two. Mathematically, these are just multiplication tables. In the first two tables one finds 81 pairs of numbers from 1 to 9, and in the second two are multiplication tables of these pairs. Therefore, there is nothing special about them.

However, at the bottom of these diagrams are comments that need attention. Below the first it says: "[In each cell], on top is the Gang number (i.e., division), and

$+$	0	1	2	3	4	5	6	7	8	9
0	0	1	2	3	4	5	6	7	8	9
1	1	2	3	4	0	9	5	6	7	8
2	2	3	4	0	1	8	9	5	6	7
3	3	4	0	1	2	7	8	9	5	6
4	4	0	1	2	3	6	7	8	9	5
5	5	9	8	7	6	3	2	1	0	4
6	6	5	9	8	7	2	1	0	4	3
7	7	6	5	9	8	1	0	4	3	2
8	8	7	6	5	9	0	4	3	2	1
9	9	8	7	6	5	4	3	2	1	0

$0 = (0, 0)$ $5 = (1, 4)$
$1 = (0, 1)$ $6 = (1, 3)$
$2 = (0, 2)$ $7 = (1, 2)$
$3 = (0, 3)$ $8 = (1, 1)$
$4 = (0, 4)$ $9 = (1, 0)$

For example,
$7 + 8 = (2, 3) \equiv (0, 3) = 3 \mod (2, 5)$

Fig. 11 Similar to Fig. 10, we give the 10 × 10 matrix of Choi Seog-jeong as a Cayley table of the group $Z/2 \oplus Z/5$. On the right is the assignment of its elements to Choi's symbols {0, 1, … 9} and an example of addition 7 + 8 = 3

on the bottom is the Mu number (i.e., subdivision)" 上爲綱數，下爲目數. The method of using *gang-mu* 綱目 is said to have originated from Zhu Xi's book *Zizhi tongjian gangmu* 資治通鑑綱目 (1172), which is an abbreviated version of Sima Guang's (1019–1086) original history *Zizhi tongjian* 資治通鑑 (1084). *Gangmu* literally means dividing into "chapters and sections" or "divisions and subdivisions." Therefore, Choi Seog-jeong's comment may simply mean "the first number and the second number." But below the second diagram of symbols his comment reads:

> 右爲目數左爲綱數，夫綱而妻目也。此圖卽蔡氏範數方圖。
>
> [In each cell] on the right is the Mu number, on the left is the Gang number and a husband is Gang and [his] wife is Mu. This diagram is Cai's "Fanshu fangtu" 範數方圖.

According to Cho (2017), this diagram is the "Jiujiu fangshu tu" 九九方數圖 from Cai Shen's *Hongfan huangji neipian* 洪範皇極內篇, and when following his *shicao* 蓍草 method of fortune telling, Cai used the terms *gang* and *mu* for the first and second numbers drawn, respectively. In fact, in Cai's book there are several diagrams with the 81 pairs listed with fortune telling comments (Fig. 12).

Therefore it is reasonable to assume that Choi Seog-jeong consulted Cai's book for his first two diagrams.

Of course, the second two diagrams we do not find in Cai's book. Also, Korea did not have the multiplication rule as a table before Choi Seog-jeong. According to Hong (2006), this diagram comes from the *Tongwen suanzhi*, and it seems reasonable to assume that Choi Seog-jeong first found the table in this book (Fig. 13).

This means that in the beginning of the *Gusu-ryag* Choi Seog-jeong mentions the two tables as important, one from the East and the other from the West. Looking at this, we can almost feel the thrill he felt when he found these two tables from opposite ends of the world and the products of completely different contexts but matching perfectly to complete a theory. He might have interpreted this as the unification of East and West, or at least the agreement of the ultimate theories of both sides.

This might have been reason for Choi to believe that Western mathematics can be categorized and interpreted using Eastern philosophy, especially by those philosophies heavily related to numbers like Cai's. And this might have further driven him to do research to find new tables or diagrams obeying the rules of Cai's philosophy, believing that such findings would be important in mathematics. In fact, these diagrams are one of the bases for his diagrams in the last part of his book.

7.2 *The Four Symbols in the Four Operations*

The calculation methods in Part 2 ("Eulpyeon") are from the *Tongwen suanzhi* (Hong 2013, 2006).

In Part 2 are explained the four basic operations and their variations. Choi Seog-jeong tried to categorize the calculation methods into 4 or 8 categories as explained above.

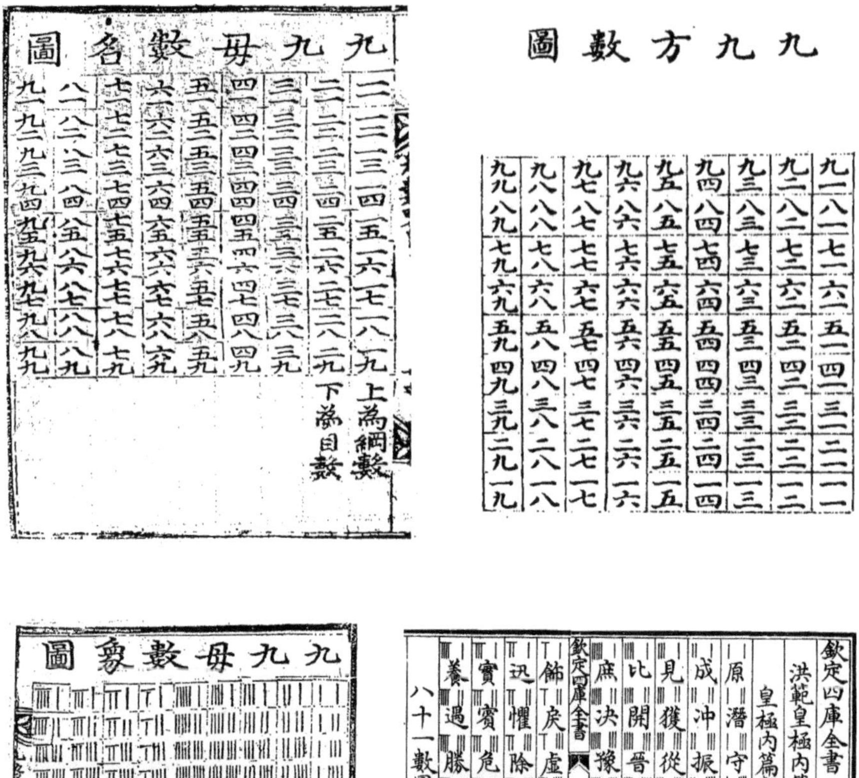

Fig. 12 Nine-nine multiplication table ("Gugu mosu myeongdo" 九九母數名圖) from the *Gusu-ryag* (**top left**) and the Nine-nine Table ("Jiujiu fangshu tu" 九九方數圖) from Cai Shen's book (**top right**) as compared to similar tables in counting rods symbols from the *Gusu-ryag* (**bottom left**) and Cai Shen (**bottom right**). Choi comments in the bottom left table that this is "Cai's square" (*fanshu fangtu* 範數方圖). Image source: Kim (1985)

The problems were how to compute the fourth number from three given numbers so that the two pairs have the same ratios and variations thereof. This is a simple "rule of three" problem, but the *Tongwen suanzhi* labels the four numbers as "Ratio 1," "Ratio 2," etc. (1率, 2率, 3率, 4率). Choi must have been thrilled to interpret them as the Four Symbols, and he names them accordingly the "sun, moon, stars,

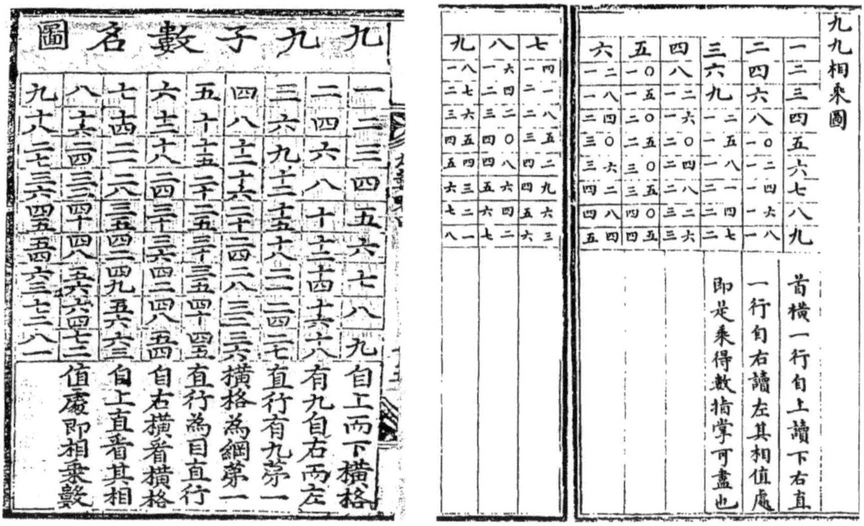

Fig. 13 The multiplication tables in the *Gusu-ryag* (**left**) and *Tongwen suanzhi* (**right**). Image source: Kim (1985) (left) and Guo (1993) (right).A for right

and planets" 日月星辰, which is not in the *Tongwen suanzhi*. He must have considered this as a structure of Oriental philosophy projected in Western mathematics.

7.3 Tools from East and West: Napier's Rods and Abacus

In the appendix to his work ("Jeongpyeon"), Choi Seog-jeong briefly explains two tools of computation. One is the Chinese abacus, and the other are the so-called Napier's rods. Here he mentions that the abacus was used by Chinese officials in Joseon, which implies that it was not widely used in Joseon.

On the other hand, Napier's rods were probably new to him, and he refers to the book *Chousuan* by Jaques Rho (Luo Yagu 羅雅谷, 1593–1638).[9] These two tools are rather limited in their ability compared to counting rods. His comment on the abacus is that it is not wise to use it in place of counting rods. He did not seem to have seized the usefulness of Napier's rods (Hong 2006), and he comments that Napier's rods are very limited. In spite of Choi Seog-jeong's depreciations, he thought that they were worth mentioning, and the first encounter with West must have made a strong impression on him. It is interesting that he includes the two tools side by side, just like he juxtaposed the four diagrams (or multiplication tables). Maybe his intention was to emphasize the similarity of tools from East and West.

[9] This book is Rho's editing of Napier's (1550–1617) original *Rabdologiae, seu numerationis per virgulas* (1617) and is included in the *Xiyang xinfa lishu* (1645).

8 Conclusion

Choi Seog-jeong lived as an official all his life. Starting as a history official, went into the military administration up until he eventually became a Minster (Left, Right, and Prime). He liked mathematics and studied astronomy even if he was busy with politics, his official duties, and studying the Confucian classics.

In this paper, we went through Choi's *Gusu-ryag* and took a rough look at several points therein. As is well acknowledged by recent scholarship on the book, it was written in paradigm of Confucian philosophy, especially that of Shao Yong and Cai Shen (Jung 2009; Cho 2017). Following Feng Youlan 馮友蘭 (1894–1990), we may argue that, if many Joseon scholars took after Shao Yong, "it was probably not a coincidence" 亦非無故 (Feng 1934). But here we want to point out the opposite.

New Ideas and Findings of Humankind Humankind has advanced its civilization through learning from its environment. The most basic form of learning is pattern recognition. Patterns tell us about inherent rules and/or structures, and they usually lead us to meanings. In statistics, it is said that this learning-through-patterns method causes two types of errors (Wheelan 2013).

The first is called *false pattern*, and it is an error by which we wrongly think there is a pattern. An example is believing that there is a face in the moon. The second is called *false meaning*, which is an error by which we assign a wrong meaning to a pattern. An example is that, after one sees a comet regularly (though not necessarily at fixed intervals), one wrongly attributes to it the bad things that might happen after seeing it, believing that this comet brings disaster.

Such errors are inherent in the pattern-to-meaning process, and to avoid them one must understand the whole structure or see the big picture. This big picture is one thing's relationship with others, and in mathematics it is knowing many examples and counter examples related to it.[10]

It is not hard to see that *Choi Seog-jeong's findings are good patterns, but his error was assigning false meanings to them.*

In the history of mathematics, many inventions and discoveries did not turn out to be successes of the proposed theories. On the contrary, those discoveries they thought useful may be shown by later studies to open a door to a new field. This is usually the case with a new mathematical creation.

In mathematical research, one has a goal of finding something or understanding something. One then sets up a model that explains that thing and tries to find rules (pattern) therein. If it succeeds, the rules become a theory (meaning) that supports the model. But whether or not it turns out to be a success, those rules and examples that are discovered in the process have meaningful mathematical properties and structures in themselves. And when their time comes, they might blossom in totally different ways. One can name many cases like this in the history of science and

[10] This learning-through-patterns method is still persistent in modern science and is ever more important in the field of data science with similar inherent errors pertaining to it.

mathematics, but here we name only one case: Sophus Lie had an idea of classifying solutions of differential equations according to the group actions that leave the equation invariant. Lie was applying the idea of Galois' theory to his problems. His idea did not work out well, but it opened a whole new world for the theory of Lie groups and its applications.

In the case of Choi Seog-jeong, his diagrams represent some relationships in the theory of change (*Yixue* 易學), that is they show some abstractions and/or variations following the laws of change. His intention to clarify mathematics through this idea did not result in success and instead led Joseon scholars into false beliefs. However, the 9×9 and 10×10 diagrams are important discoveries in some other directions.

We can think of this happening as a typical example of mathematical discoveries, as explained above. The laws of change have many sequences (transformations) and symmetries. These can be interpreted as mathematical symmetries or other kinds of mathematical rules, and this kind of combinatorial thought contributed to Choi Seog-jeong's discovery of the diagrams.

The ideas of *setting-up of a model and finding rules in it* is also important in today's research and education. We value creativity, and creativity seems to reside in trying to find new examples and possible rules in them. Choi Seog-jeong's theory may not be a successful one, but his attempt can be praised as having succeeded in the sense that it led Oriental philosophy and mathematics to find a clue to an important field of mathematics. We value this aspect of his endeavor.

Bibliography

Adler, Allan, and Shuo-Yen Robert Li. 1977. Magic Cubes and Prouhet Sequences. *The American Mathematical Monthly* 84(8): pp. 618–627.

Carter, Nathan C. 2009. *Visual Group Theory*, vol. 32. Washington: The Mathematical Association of America.

Cho Hie-Young 조희영. 2017. 『구수략(九數略)』, 송대 도서상수학으로 짜인 조선수학서 - 인문학적 지평에서 (*GuSuRyak*, Joseon Dynasty Math Book Consisting of Song Dynasty Do Seo Xiang Shu: The Horizon of Mathematical Humanities). 정신문화연구 (*Korean Studies Quarterly*) 40(1):105–132.

Feng Youlan 馮友蘭. 1934. *Zhongguo zhexue shi* 中國哲學史 (*History of Chinese Philosophy*). Shanghai: Shangwu yinshuguan.

Guo Shuchun 郭書春, ed. 1993. *Zhongguo kexue jishu dianji tonghui: shuxue juan* 中國科學技術典籍通彙·數學卷. Zhengzhou: Henan jiaoyu chubanshe.

S. G. Hahn and Y.-Y. Oh. 최 석정과 그의 마방진 (Choi Seok-Jeong and his magic squares). *Journal of the Korea Society of Mathematical Education Series A*, 32(3):201–219, 1993.

S. S. Hong. 2013. 崔錫鼎의 九數略 (Gusu-ryag of Choi Seog-jeong). *Journal for History of Mathematics*, 26(2–3): 1–6.

Hong Young-Hee 홍영희. 2006. 조선(朝鮮) 산학(算學)과 수리정온(數理精蘊) (Mathematics of Chosun Dynasty and *Shù lǐ jīng yùn* (數理精蘊)). 한국수학사학회지 (*Journal for History of Mathematics*) 19(2): 25–46.

Jin Yuzi 金玉子. 默思集算法과 17세기 朝鮮 算學(*Muk Sa Jib San Beob* and Mathematics of 17th Century Chosun). Ph.D. thesis (Korea University), 2009.

Jin Yuzi 金玉子 and Kim Young-Wook 김영욱. 2009. 17세기 조선 산학(朝鮮 算學)과 默思集算法(Chosun mathematics in the 17th Century and Muk Sa Jib San Beob). 한국수학사학회지 (*Journal for History of Mathematics*) 22(4): 15–28.

Jung Hae-Nam 정해남. 2009. 수학철학적 관점에서 본 <구수략> (<GuSuRyak> on Perspective of Philosophy of Mathematics). 한국수학사학회지 (*Journal for History of Mathematics*) 22(4): 67–82.

Kim Sung-Sook 김성숙 and Khang Mee-Kyung 강미경. 2010. 최석정의 직교라틴방진 (Orthogonal Latin Squares of Choi Seok-Jeong). 한국수학사학회지 (*Journal for History of Mathematics*) 23(3): 21–31.

Kim Yong-un金容雲, ed. 1985. *Hanguk gwahakgisulsa jaryo daegye: suhak pyeon* 韓國科學技術史資料大系·數學篇. Seoul: Yŏgang ch'ulp'ansa.

Kwon Gyunuk 권균욱, Park Sang Hu 박상후, Song Yun Min 송윤민, Choi Seong Woong 최성웅, and Park Poo-Sung 박부성. 2014. 합의 범위를 이용한 지수귀문도 해의 탐구 (A Study on Solutions of Jisuguimundo Using the Range of Magic Sums). 한국수학사학회지 (*Journal for History of Mathematics*) 27(2): 111–125.

Lee Kyung-Eon 이경언. 2010. 정사각형 형태가 아닌 마방진에 대한 고찰 (A Study on Various Non-Regular Magic Square). 한국수학교육학회지시리즈E: 수학교육논문집 (*Journal of the Korean Society of Mathematics Education Series E: Communications of Mathematical Education*) 24(1): 195–220.

Li Yan 李儼 and Du Shiran 杜石然. 1987. *Chinese Mathematics: A Concise History*. Oxford: Oxford University Press.

Lih Ko-Wei 李國偉. 2010. A Remarkable Euler Square before Euler. *Mathematics Magazine* 83(3): 163–167.

Park Kyo Sik 박교식. 2018. 지수귀문도를 만드는 근사적 방법 (An Approximate Method to Make *Jisuguimundo*). 한국수학사학회지 (*Journal for History of Mathematics*) 31(4): 183–196, 2018.

H.-Y. Song. 최석정 선생, 오일러를 최소 61년 앞서 직교라틴방진을 만들다 (Choi suk-jung's euler square is earlier than Euler by 61 years). Information and Communications Magazine (Korean), 30(10):101–108, 2013.

Wheelan, Charles J. 2013. *Naked Statistics: Stripping the Dread from the Data*. New York: W.W. Norton & Company.

Solar Activities and Climate Change During the Last Millennium as Evidenced in Korean Chronicles

Yang Hong-Jin

Abstract Korean historical chronicles provide a two-thousand-year observational record of natural phenomena, including astronomical and meteorological events. In this paper, we use their records of sunspots and frost to examine the correlation of between solar activity and climate change in Korea. In total, we have 42 sunspot records for the Goryeo Dynasty (AD 918–1392) and 13 for the Joseon Dynasty (AD 1392–1910). Those for the Goryeo Dynasty show a periodicity in good agreement with the well-known 11-year cycle of solar activity, and those for the Joseon Dynasty suggest that solar activity decreased over the following ~500 years. In order to examine the long-period variation of solar activity, we supplement our data with Chinese sunspot records from the same period, and we find a new ~240-yr period of solar activity via power spectral analysis. The Korean chronicles also contain ~700 frost records spanning the last thousand years, and we investigate these records for signs of cooling indicative of climate change. These records point to a ~240-yr cooling period, which is well in accord with that of solar activity. From this, we argue that solar activity has decreased over the last one thousand years and has a long-term variation of ~240 years.

Keywords Korean chronicles · Sunspot · Frost · Climate change · Long-period solar activity · Power spectral analysis

This paper presents a synopsis of the author's findings in the original, larger paper Yang et al. (2019).

Yang Hong-Jin (✉)
Korea Astronomy and Space Science Institute, Daejeon, South Korea
e-mail: hjyang@kasi.re.kr

D. P. Morgan, T. Quan (eds.), *Mathematics and Astronomy in the Ancient World*,
https://doi.org/10.1007/978-3-031-93809-2_8

1 Introduction

Korea has a long history of astronomical and meteorological observation spanning some two thousand years. Indeed, the historical chronicles *Samguksagi* 三國史記, *Goryeosa* 高麗史, and *Joseon wangjo sillok* 朝鮮王朝實錄 (hereafter, *Sillok*) contain some 25,000 observational reports of such phenomena as eclipses, sunspots, meteors, meteor showers, meteorites, novae, comets, planetary movements, and aurorae (Park 2008). Moreover, astrophysical analysis has confirmed most of these reports are reliable, in that they correspond with observable historical phenomenon that were seen elsewhere and/or can now be calculated backward (Park and Na 1994; Yang et al. 1998, 2005a,b; Lee et al. 2004, 2009).

Since the scientific observation of such phenomena only began several centuries ago, such historical records are important materials for modern astronomical research. For example, the analysis of ancient sunspot records has revealed long-term variations in solar activity. Though nearly invisible to the naked eye, there are 42 unambiguous reports of sunspots in the *Goryeosa*, which covers AD 1151–1387, and another 13 in the *Sillok*, which covers AD 1402–1743. Alone, these two chronicles cover the better part of a thousand years, and Yang et al. (1998) and Lee et al. (2004) have shown their sunspot and aurora records to reflect the well-known 11-year periodicity of solar activity.

Solar activity is a major factor of global climate change, since solar radiation has a significant influence on the earth's temperature. Korean chronicles document its 11-year periodicity going further back than is possible in the West, where the observation of sunspots did not begin in earnest until the seventeenth century. Korean records are admittedly too few in number to speak to the long-term variation of solar activity in and of themselves, but they contain a variety of meteorological observations made over two the last thousand years that, in conjunction with the sunspot records, provide invaluable data for understanding long-term cycles and trends.

The past few decades have seen a growing interest in climate change, which scientists attribute to such factors as solar radiation, volcanic activity, topography, the variation of the earth's orbit, and CO_2. The world may currently be focused on CO_2, but we know that, for most of history, volcanic eruptions and solar activity were the principal factors at play (Crowley 2000). Among the many studies on long-term climate change at the scale of human history, several have brought historical observations to bear on what we can extrapolate from the current day. Following Dansgaard's (1969) study of Greenland ice core data, for example, Zhu (1973) has shown historical Chinese meteorological records to be in good agreement with Dansgaard's results for the past two thousand years. Such studies confirm that pre-modern meteorological records might provide useful indicators of global climate change, but such studies are nevertheless rare. The reason for this is twofold, particularly as concerns Western-language sources: the lack of systematicity and contiguity in the pre-modern observational record, and the difficulty of confirming the reliability of individual reports. In this regard, Korea and China have left us with superior records with which we may complement the instrument-based measures of the modern period (Liu et al. 2011).

In this paper, we will study the long-term variation in solar activity and global climate change over the last thousand years using Korean astronomical and meteorological records. For this purpose, we will begin in Sect. 2 with an analysis of historical sunspot and frost records in Korean chronicles using the Fourier spectral analysis method. Since the Chinese sunspot records are more sensitive to the long-term variation of solar activity, we will combine Korean and Chinese sunspots records in this analysis. As the temporal variation of frost-free periods can provide the information on climate change, we will also examine the relationship between climate change and solar activity by comparing temporal variations in sunspot and frost records. We will present our results in Sect. 3, together with discussion, followed by a summary and conclusions in Sect. 4.

2 Data and Methods

2.1 Historical Korean and Chinese Sunspot Records

Records of sunspots in East Asian chronicles go back prior to the Common Era. The earliest was recorded in 28 BC, in Han Dynasty China. In Korea, the *Samguksagi* contains a few early observations of what would seem to be sunspots in the Three Kingdoms period (57 BC–AD 935). By means of comparison, the observation of sunspots in the West only began in earnest from 1611 on, with but two earlier observations dating to fourteenth-century Russia (Park 2008).

For the Goryeo Dynasty (AD 918–1392), the *Goryeosa* contains 42 unambiguous reports of sunspots, starting from AD 1151. The first record states that "There is a black spot on the sun, and it is the size of an egg."[1] In the Korean chronicles, the size of a sunspot is reported as being that of a plum, egg, peach, or pear in an effort to make a system of consistent measurements.

In Korea, the Goryeo Dynasty was eventually replaced by Joseon Dynasty (1392–1910), both of which lasted ~500 years. In the political transition, the state astronomical office and its staff simply switched allegiances, carrying on with their observational duties as they had under the previous ruling family. The result is that the Korean observational record maintains a strong systemacity and contiguity across the historical chronicles of the two dynasties, AD 918–1910. Hence, it can be argued that Korean historical observations are generally more reliable than those of other countries.

Yang et al. (1998) and Lee et al. (2004) have studied the sunspot and aurora records of the Goryeo Dynasty (918–1392) using Fourier spectral analysis, and they

[1] *Goryeosa* 高麗史, Kim Jong-seo金宗瑞 (1383–1453) et al. (ed.), 1451 (1451 woodblock print edition), v. 1, p. 16.

have found that they align with the aforementioned 11-yr periodicity of solar activity (Fig. 1).

Figure 1 shows the temporal distributions of sunspot and aurora records (a, c) in the *Goryeosa*, the representative history of the Goryeo Dynasty, and power spectra measured from these records (b, d). It is notable that the power spectra of both sunspot and aurora records clearly show the well-known 11-yr period of solar cycle. The dotted and dashed lines in Fig. 1b and d indicate the confidence levels of the upper 0.14% and 0.23% estimated from the similar analysis of random catalogues generated by the Monte-Carlo method. As shown in the temporal distributions, all the sunspots and aurorae in the *Goryeosa* are divided into five classes according to their size and intensity (Yang et al. 1998).

We collected the records of sunspots in Chinese chronicles between AD 918 and 1910, which corresponds to the Korean Goryeo and Joseon Dynasties, finding a total of 134.[2] Chinese sunspot records are generally more numerous than Korean records. Although the distribution of Chinese observations shows similar behaviour from the early tenth to the early sixteenth century, the number of sunspot records becomes too large after the sixteenth century and does not match with the solar activity minimum. This is probably due to the fact that the sunspot records were duplicated in many different Chinese chronicles. The number and chronological span of these records aid in finding long-period signals, but they do not contain

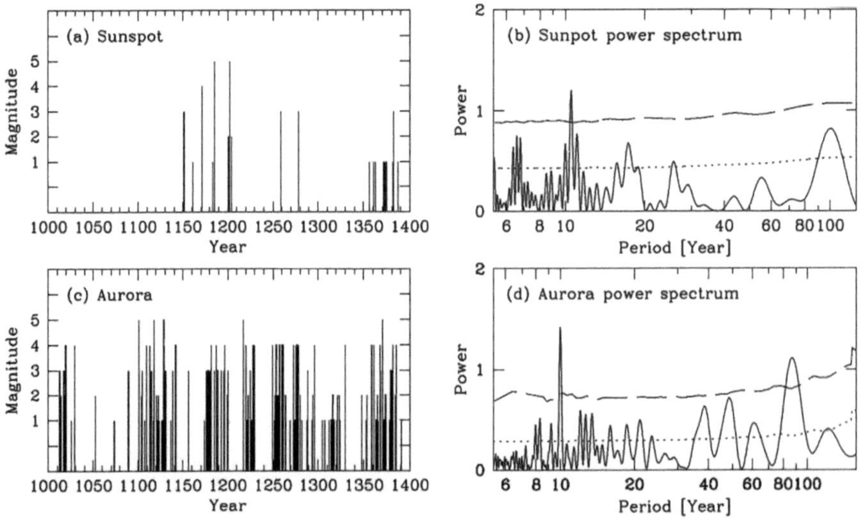

Fig. 1 Temporal distributions (**a**, **c**) and power spectra (**b**, **d**) of sunspots and aurorae under the Goryeo Dynasty (918–1392). Magnitude is divided into five classes according to size and intensity. The dotted and dashed lines in panels (**b**) and (**d**) indicate the confidence levels of upper 0.14 and 0.23% estimated via Monte-Carlo analysis. In the power spectra, the normalization is arbitrary. Source: Yang et al. (1998)

[2] Source: Beijing tianwentai (1988). Note that the actual period covered in Chinese sources is AD 927–1905.

reliable short-period information reflecting the 11-year periodicity of solar activity. Therefore, in this study we use the Chinese sunspot records for the purpose of finding the long periodicity of solar activity, relying on Korean records to identify the relative changes in the number of sunspot records.

2.2 Historical Korean Frost Records

Combined with sunspots, historical meteorological records can provide useful information for understanding climate change. The Korean chronicles contain observations of various meteorological phenomena such as drought, frost, flood, rain, hail, and snow over the last two thousand years. In order to examine historical temperature variations, we shall use the frost records.

The *Samguksagi*, *Goryeosa*, and *Sillok* contain 38, 48, and 667 frost records, respectively. Most historical frost records are dated, which allows us to determine the length of the frost-free period, by year, and that allows us to estimate the mean temperature for said year. To this end, we examined records for the first (autumn) and last (spring) frost of each year, and since there may be missing records, we shall look at the overall trends in the long-term data.

In the analysis of frost records, the observation site is an important factor because temperature varies with latitude. As the capital of Goryeo was located at higher latitude than that of Joseon, we thus examined their records separately. Lastly, we excluded the frost records of the Three Kingdoms, because they are small in number and, thus, ill-suited for investigating climate change.

2.3 Data Analysis

Our chronicles give dates as per the contemporary luni-solar civil calendar. We have converted these dates using the *Korean Chronological Tables* (Ahn et al. 1999, 2000a, 2000b; Han 2001). For the luni-solar conversions, we use Julian calendar for dates prior to Oct. 4, 1582, and the Gregorian calendar after Oct. 15, 1582. In this analysis, we only use the Goryeo and Joseon records, which provide the specific year, month, and date; we exclude those from the Three Kingdoms, since they omit the date and are small in number. The same conversion is made for the Chinese historical sunspot records, and for those that only give the year, we set the date as January 1. Setting the date in this way does not affect our result for the long-term variation of solar activity.

One notes that, due to political and environmental factors, the number density of historical records is not uniform. Besides, some astronomical phenomena might be occluded by clouds. Therefore, it is generally necessary to correct the effect of the non-uniformity of observations on the data analysis of astronomical records. However, in this work, we do not consider such bias in our Fourier spectral analysis of the sunspot records, because the 11-year periodicity appears in the power spectrum of sunspot records of the Goryeo Dynasty without the bias correction. Moreover, the purpose of this study is to identify long-term changes in solar activity, so the effect of the non-uniform number density of historical records is not important.

We regard the temporal distribution of sunspot records as the point process, and express the number density field of the records $\delta(t)$ as the sum of one-dimensional delta functions. For N data points located at t_i's ($i = 1, \cdots, N$) during the time span T, the density field can be decomposed into waves in the Fourier space as

$$\delta(t) = \frac{T}{\sum w_j} \sum_j w_j \delta^{(1)}(t - t_j) = \sum_{k'} \delta_{k'} e^{-ik't}, \quad (1)$$

where w_j is the weight for the jth data point and δ_k is the Fourier transform of the number density field and can be expressed as

$$\delta_k = \frac{1}{\sum w_j} \sum_{j \in T} w_j e^{ikt_j}. \quad (2)$$

The wavenumber k is related to the period by $\lambda = 2\pi/k$. The power spectrum of the temporal distribution of the discrete points is given by the variance of the Fourier mode,

$$P(k) = |\delta_k|^2. \quad (3)$$

For more detailed description, see Yang et al. (1998). In this study, we do not correct for the contribution due to the Poisson noise arising from the discrete data points because we are interested in finding the periodic signals, not the actual normalization of the power spectrum.

In order to examine the long-term variations of solar activity in detail, we add the Chinese sunspot records (134 data points) to the Korean records (55 data points) and analyze the combined records. In the power spectrum estimation, each data point is weighted by the size of the sunspot, and the ratio of number density of Chinese records relative to that of Korean records is multiplied to the weights of the Korean data points in order to equalize the number density of both data sets.

To estimate the statistical significance of the periodic signals in the measured power spectrum, we generated 10,000 catalogs that include the random data sets with the same number of data points (N = 189) as in the Korean-Chinese sunspot records within the same time span (AD 918–1910) and analyze them in the same way as the observation records are analyzed. The level of statistical significance at each wavenumber k is determined from 10,000 measured powers of the random catalogs. This Monte Carlo test enables us to assess the reliability of the historical records.

3 Results and Discussions

3.1 Historical Sunspot Records and Solar Activity

The *Sillok*, which covers the Joseon Dynasty, contains ~20,000 astronomical observations. Although the *Sillok* has many more records than the *Goryeosa* (~5000), it has only 13 unambiguous records of sunspots (Chunchugwan 1391–1863). As is well known, there were the solar minimums of Spörer, Maunder, and Dalton, and they caused the Little Ice Age (LIA) from the sixteenth to nineteenth century (Michael 2003). That is clearly the reason for the lack of sunspot records during the Joseon Dynasty. The temporal distribution of sunspot records shows that the solar activity in the sixteenth to seventeenth century was lower than those in the previous two long cycles, which strongly suggests that the overall solar activity has declined for about 600 years (see the histogram of Korean sunspot records in Fig. 5).

Figure 2 shows the power spectrum measured from the Sino-Korean sunspot records. The dotted and dashed curves denote 3 and 4 sigma confidence limits, respectively, estimated from the Monte-Carlo test. A few significant periodicities are seen in the power spectrum of these records. In Fig. 2, the most notable signal is the peak around a 240-yr periodicity of solar activity, with a second and third peak at 60 and 11 yr.

Similarly, Vaquero et al. (2002) reported a notable 250-yr solar cycle using the spectral analysis of naked-eye observations of 240 sunspots from 165 BC to AD 1918, based on the catalogues of Yau and Stephenson (1988) and Wittmann and Xu (1987). They also reported the solar cycles of 115, 85, 60, and 11 years. Although some Korean historical sunspots are omitted in these catalogues, the result of Vaquero et al. (2002) is similar to our result.

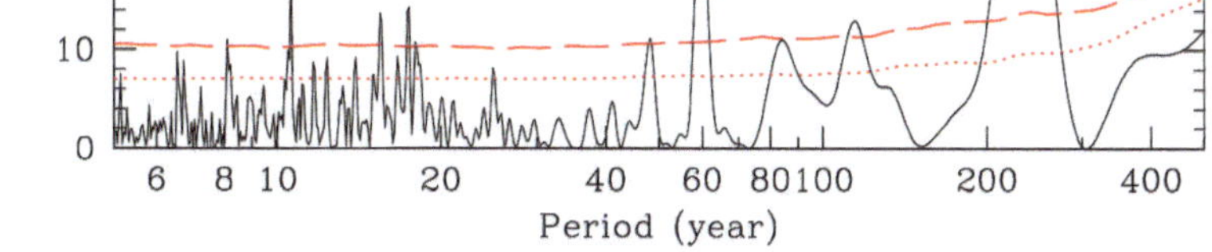

Fig. 2 Power spectrum of Korean and Chinese sunspot records over the last millennium. The three most powerful signals correspond to the periods of 240, 60, and 11 years. The 3 sigma and 4 sigma levels of statistical significance estimated via Monte Carlo analysis are shown as dotted and dashed curves, respectively

Our results from the analysis of historical sunspot records indicate the following facts. First, solar activity has decreased during the last one thousand years. Second, the long-term period of 240-yr solar activity has been verified with high significance from the power spectrum of historical Sino-Korean sunspot records.

3.2 *Historical Frost Records and Temperature Variation*

To test whether the frost records reflect temperature at the observation site, we analyzed frost records collected by the Korea Meteorological Administration (KMA) from 1992 to 2011 (KMA 2012). We collected the annual variation of the last (spring) and the first (autumn) dates of frost observation in two Korean sites, Cheorwon and Cheongju. Geographically, Cheorwon is located in a higher latitude, 1.6° north of Cheongju, and at a higher altitude above the sea level. We found that the mean temperature of Cheorwon is lower than that of Cheongju, and that the colder city, Cheorwon, has a shorter frost-free period, which demonstrates that frost records are a good indicator of mean temperature.

In order to assess the correlation between solar activity and climate change, we analyzed 753 historical frost records as indicators of mean temperature that reflect climate change. Figure 3 shows the last and first frost records for the Goryeo (triangles) and the Joseon Dynasty (circles). Linear regression lines are shown separately for each dynasty, as their capitals were at different latitudes, with Gaeseong (Goryeo) to the north of Seoul (Joseon). Based on the linear fits to the frost observations, especially from the fit to the last frost observations in spring, we deduce that

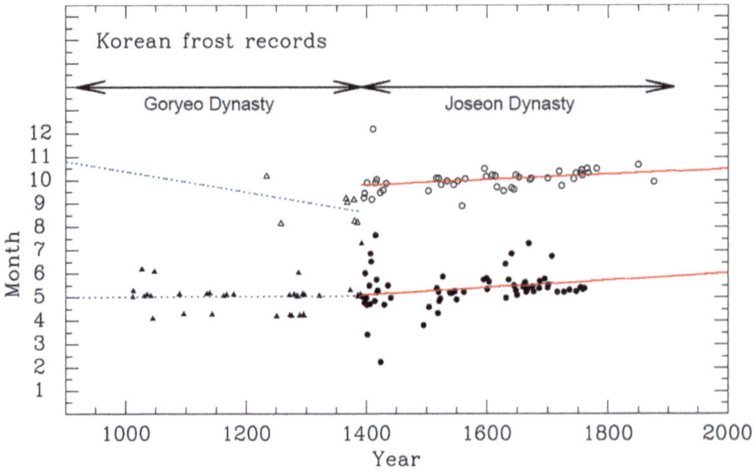

Fig. 3 Temporal distribution of the first and last frost records of the *Goryeosa* and *Sillok* (918–1392). Triangles and circles denote the records of the Goryeo Dynasty (observed at Gaeseong) and Joseon Dynasty (at Seoul), respectively. Filled symbols represent the last frost records in spring, and open symbols the first records in autumn

the temperature cooled over the Joseon Dynasty. For the Goryeo Dynasty, such behavior is not clearly seen, and it is difficult to speak to climate change over this period due to the lack of observational records. The tendency in temperature variation is similar to than in solar activity (i.e., the long-term decline in the number of sunspots) as seen in historical sunspot records.

Korean frost records show that the temperature (i.e., the length of the frost-free period) gradually decreased from the Goryeo to Joseon Dynasties. In Fig. 3, one sees two notable cold epochs in the early fifteenth and late seventeenth centuries. We surveyed the frost records of the Joseon Dynasty and drew distributions of the frost records observed at various regions. As shown in Fig. 4, the two cold epochs also appear in the plots of frost observations in four provinces of Korea. These frost records also roughly indicate that the temperature decreased over the Joseon Dynasty. This is particularly clear in the distribution of the numerous last (spring) frost records. On the other hand, few the first (autumn) frost records demonstrate behaviors inconsistent with the spring records in some areas, including Seoul. We think that a more careful follow-up study is needed for this issue. Besides, the cold period influenced by the solar cycle is also evident in other studies on the weather of the Joseon Dynasty. For example, based on several kinds of meteorological records, Bae (2004) reports that the temperature of the Korean Peninsula was unusually low during the eighteenth century.

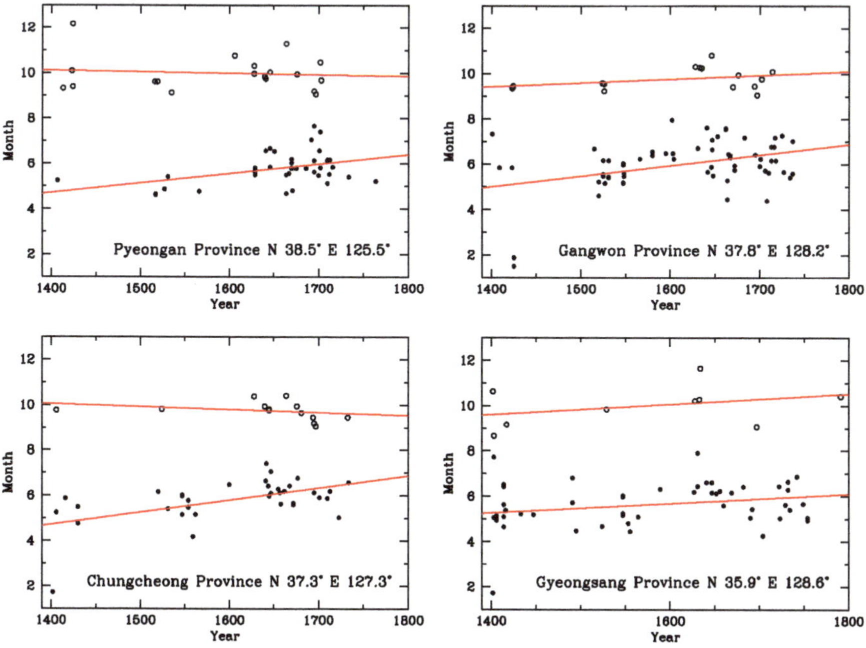

Fig. 4 Temporal distributions of frost records observed in four provinces of the Joseon Dynasty (1392–1910). The four provinces are Pyeongan, Gangwon, Chungcheong, and Gyeongsang, and their representative geographical locations are N38.5° E125.5°, N37.8° E128.2°, N37.3° E127.3°, and N35.9° E128.6°, respectively

3.3 Correlation Between Solar Activity and Temperature Variation

To investigate the correlation between solar activity and the temperature change, we have compared the temporal distributions of frost and sunspot records in Fig. 5. We can see that the two cold epochs correspond to periods of the minimum solar activity. The sinusoidal curve along with the histogram shows the 240-yr cycle of solar activity as derived from the power spectral analysis.

Korean frost records indicate that the temperature has been cooling for the last thousand years, and this matches the variation in solar activity. We also find that the transient cooling periods around the fifteenth and seventeenth centuries from the frost records are in good agreement with the solar minima of a 240-yr cycle.

Since the Little Ice Age and the solar minimums of Spörer, Maunder and Dalton in the fourteenth century, Korea's sunspot records are have been few in number. Thus, in order to study the long-term variation of solar activity, we supplemented these with Chinese records and found a long-term periodicity in solar activity of around 240 years. We also examined the change of the frost-free period in the Korean frost records and found that the temperature gradually decreased from AD 1200 to 1800. This result coincides with the trend of solar activity as seen in the sunspot number distribution over the same period, and it is similar to that of Chinese historical records and to the climate change reconstructed from the analysis of Greenland ice core data (Dansgaard et al. 1969; Zhu 1973).

Turning to frost records, we collected data for several Korean regions to perform a detailed examination of temperature variation from the fourteenth century on. We have identified two periods of abrupt cooling in these records: the early fifteenth and late seventeenth centuries. Comparing this to our results from the sunspot records, we found that the two periods are closely related to the 240-yr cycle in solar activity, which demonstrates that the long-term temperature variations over the last thousand years have been directly affected by solar activity. Studies of medieval climate change also point to cooling at the same time. Bard et al. (1997) report cooling periods around the fifteenth and early eighteenth centuries based on the ^{10}Be measurements of South Pole ice and Δ14C values of American-European tree rings. Likewise, Liu et al. (2011) show that there was an extremely cool epoch around the seventeenth century from the analysis of Tibetan tree-rings. In Fig. 5, we also found that there is a slight time delay between the temperature change and the solar activity cycle, which merits future investigation.

4 Summary and Conclusions

The Korean historical chronicles have left us with a wealth of astronomical and meteorological observations spanning the past two thousand years. These are factual records, as examination has revealed, and they are of great use to modern research in astronomy and meteorology. Among other things, their historical sunspot records

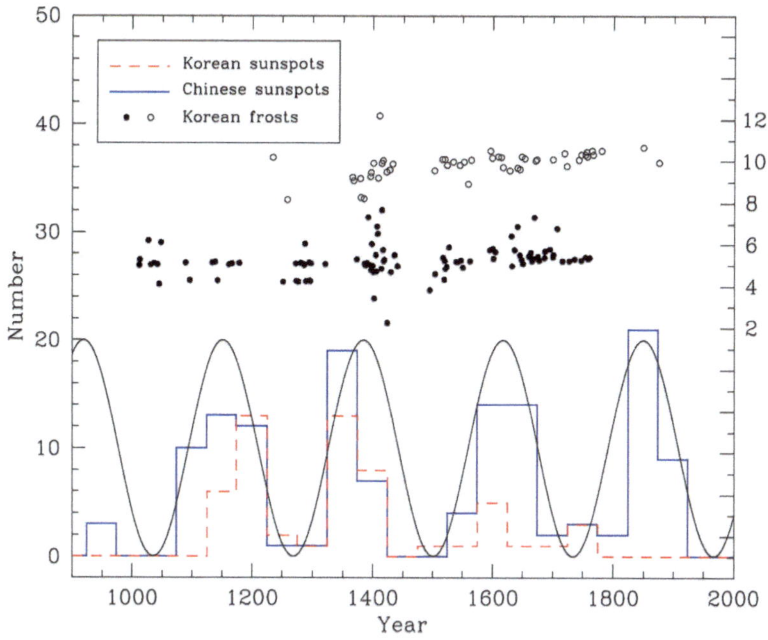

Fig. 5 Histogram of Korean (AD 1151–1743; red dashed) and Chinese sunspot records (AD 927–1905; blue solid curves). For comparison, a sinusoidal curve with a period of ~240 yr. is drawn on the histograms, together with the frost records of the Goryeo and Joseon Dynasties presented in Fig. 3, are shown above the histogram of sunspot records

provide useful information for understanding the long-term variation of solar activity and climate change. In particular, these records directly reveal the level of solar activity at the time of observation, because they classify sunspots by size, and size is proportional to the strength of solar activity. The distribution of Korean sunspot records clearly shows that solar activity decreased over the past one thousand years (see Fig. 5). Historical frost records in Korean chronicles are also useful for studying the long-term variation in temperature, because the recorded dates of first and last frost is show to reflect the mean temperature at the site of observation.

In this paper, we analyzed these records to study the long-term variation in solar activity and climate change in historical times. Two important features were identified in the Korean sunspot and frost records. First, over the last millennium, they show both solar activity and temperature to have gradually decreased, suggesting that long-term solar activity has been at the root of climate change for over a thousand years. Second, in the frost records, there are two periods during which the temperature decreased sharply, in good agreement with the 240-yr cycle of solar activity. This also confirms that temperature variations over the past one thousand years were directly affected by solar activity. In addition, the correlated long-term change of 240 yr. in the sunspot and frost records suggests that the cooling around early fifteenth and late seventeenth centuries was due to the variation of solar activity.

Bibliography

Ahn, Y.S., Lee, Y.B., Lee, Y.S., and Han, B. S. 2000a. *Chronological Table of the Three Kingdoms Period* (三國時代年曆表). Daejeon: Korea Astronomy & Space Science Institute.

Ahn, Y.S., Shim, K.J., Han, B.S., and Song, D.J. 2000b. *Chronological Table of the Joseon Dynasty* (朝鮮時代年曆表). Daejeon: Korea Astronomy & Space Science Institute.

Ahn, Y.S., Shim, K.J., Song, D.J., Han, B.S., and Yang, H.-J. 1999. *Chronological Table of the Goryeo Dynasty* (高麗時代年曆表). Daejeon: Korea Astronomy & Space Science Institute.

Bae, J.H. 2004. The Abnormal Climate and Agricultural Conditions of the Samcheok Area in the late 18th century. *The Daegu Historical Review* 大邱史學 75: 99–147.

Bard, E., Raisbeck, G.M., Yiou, F., and Jouzel, J. 1997. Solar Modulation of Cosmogenic Nuclide Production Over the Last Millennium: Comparison between ^{14}C and ^{10}Be Records. *Earth and Planetary Science Letters* 150.3–4: 453–462.

Beijing observatory 北京天文台 (ed.). 1988. *Zhongguo gudai tianxiang jilu zongji* 中國古代天象記錄總集 (General Compilation of Chinese Ancient Astronomical Records). Nanjing: Jiangsu kexue jishu chubanshe.

Crowley, T.J. 2000. Causes of Climate Change over the Past 1000 Years. *Science* 289.5477: 270–277.

Dansgaard, W. et al. 1969. One Thousand Centuries of Climatic Record from cAMP Century on the Greenland Ice Sheet. *Science* 166.3903: 377–380.

Han, B.S. 2001. *Korean Calendar Conversion Table* (韓國年曆大典). Daegu: Yeongnam Univ. Press.

KMA, 1992–2011, Annual Climatological Report, Korea Meteorological Administration (KMA), Seoul: Korea Meteorological Administration (in Korean).

Lee, E.H., Ahn, Y.S., Yang, H.-J., and Chen, K.Y. 2004. The Sunspot and Auroral Activity Cycle Derived from Korean Historical Records of the 11th–18th Century. *Solar Physics* 224: 373–386.

Lee, K.-W., Yang, H.-J., and Park, M.-G. 2009. Orbital Elements of Comet C/1490 Y1 and the Quadrantid Shower. *Monthly Notices of the Royal Astronomical Society* 400: 1389–1393.

Liu, Y., Cai, Q.F., Song, H.M. et al. 2011. Amplitudes, Rates, Periodicities and Causes of Temperature Variations in the Past 2485 Years and Future Trends Over the Central-eastern Tibetan Plateau. *Chinese Science Bulletin* 56.28–29: 2986–2994.

Michael, E.M. 2003. Little Ice Age. In: *Encyclopedia of Global Environmental Change, Volume 1, The Earth System: Physical and Chemical Dimensions of Global Environmental Change*, ed. M.C. Mac Cracken and J.S. Perry, 504–509. Chichester: John Wiley & Sons (Retrieved 17 November 2012).

Park, C. 2008. *Astronomy: Traditional Korean Science*. Seoul: Ewha Womans Univ. Press.

Park, C., and Na, D. 1994. Confirmation and Historical Consequences of Astronomical Records in *Samguksagi*. *Journal of Korean History of Science Society* 16(2): 167–202.

Vaquero, J.M., Gallego, M.C., and Garcia, J.A. 2002. A 250-year Cycle in Naked-eye Observations of Sunspots. *Geophysical Research Letter* 29(20): 58-1–58-4.

Wittmann, A.D., and Xu, Z.T. 1987. A Catalogue of Sunspot Observations from 165 BC to AD 1684. *Astronomy and Astrophysics Supplement Series* 70(1): 83–94.

Yang, H.-J., Park, C., and Park, M.-G. 1998. Evidence for the Solar Cycle in the Sunspot and Aurora Records of the Goryeo Dynasty. *Publications of the Korean Astronomical Society* 13(1): 181–208.

Yang, H.-J., Park, C., and Park, M.-G. 2005a. Korean Nova Records in A.D. 1073 and A.D. 1074: R Aquarii. *Astronomy & Astrophysics* 435(1): 207–214.

Yang, H.-J., Park, M.-G., Cho, S.-H., and Park, C. 2005b. Analysis of Historical Meteor and Meteor Shower Records: Korea, China, and Japan. *Icarus* 175(1): 215–225.

Yau, K.K.C., and Stephenson, F.R. 1988. A Revised Catalogue of Far Eastern Observations of Sunspots (165 BC to AD 1918). *The Quarterly Journal of the Royal Astronomical Society* 29: 175–197.

Zhu, K. 1973. A Preliminary Study on the Climate Changes Since the Last 5000 Years in China. *Science in China* 2: 168–189.

Part III
Japan

Japanese Mathematics from the Seventh to Sixteenth Century

Ueno Kenji

Abstract In ancient times, the Japanese learned mathematics from China and Korea. In the middle of the seventh century, the Japanese court had started to introduce administrative organization, including in education, which modelled the Chinese system. This included mathematical education. All of the major mathematical textbooks were brought to Japan, but it is not clear how the Japanese learned and taught mathematics. While the Japanese would not develop their own mathematics until the Edo period (1602–1868), basic mathematics were nevertheless learned and used as skill. In the present article, we will report a few facts on Japanese mathematics from the seventh to the sixteenth century. We follow mainly to the description of Fujiwara (1954–1960), adding a few recently discovered facts.

Keywords Sanhakase · *Man'yōshū* · Multiplication table · *Kuchizusami* · Counting rods · Abacus · *Jinkōki*

1 The Yōrō Code and Early Mathematics Education

In the first official history of Japan, the *Nihon Shoki* 日本書記 of 720 CE, we find several records related to mathematics. For example, in the fifteenth year of Emperor Kinmei 欽明 (553 CE), we read that a master of the calendar (*rekihakase* 曆博士)

The present work was done when the author visited the Institute for Advanced Study in History of Science, Northwest University. The author thanks Professor Qu Anjing and the Institute for their hospitality and excellent working conditions.

Ueno Kenji (✉)
Seki Kowa Institute of Mathematics, Yokkaichi University, Yokkaichi, Japan

D. P. Morgan, T. Quan (eds.), *Mathematics and Astronomy in the Ancient World*,
https://doi.org/10.1007/978-3-031-93809-2_9

and a master of the *I Ching* (*ekihakase* 易博士) visited Japan from Baekje 百済.[1] Also, in the tenth year of Suiko 推古 (602 CE), we read that Kanroku 観勒, a monk from Baekje, visited Japan carrying books on calendrics, astronomy, geography, and *dunjia* 遁甲. The record states that a few students studied these books under Kanroku's supervision and got good results.[2]

Over the course of the seventh century, the Japanese court set about creating an effective administrative apparatus on the Chinese model, the first result of which was the Taihō Code (*Taihō-ritsuryō* 大宝律令) of 703. The Taihō Code does not survive, but we do have the revised edition of 718, named the Yōrō Code (*Yōrō-ritsuryō* 養老律令). As concerns the Office of Education, or Daigakuryō 大學寮, the Yōrō Code stipulates there be two masters of mathematics (*san-hakase* 算博士) with thirty students. The *Ryō no gige* 令義解 commentary of 833 further elaborates on how the students were to be tested. Namely, the thirty students were divided into two classes, one of which had to learn the *Jiuzhang suanshu* 九章算術, *Haidao suanjing* 海島算經, *Zhoubi suanjing* 周髀算經, *Wucao suanjing* 五曹算經, *Kyūshi* 九司, *Sunzi suanjing* 孫子算經, and *Sankai chongcha* 三開重差, while the other fifteen students had to learn the *Zhuishu* 綴術 and *Liuzhang* 六章.[3]

The Office of Education was modelled more or less faithfully on the Tang equivalent, but the textbooks *Kyūshi*, *Sankai chongcha*, and *Liuzhang* were added. Indeed, these texts appear nowhere in Chinese sources, but they are cited in the *Samguk sagi* 三國史記, the history of Baekje, Silla, and Goguryc, the three kingdoms in Korea (Fig. 1).

2 *Nihonkoku genzai sho mokuroku* 日本國見在書目録

All the textbooks cited in the *Ryō no gige* appear in Fujiwara no Sukeyo's 藤原佐世 *Nihonkoku genzai sho mokuroku* 日本國見在書目録 (*Catalog of Extant Texts in the Country of in Japan*) of 875. The *Nihonkoku genzai sho mokuroku* records a *Kyūshi* in 5 volumes and a second work entitled *Kyūshi sanjutsu* 九司算術 (*Kyūshi*

[1] 二月、百濟遣下部杆率將軍三貴・上部奈率物部烏等乞、救兵。仍貢德率東城子莫古、代前番奈率東城子言。五經博士王柳貴、代固德馬丁安。僧曇慧等九人、代僧道深等七人。別奉勅、貢易博士施德王道良・曆博士固德王保孫・醫博士奈率王有悛陀・採藥師施德潘量豐・固德丁有陀・樂人施德三斤・季德己麻次・季德進奴・對德進陀。皆、依請代之。三月丁亥朔、百濟使人中部木劦施德文次等、罷歸。夏五月丙戌朔戊子、內臣、率舟師詣于百濟 (*Nihon Shoki*, 19. Koten bungaku taikei vo.68 "Nihon shoki" ge, p.109).

[2] 冬十月、百済僧観勒来之。仍貢暦本及天文地理書、并遁甲方術之書也。是時、選書生三四人、以俾学習於観勒矣。皆学以成業 (*Nihon Shoki*, 22. Koten bungaku taikei vol.68 "Nihon shoki" ge, p.179).

[3] 其算学生辯明術理然後為通。試九章三條、海島、周髀、五曹、九司、孫子、三開重差各一條、試翕然通為甲、通六為乙。若落九章者雖通六猶為不第。其試綴術、六章者推前。綴術六條、六章三條(謂若之級商與綴術、及六章與海島等六経、願受試者亦同令聴也)試翕然通為甲、通六為乙、若落経者(謂六章忽不通者也)雖通六猶為不第 (*Ryō no gige*, 3. fol. .28–29, "Gakuryō" 学令).

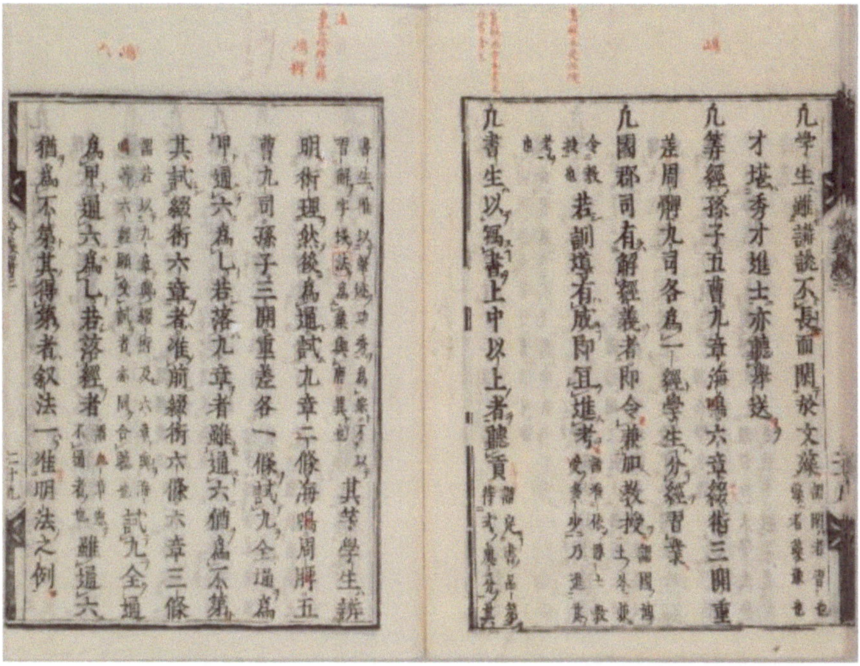

Fig. 1 List of mathematical textbooks stipulated in the *Ryō no gige*. National Institute of Japanese Literature 国文学研究資料館 (http://codh.rois.ac.jp/iiif/iiif-curation-viewer/index.html?pages=200020588&pos=171&lang=ja)

Mathematical Procedures) 1 volume. Interestingly, the catalog also attributes the *Liuzhang* to one "Mr. Gao" 高氏, as does the *Yōrō gakuryō shūge* 養老学令集解 "Sankeijō" 算経條 for the *Sankai chongcha*. "Mr Gao" would appear to be Gao Yun 高允 (390–487) of the Northern Wei (Fig. 2).[4]

3 The *Man'yōshū* 万葉集 and the Multiplication Table

Other evidence that Japanese studied basic mathematics is found in the *Man'yōshū* 万葉集. Compiled in the eighth century, the *Man'yōshū* is the oldest extant collection of Japanese poetry, containing more than 4500 poems going back to the sixth century, many of which are anonymous or misattributed (usually to well-known poets). At the time, the Japanese had not yet developed their own syllabary, so they used Chinese characters in a variety of functions: in their usual logographic sense, to represent Japanese syllables phonetically, and sometimes in a combination of

[4] Feng and Li (2000, 2001), Osada (2018), and Sun (2015).

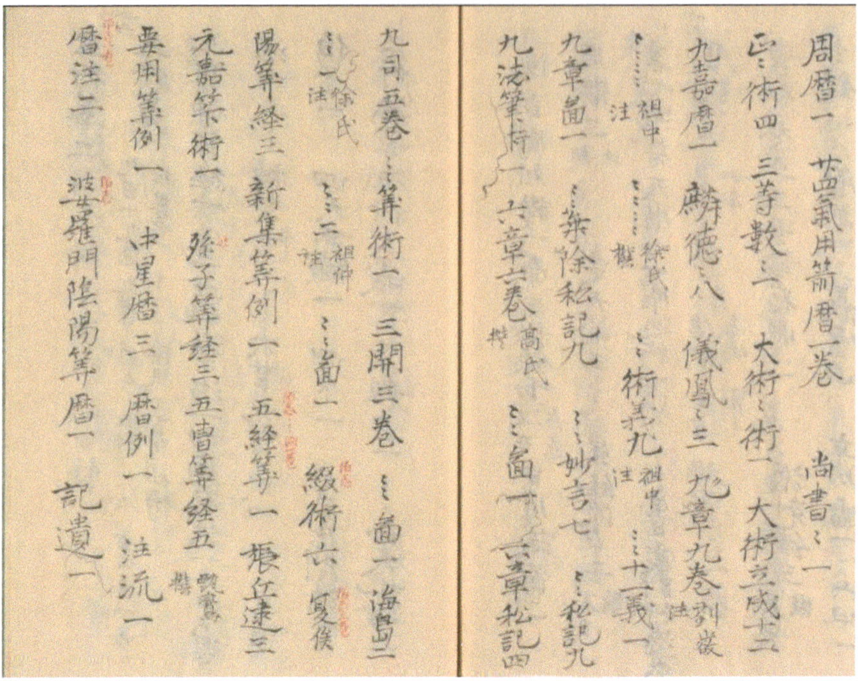

Fig. 2 *Nihonkoku genzai sho mokuroku,* National Diet Library 日本国会図書館, DOI 10.11501/2540620, https://dl.ndl.go.jp/pid/2540620/1/34

these functions. Like any good poets, these authors play with the language, and sometimes their sense of humor involves mathematics. For example, in Japanese, the Chinese expression "four fours, sixteen" 四四十六 (i.e., 4 × 4 = 16) is pronounced *shi shi jūroku*, and we find examples in the *Man'yōshū* where 十六 (*jūroku*) is read *shi shi* ("four fours"). For example, in poem 239, by Kakinomoto no Hitomaro 柿本人麻呂, we can find the expression "even deer and wild boars" 十六社者 read *shi shi koso ba* 獣(しし)こそば. Similarly, since "nine nines, eighty-one" 九九八十一 (*ku ku hachijūichi*), 八十一 (*hachijūichi*) was sometimes read as *ku ku*, hence "is tying up" 八十一里喚鶏 becomes *kuku ri tsutsu* 括りつつ, which sounds like a chicken's cluck. Also, *Man'yōshū* contains the following expressions:

- "Please answer I don't know (his name)" 不知二五寸許瀬 (*isa towo kikose* いさとを聞こせ), where 十 (10) is read *towo*, from" two times tow is ten", nigoju 二五十 (2 × 5 = 10).
- "Thus succesive emperor have ruled" 如是二々知三 (*kaku shi rasamu* かく知らすむ), where 二々 (two twos) is read *shi* (4), from "two times two is four" ninin ga shi 二二如四 (2 × 2 = 4).
- "I lost my reason for living" 生友奈重二 (*ikeri tomo na shi* 生けるともなし), where 重二 (double two) is likewise read *shi* (4).

These examples show a certain familiarity with the multiplication table in eighth-century Japan, at least among poets.

The *Man'yōshū* also contains a poem written by a mathematician. For background, the Dazaifu 太宰府 was the imperial office governing Kyūshū and responsible for hosting foreign dignitaries from China and Korea. In 730, the head of the Dazaifu, Ōtomo no Tabito 大友旅人 (665–731), held a plum blossoms party at which the guests were requested to write poems praising plum blossoms, thirty-two of which are preserved in volume 5 of the *Man'yōshū*. Among these is a poem composed by "Master of Mathematics Shishi no Ōmichi" 算師志氏大道, who it seems had a position at the Dazaifu (Fig. 3).

As illustrated in Fig. 4, one also finds evidence of the official study of the multiplication table in the Nara period (710–794) from recently excavated *mokkan* 木簡, manuscripts written on wood strips.

4 *Kuchizusami* 口遊

The *Kuchizusami* is a primer compiled in 970 for Fujiwara Tamemitsu's 藤原為光 (942–992) seven-year-old son, Matsuo 松雄.

The *Kuchizusami* also contains a number of mathematical problems. One is the problem of the pregnant woman, similar to that in the *Sunzi suanjing*.[5] Another concerns a bundle of bamboo and has no precedent in the *Jiuzhang suanjing*, *Haidao suanjing*, *Zhoubi suanjing*, *Wucao suanjing*, or *Sunzi suanjing*.[6] The term 'bundle of bamboo' 竹束 does appear in the *Shin Sarugaku ki*, however, so the *Sankai chongcha* or *Liuzhang* might well have contained a similar problem (Fig. 5).

5 *Sangi* (Counting rods) and *Soroban* (Abacus)

Before the abacus was brought to Japan, the Japanese used *sangi* 算木 (counting rods) for calculations, which had been introduced from China. Since few knew how to use them, many Japanese believed that counting rods had magical powers. This

---

[5] 産婦令妊婦可生子知男女法

術曰置婦女年數(自生年至妊年)加十二神爲實可除天一地二人三四時五行六神七皇八風九宮殘一三五七(爲陽男也)二四六八(爲陰女也)一説以九除也(今案同法也)口傳曰若自去年妊者可加定筭三加婦女之年也.

The last problem of *Sunzi suanjing* is the following.

今有孕婦行年二十九難九月未知所生.答曰生男.

術曰置四十九加難月減行年所餘以天除一地除二人除三四時除四五行除五六律除六七星除七八風除八九州除九其不盡者奇則爲男耦則爲女.

[6] 今有竹束周員廿一問惣数幾曰四十八.術曰置周員加三筭自乘得五百七十六以十二除得四十八.口傳曰不盡法半已上者取一從惣數以六爲法半.

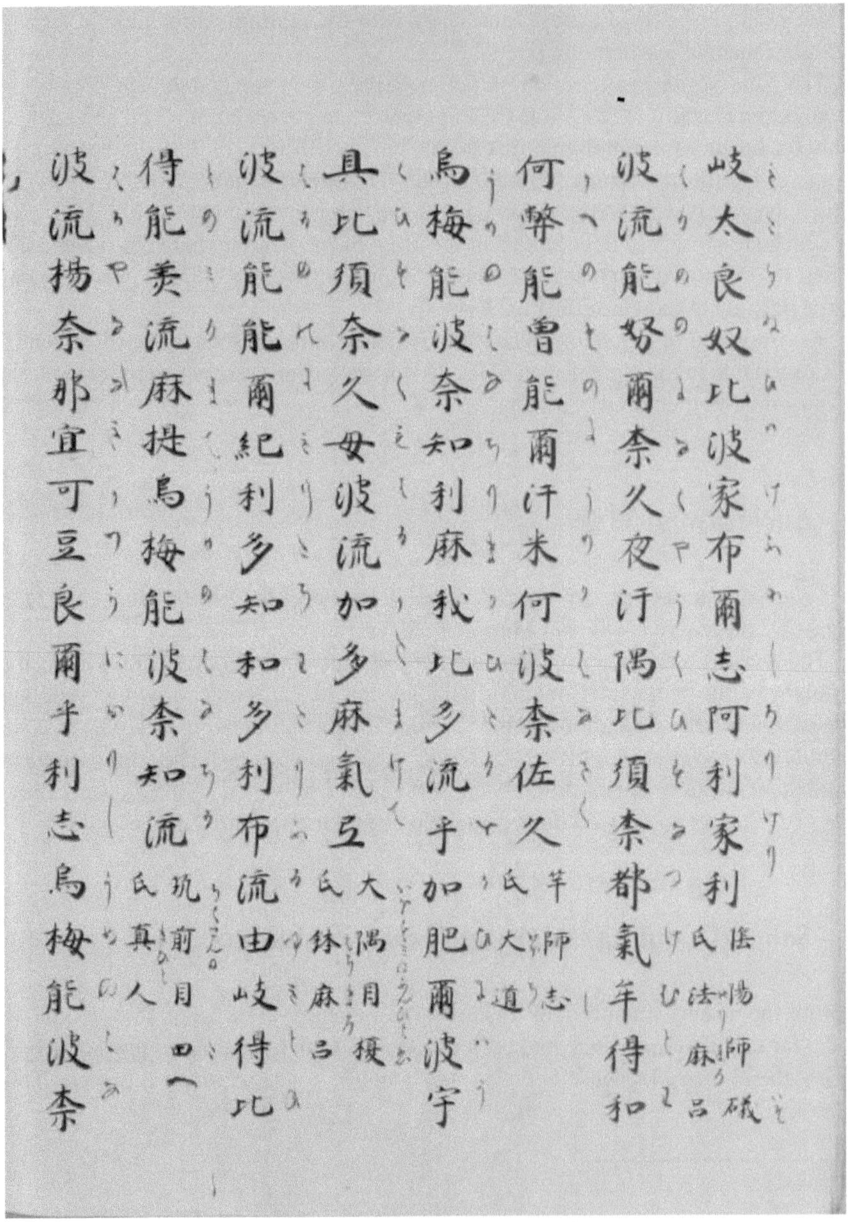

Fig. 3 The second and third lines from the right we read a poem by "Master of Mathematics Shishi no Ōmichi" 算師志氏大道 in *Man'yōshū*. His name is written at the last part of the third line. "In a spring field Japanese nightingale are singing. To invite them to my garden, plum trees are full of blossoms" (*haruno no ni nakuya uguisu natsukemu to wagahe no sononi ume ga hana saku* 春の野に鳴くや鶯なつけむと我が家の園に梅が花咲く), *Manyoshu*, vol.5, no.837, National Institute of Japanese Literature 国文学研究所, http://codh.rois.ac.jp/iiif/iiif-curation-viewer/index.html?pages=200015542/252&lang=ja]

Fig. 4 *Mokkan* multiplication table found in Nara, reading "[three] nines, twenty-seven; [two] nines, eighteen; one nine equals nine" □九廿七 □ 九十八 一九如九. Nara National Research Institute for Cultural Properties 奈良文化財研究所. 6AAFJE270001251

belief is reflected in several stories in the *Konjaku Monogatari Shu* 今昔物語集,[7] such as story 22 of volume 24, in which laughing was controlled by counting rods. The last part of the story is as follows. "A night for Koshinmachi[8] court ladies were gathered in a room of the monk Takashina no Toshihira's house but had nothing to do. One of them asked Toshihira to talk a funny story. But he refused it so that many court ladies asked him to perform a funny dance. Instead he brought counting rods and arranged them in a certain shape on a floor, and asked court ladies to laugh. They started laughing and soon they found that they could not stop laughing. They begged to Toshihira to stop laughing. When he broke the arrangement of counting rods, they could stop laughing." (Shin nihon koten bungaku taikei vol. 36 "Konjaku monogatari shu" 4, p.426–427.)

[7] *Konjaku Monogatari Shu*今昔物語集 is also known as *Konjaku Monogatari*今昔物語. It is a collection of more than one thousand tales written down the late Heian period (twelfth century). The entire collection is in 31 volumes, of which 28 remain today.

[8] *Onmyodo* tells that on the eve of 57th day of the sexagenary cycle night three worms escape from the body of each person to tell his or her evil deeds to the Lord of Heaven so that he or she will be shortened his or her life or after death he or she will go to Hell. Many people stayed up at the night to prevent three worms to escape from their bodies. This was called *Koshinmachi* 庚申待.

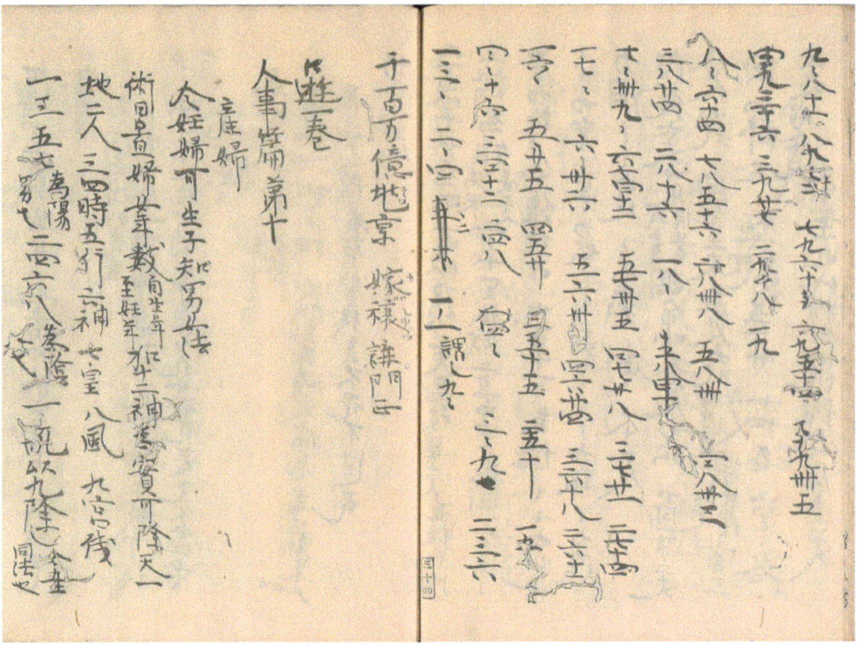

Fig. 5 Multiplication table in the *Kuchizusami*. Note that "two fours, eight" (*ni shi ga hachi*) 二四八is used instead of "two fours equals eight" 二四如八, commonly used in Chaina. In the left page the problem of the pregnant woman is written. National Diet Library 日本国会図書館, DOI 10.11501/2540620, https://dl.ndl.go.jp/pid/2540624/1/37

One sees this also in the sixteenth- and seventeenth-century Nara picture books (*Nara ehon* 奈良絵本), of which *Tanabata* 七夕, for example, illustrates counting rods being used to count the number of grains of rice in a giant granary (Fig. 6).

It is not known when the Chinese abacus first arrived in Japan. In the fifteenth and sixteenth centuries, many Japanese travelled to other Asian countries for trade, and it is quite likely they brought the abacus back with them. The Japanese–Portuguese Dictionary published by the Jesuits in Nagasaki in 1603 and its supplement of 1604 feature entries for both *soroban* 算盤 and *san* 算. *San* is defined as "certain pieces of wood used to throw lots or to calculate. Originally it is called *sangi*." The *soroban* (abacus) is described as a common tool for calculation in China and Japan. This shows that the abacus and counting rods coexisted before the former replaced the latter as the standard tool for calculation in the seventeenth century. Part of what facilitated this transition was the appearance of good textbooks for the abacus, foremost amongst which was Yoshida Mitsuyoshi's 吉田光由 *Jinkōki* 塵劫記 of 1627.[9]

[9] See for example Horiuchi (2010) and Ueno (2017, 2023).

Fig. 6 Nara Ehon *Tanabata*. A red demon uses counting rods. Graduate School of Literature, Kyoto University

6 Conclusion

The Japanese learned mathematics from China and Korea. They learned it as skill, and it was a rare skill at that. The advanced mathematics in China was not required by Japanese society at that time, so there was little opportunity for Japan to develop its own mathematics.

By at least the thirteenth century, however, big temples in Kyoto bought extensive collections of books published China and Korea, several of which survive to our day. Among them, we find mathematical books. For example, Tofukuji Temple has a copy of the *Suanxue qimeng* 算學啓蒙 (1299) used to publish the Japanese edition of 1658. Also, it is said that Sawaguchi Kazuyuki 澤口一之 studied "Suanxue qimeng" possessed by the temple Fujian 不二庵belonging to the Tofukuji Temple 東福寺and understood *tianyuanshu* 天元術 algebra by himself. There is as yet no through study of the books in these collections of the big temples in Kyoto, nor on the mathematical knowledge of monks that lived in these temples in fifteenth and sixteenth centuries. Such a study would shed new light on how mathematicians in the early Edo period (1603–1868) inherited mathematical knowledge from the past.

In any case, before the seventeenth century only basic mathematics were used and preserved in Japan, which served as a definite basis to support a new development in the Edo period.

Bibliography

Primary Sources

Hoyaku Nippo jisho 邦訳日葡辞書, edited and translated by Doi Tadao et al, 1980, Tokyo, Iwanami Shoten (Modern Japanese translation of *Vocabvlario da Lingoa de Iapam* compiled by Jesuit missionaries and published in Nagasaki, Japan, in 1603 and 1604).

Konjaku mongarari shu 今昔物語集; modern printed version, in *Shin Nihon koten bungaku taikei* 新日本古典文学大系 vol. 33–37, Tokyo: Iwanami Shoten, 1999–1996 (5 vols.).

Nihon Shoki 日本書記, Prince Toneri 舎人親王 et al., 720; modern printed version, in *Nihon koten bungaku taikei*日本古典文学大系 vol. 67, 68, Tokyo: Iwanami Shoten, 1965–1967 (2 vols.).

*Nihonkoku genzaisho mokuroku*日本国見在書目録, Fujiwara no Sukeyo 藤原佐理世, 891; handwritten manuscript dated 1835, National Diet Library 日本国会図書館, DOI 10.11501/2540620, https://www.dl.ndl.go.jp/api/iiif/2540620/manifest.json

Ryō no gige 令義解, Kiyohara no Natsuno 清原夏野, Ono no Takamura 小野篁 et al., 833; woodblock print, published by Yoshida Siroemon 吉田四郎右衛門, Kyoto, 1650 (National Institute of Japanese Literature, http://codh.rois.ac.jp/iiif/iiif-curation-viewer/index.html?pages=200020588&lang=ja) (10 vols.).

Samguk sagi 三國史記, Kim Bu-sik 金富軾, 1145; modern printed version from The Society of Historical Studies of Choson 朝鮮史学会, 1928, (National Diet Library 日本国会図書館 DOI 10.11501/1920811(https://dl.ndl.go.jp/info:ndljp/pid/1920811).

Shin sarugaku ki 新猿楽記, Fujiwara no Akihira 藤原明衡; eleventh-century; handwritten manuscript, National Institute of Japanese Literature (http://codh.rois.ac.jp/iiif/iiif-curation-viewer/index.html?pages=200018583&lang=ja).

Secondary Sources

Feng Lisheng 馮立昇 and Li Di 李迪. 2001. Gao Yun and the Two Mathematical Works Liuzhang and Sankai, in Proceedings of the Fourth International Symposium on the History of Mathematics and Mathematical Education Using Chinese Characters, Maebashi 1999. Maebashi: Maebashi Institute of Technology, 41–50.

Feng Lisheng 馮立昇 and Li Di 李迪. 2000. *Liuzhang*, *sankai* xintan 『六章』、『三開』新探 (On the two Mathematical Works "*Liuzhang*" and "*Sankai*"). *Xibei daxue xuebao (ziran kexue ban)* 西北大學學報 (自然科學版) 30.1: 89–92.

Fujiwara Matsusaburō 藤原松三郎. 1954–1960. *Meijizen Nihon sūgakushi* 明治前日本数学史 (History of Japanese Mathematics before the Meiji), 5 vols. Tōkyō: Iwanami Shoten.

Horiuchi, Annick. 2010. *Japanese Mathematics in the Edo period (1600–1868)*. Basel: Birkhäuser.

Osada Naoki 長田直樹. 2018. Daigakuryō sanka no kyokasho 大学寮算科の教科書 (Textbooks of Daigakuryo *sanka*). Sūgakushi kenkyū 数学史研究 229: 23–39.

Sun Meng 孫猛. 2015. *Riben guo jianzai shu mulu xiangkao*. Shanghai: Shanghai guji chubanshe.

Ueno Kenji 上野健爾. 2017. *Wasan eno izanai: sūgaku o tanoshinda Edo jidai* 和算への誘い:数学を楽しんだ江戸時代 (Invitation to Japanese Mathematics, the Edo Period in which many people enjoyed mathematics). Tōkyō: Heibonsha.

Ueno Kenji 上野健爾. 2023. Seki Takakazu no Sugaku 関孝和の数学 (Mathematics of Seki Takakazu). In Seki Takakazu Zenshu 関孝和全集 (Collected Works of Seki Takakazu), 3vols, ed. by Ueno Kenji, Ogawa Tsukane 小川束, Kobayashi Tatsuhiko 小林龍彦 & Satô Kenichi 佐藤賢一. Tōkyō: Iwanami Shoten..

Sunrise and Sunset Times of the Chinese *Chóngxiū-Dàmíng* Calendar 重修大明曆

Sôma Mitsuru

Abstract Choi et al. (*Advances in Space Research* 62, 3237–3248) studied the sunrise and sunset times of the *Chóngxiū-Dàmínglì*, but they made a small mistake in deriving the formula for them, which caused a maximum error of 1.73 seconds in the current unit of time. They also suggested that the sunrise and sunset times of the *Chóngxiū-Dàmínglì* were based on observations at the capital of the Northern Song dynasty, but it turns out that they were based on the same observations as the older Chinese calendar *Xuānmínglì*.

Keywords Calendar · *Chóngxiū-Dàmínglì* · Sunrise and sunset times

The *Chóngxiū-Dàmínglì* 重修大明曆 was used in China from 1182 to 1280, under the dynasties of the Jīn 金 (1115–1234) and Yuán 元 (1271–1368). It was also used in Korea during the reign of King Sejong 世宋 (r. 1418–1450) of the Joseon dynasty 朝鮮 (1392–1897). Choi et al. (2018) studied the sunrise and sunset times of the *Chóngxiū-Dàmínglì*, but they made a mistake in deriving the formula for sunrise and sunset times, which caused small errors in their calculations. Namely, they derived the formula for the sunrise time $t_i^R(n)$ on the *n*th day in the *i*th solar term as

$$t_i^R(n) = T_i^R + nC_i^I + \frac{n(n-1)}{2}A_i^I + \frac{n(n-1)(n+1)}{6}B_i$$

where T_i^R etc. are given in Appendix A.1 of Choi et al. (2018). However, it can be seen that the correct formula is

$$t_i^R(n) = T_i^R + nC_i^I + \frac{n(n-1)}{2}A_i^I + \frac{n(n-1)(n-2)}{6}B_i$$

It should be noted that, although they made a mistake in the formula, the errors in the times of sunrise and sunset due to this mistake are not significant, and the

Sôma Mitsuru (✉)
National Astronomical Observatory of Japan, Mitaka, Tokyo, Japan

D. P. Morgan, T. Quan (eds.), *Mathematics and Astronomy in the Ancient World*,
https://doi.org/10.1007/978-3-031-93809-2_10

maximum error is 0.1050 *fen* 分, where 1 day equals 5230 *fen*. This error of 0.1050 *fen* corresponds to 1.73 seconds in the current unit of time.

Choi et al. (2018) suggested that the sunrise and sunset times of the *Chóngxiū-Dàmínglì* were based on observations made at Kaifeng, the capital of the Northern Song dynasty (960–1127), but we can see that there are systematic differences between the sunrise and sunset times of the *Chóngxiū-Dàmínglì* and those at Kaifeng. Indeed, it turns out that the sunrise and sunset times of the *Chóngxiū-Dàmínglì* were based on the same observations as the older Chinese calendar, *Xuānmínglì* 宣明暦, which was used in 822–892 in China and in 862–1684 in Japan. It can also be seen that the sunrise and sunset times of the *Chóngxiū-Dàmínglì* are very close to those given in the Japanese calendar called *Guchûreki* 具注暦, which was printed in the tenth century.

The details about the results mentioned here will be published elsewhere.

Bibliography

Choi, G.-E., Lee, K.-W., and Mihn, B.-H. 2018. Models of solar and lunar motions in the Chinese *Chongxiu-Daming* calendar. Advances in Space Research 62(11): 3237–3248.

Part IV
West and East from Asia

Three Texts Concerning Planetary Periods from Babylon

John Steele

Abstract This chapter presents editions and translations of three cuneiform tablets from Babylon which contain statements of planetary periods. One of these texts is an astronomical procedure text which presents instructions for how to predict future planetary phenomena from earlier observations. The other two tablets, however, are primarily astrological tablets: one tablet contains sections of material dealing with predicting the number and general characteristics of children to be born, material relating to astral medicine, and associations between the signs of the zodiac and various different kinds of death of an individual, along with the statements of planetary periods; the other tablet contains various astrological associations along with the statements of planetary periods. Comparing these and other previously published tablets which contain planetary periods I show that there is a difference in the way that the periods are presented in astronomical and astrological texts and discuss the importance of considering the periods given in primarily astrological texts in the context of astrology rather than predictive astronomy.

Keywords Babylonian astronomy · Babylonian astrology · Cuneiform texts · Planetary periods

1 Introduction

Our knowledge of Babylonian astronomy and astrology is based upon approximately five thousand cuneiform tablets containing astronomical and astrological texts which have been recovered from sites in Babylonia and almost all of which date to the so-called Late Babylonian period (ca. 750 BC to AD 100). Roughly half of these tablets have been studied in detail and published by modern scholars, providing us with a detailed, but not complete, picture of Babylonian astronomical and

J. Steele (✉)
Brown University, Providence, RI, USA
e-mail: john_steele@brown.edu

D. P. Morgan, T. Quan (eds.), *Mathematics and Astronomy in the Ancient World*,
https://doi.org/10.1007/978-3-031-93809-2_11

astrological practice. This practice included a diverse range of activities including the regular and systematic observation of selected astronomical phenomena and their recording in texts dubbed by modern scholars 'Astronomical Diaries' (Sachs 1974), the prediction of future astronomical phenomena by applying lunar and planetary periods to earlier observations in what has become known as 'goal-year astronomy' (Brack-Bernsen 1999; Steele 2000, 2011), and the development of mathematical astronomy which computes the same types of astronomical phenomena using purely arithmetic means (Neugebauer 1955; Ossendrijver 2012). In addition, it encompassed the continuation of early traditions of 'schematic astronomy' which use simple mathematical schemes to model astronomical phenomena such as the variable length of day and night, the duration of visibility of the moon throughout the month, and the length of the shadow cast by a gnomon (Steele 2021), along with the development of new forms of astrology such as horoscopic astrology (Rochberg 1998) and astral medicine (Heeßel 2008). Whilst new discoveries continue to be made through the study of previously unpublished material which allow us to refine the details of these diverse practices, this overall picture of Babylonian astronomy and astrology in the Late Babylonian period is unlikely to change.

At the heart of both goal-year astronomy and mathematical astronomy lie lunar and planetary periods: a number of years after which a lunar or planetary phenomenon repeat on roughly the same day in the Babylonian calendar. Two types of planetary phenomena were of concern in Late Babylonian astronomy: the first and last appearances, stations, and acronychal risings of the planets, collectively known as 'synodic phenomena', and the passage of the planets past a group of standard reference stars known to modern scholars as 'Normal Stars'. The use of periods within goal-year and mathematical astronomy has been studied by several earlier scholars including Kugler (1907), Aaboe (1964), Neugebauer and Sachs (1967), Steele (2011) and Rochberg (2014). Only recently, however, has attention been drawn to the use of planetary periods within astrology. Hunger (1976) discussed the appearance of some planetary periods in texts concerning the astrological prediction of weather and Ossendrijver (2019) has discussed several astrological procedures for predicting the rise and fall of the market in Babylon which make use of planetary periods. (For the present purposes, astronomical and astrological prediction can be distinguished by the realm of what is being predicted or, on occasion, retrocalculated: the celestial in the former case and here on Earth in the latter (weather phenomena are ambiguous, usually taking place from a Babylonian perspective in the celestial realm but directly affecting the terrestrial realm)).

Several tablets are known which preserve texts containing statements of planetary periods. These texts range from astronomical procedure texts which explain how to compute astronomical phenomena using periods to simple statements of planetary periods. These tablets include:

- BM 41004 (Neugebauer and Sachs 1967 Text E) Rev. 1–17. This tablet contains sections presenting planetary periods for each of the five planets along with day and position corrections for these periods. The periods are said to be used for computing the passages of planets past the Normal Stars. The rest of the tablet contains sections concerning the latitudinal motion of the moon.

- BM 76488 (Ossendrijver 2017). This badly damaged tablet contains procedures for computing planetary phenomena and planetary conjunctions using periods.
- AO 6455 (= TU 11) (Brack-Bernsen and Hunger 2002). Alongside a number of sections devoted to computing the visibility of the moon and the length of the month and the time of eclipses as well as various astrological material, this tablet includes two sections which give planetary periods in the context of predicting the weather.
- AO 6488 (= TU 20) (Hunger 1976). This tablet lists several planetary periods in the context of predicting the weather.

In this chapter, I edit and discuss three further cuneiform tablets which contain lists of planetary periods. Two of these tablets, BM 45728 and BM 35402, were published by Kugler (1907: 45–53) and have been discussed by Neugebauer and Sachs (1967: 206–207), Hunger and Pingree (1999: 203). Britton (2002: 59–61) reedited and discussed BM 45728, although his edition contains several errors. The third tablet I edit here, BM 40113, was discussed briefly by Neugebauer and Sachs (1967: 206) but has not previously been published.

Of the various tablets containing planetary periods, three are primarily concerned with astrology: most of BM 35402 contains birth omens, one side of BM 40113 contains astrological associations of the months of the year with omen apodoses, and AO 6488 contains astrological procedures for predicting weather. Nevertheless, most previous studies of the planetary periods contained in these tablets have taken the planetary periods as embedded astronomical knowledge, ignoring the role they may play in the astrological material contained on the rest of the tablet, and simply studying them in comparison with other astronomical sources. In order to fully understand these texts, however, and indeed the general topic of planetary periods, it is important to try to go beyond studying the astrological texts simply from the viewpoint as repositories of embedded astronomical knowledge. Instead, in addition to comparing the periods found in these texts with those given in astronomical texts, we should also try to understand what roles these periods played within the astrology which is the subject of the tablet. I will return to this question at the end of this paper.

2 An Introduction to Planetary Periods

To illustrate what is meant by a planetary period, let us consider the simple example of Venus. In Table 1 I give the dates in the Babylonian calendar for the first and last appearances of Venus in the east and the west computed using modern theory for the first twelve years of the Seleucid Era (311–299 BC). Inspecting this table we can see that the entries in the first line have very similar dates to the entries in the sixth line: Month V day 10 for first appearance in the east in SE 1 compared to Month V day 7

Table 1 Computed dates in the Babylonian calendar of the first and last appearances of Venus in the east and west for the first twelve years of the Seleucid Era (311–309 BC)

First appearance in the east	Last appearance in the east	First appearance in the west	Last appearance in the west
SE 1 V 10	SE 2 I 10	SE 2 IV 3	SE 2 XII 27
SE 2 XII_2 2	SE 3 VIII 25	SE 3 X 30	SE 4 VII 9
SE 4 VII 27	SE 5 IV 18	SE 5 VI 27	SE 6 II 11
SE 6 II 23	SE 6 X 29	SE 7 I 14	SE 7 X 10
SE 7 X 14	SE 8 VII 5	SE 8 IX 16	SE 9 IV 18
SE 9 V 7	SE 10 I 7	SE 10 III 28	SE 10 XII 24
SE 10 XII 28	SE 11 VIII 21	SE 11 X 26	SE 12 VII 4

Dates are taken from the tables produced by N. A. Roughton and described in Roughton (2002)

for the same phenomena in SE 9; Month I day 10 for the last appearance in the east in SE 2 compared to Month I day 7 in SE 9; Month IV day 3 for first appearance in the west in SE 2 compared to Month III day 28 in SE 10; and Month XII day 27 for last appearance in the west in SE 2 compared to Month XII day 24 in SE 10. In each case, the phenomena take place three or four days earlier in the Babylonian calendar in the entries in line six than in those in the first line. Thus, we can see that Venus's phenomena recur three or four days earlier in the Babylonian calendar after eight properly intercalated Babylonian years. If we now consider the entries in the second and the seventh lines, we find almost the same rule holds: after eight years, Venus's synodic phenomena occur either four or five days earlier in the Babylonian calendar. The same pattern would be found if we extended Table 1 to later years. Thus, we can identify an eight-year period for Venus after which its phenomena occur roughly four days earlier in the Babylonian calendar. This knowledge can be used to predict future phenomena of Venus simply by taking observations from eight years earlier, subtracting four days from the date and, where necessary, making a one-month correction to take into account the differing lengths of the Babylonian year caused by intercalation (Gray and Steele 2008, 2009). We have extensive evidence that this approach was indeed used by the Babylonian scholars to predict both synodic phenomena and passages of a planet past certain reference stars not only for Venus but for all five planets. The clearest evidence for this practice are a large number of preserved tablets containing what modern scholars call Goal-Year Texts (Hunger 2006). These texts are arranged into sections for each planet. The data in each section is taken for one period earlier than the target 'goal' year: eight years earlier for the Venus data, forty-six years earlier for the Mercury data, etc.

Within its eight-year period, Venus will exhibit each of its synodic phenomena five times. We can therefore determine the mean interval between the occurrences of two successive synodic phenomenon of the same kind (e.g. the time between one first appearance in the east and the next first appearance in the east), known as the mean synodic time, by dividing eight years, or more correctly 99 lunar months, by five to give 19.8 months. The true synodic time will vary around this mean: in Babylonian mathematical astronomy, this variation is taken into account by using step functions or (for the outer planets only) zigzag functions which maintain the correct mean value.

Ideally, a planetary period should contain an exact number of calendar years. As we have seen, Venus's eight-year period has an error of about −4 days. In principle, however, it is possible to find longer planetary periods which have no error by taking linear combinations of periods which have errors in the opposite direction. For example, it seems almost certain that two periods for Jupiter, a twelve-year period with an error of about −12 days after twelve calendar years or an error of about +5 days after twelve solar years and a seventy-one-year period which has an error of zero after seventy-one calendar years or about −6 days after seventy-one solar years, were combined in the ratio six-to-five to obtain a period of 427 years which has a zero-day error after 427 solar years (de Jong 2019).

3 BM 45728: An Astronomical Text Concerning Planetary Periods

BM 45728 (= 1881-7-6, 135) was purchased by the British Museum from the dealer Shemtob in 1879 but not registered until July 1881. It was first published by Kugler (1907: 45–48) based upon a hand-drawn copy of the tablet by Strassmaier which is published as plate II in Kugler's book. The tablet was reedited and discussed by Britton (2002: 59–61) who also included a photograph of the tablet. Unfortunately, Britton added some erroneous readings to Kugler's edition. Careful collation of the tablet by the present author in 2017 and 2018 has allowed several improvements to the earlier editions.

The upper left corner of the tablet is broken away but parts of all four edges remain. The tablet is divided into seven sections by horizontal rulings. The first six sections present short procedures for the computation of the synodic phenomena of the five planets and the star Sirius. The final section contains a colophon identifying the scribe who copied the tablet as Labaši, son of Bel-šar-ibni of the Kānik-bābi family. Unfortunately, this scribe cannot be connected to that of any other known tablets. The colophon ends with the date on which the tablet was inscribed. Unfortunately, the year number is broken and the king's name illegible and so the date of the tablet cannot be established with certainty. However, the shape of the tablet as well as the scribe's practice of writing the sign *šá* over the end of the DUMU sign suggests a late Achaemenid date (I owe this information to Christopher Walker). The tablet was therefore probably written sometime in the fourth century BC. The presence of the colophon, however, tends to suggest that the tablet is a copy of an earlier composition – at Babylon, colophons are rarely found on new compositions or texts which are the result of ongoing practice; instead, their appearance is usually confined to copies of earlier 'standard' works. Thus, the text itself may go back to early Achaemenid or even Neo-Babylonian times (ca. 600–400 BC). Britton proposed to date the tablet to either the Neo-Assyrian period (Britton 2002: 26) or year 28 of Nebuchadnezzar II's reign (577 BC) (Britton 2008: 19), but these dates were based upon an erroneous reading of second line of the colophon and so must be abandoned.

3.1 *Transliteration*

Obv.
1 [DIŠ IGI.DU$_8$.A *šá* ^{d}SAG.ME.GAR 12 MU.MEŠ] ⌈*a*⌉*-na ár-ki-ka* GUR-[*ár*]
2 [x x x x x x] *ir-bu-ú*
3 [x x x x x *ta-ṭi*]*p-pi*$_8$*-šú* MU *šá ina* MU-*k*[*a* x x]
4 [x x x x x x] ⌈1⌉7 *u*$_4$*-mu*.MEŠ *it-ti-i ta*-[x x$^{?}$]

----

5 [DIŠ IGI.DU$_8$.A *šá* ^{d}]*dele-bat* 8 MU.MEŠ *a-na ár-ki-ka* GUR-*ár*
6 [4 *u*$_4$*-mu tu-maṭ-ṭa*] 4 *u*$_4$*-mu*.MEŠ *tu-maṭ-ṭa* IGI-*mar*

----

7 [DIŠ IGI.DU$_8$.A *š*]*á* ^{d}GU$_4$.UD 6 MU-*ka a-na ár-ki-ka* GUR-*ár*
8 [x *u*$_4$*-mu*.MEŠ] *it-ti-šú ta-ṭip-pi*$_8$
9 [x *u*$_4$*-mu*.ME]Š *it-ti* IGI.DU$_8$.A *ta-ṭip-pi*$_8$ IGI.LÁ

----

10 [DIŠ IGI.DU$_8$].⌈A⌉ *šá* ^{d}*ṣal-bat-a-nu* 57$^{!}$ MU.MEŠ
11 [*ana ár-ki-k*]*a* GUR-*ár* 12 UD *it-ti-i*
12 [*ta-ṭip-pi*$_8$] ⌈1⌉2 UD *it-ti* IGI.DU$_8$.A *ta-ṭip-pi-ma* IGI.LÁ

Rev.
1 [DIŠ IG]I.DU$_8$.A *šá* ⌈^{d}SAG.UŠ 5⌉9 MU.MEŠ
2 [*ana ár-k*]*i* ⌈GUR-*ár u*$_4$*-mu a-na*⌉ *u*$_4$*-mu* IGI.LÁ

----

3 [DIŠ] ⌈IGI.DU$_8$.A *šá* MUL⌉.KAK.SI.SÁ 27 MU.⌈MEŠ⌉
4 [*ana ár*]-⌈*ki-ka* GUR⌉-*ár* UD *ana* UD ⌈IGI⌉.LÁ

----

5 [I]M ⌈*šá*⌉ ^{I}*la.ba.ši* DUMU *šá* ^{Id}EN.MAN.DÙ ⌈DUMU *šá*⌉ *ka-nik*-KÁ
6 [*pa-liḫ* ^{d}AM]AR.UTU *ina sar-tú la i-tab-bal* E^{ki} ⌈ITU.x⌉
7 [UD.x.KAM MU.x.]KAM ^{I}x x x x

3.2 *Translation*

Obv.
1–4 [¶ Appearances of Jupiter:] go back [12 years] to your rear [...] ... [... you a]dd ... of your year [...] 17 days with [...]

5–6 [¶ Appearances of] Venus: go back 8 years to your rear (and) [4 days you subtract.] 4 days you subtract and see it.

7–9 [¶ Appearances o]f Mercury: go back 6 (of) your years to your rear [x] additional [days] you add. [x] additional [day]s to its appearance you add and see it.

10–12 [¶ Appearan]ces of Mars: go back 57 (error for: 47) years [to yo]ur [rear] 12 additional days [you add]. 12 additional days to its appearance you add and see it.

Rev.

1–2 [¶ App]earances of Saturn: go back 59 years [to yo]ur [rear]. Day to day see it.

3–4 [¶] Appearances of Sirius: go back 27 years [to] your [re]ar. Day to day see it.

5–7 [Tab]let of Labaši, son of Bel-šar-ibni, son of Kānik-bābi. [He who reveres Mar]duk shall not falsely take it away. Babylon, Month [x, the xth day, the x]th [year] of …

3.3 *Critical Apparatus*

Obv.

1 For the restoration 12 MU.MEŠ, see the commentary below.
3–4 Strassmaier drew a horizontal ruling between these two lines and so Kugler, followed by Britton, divided the first section into two sections (lines 1–3 and line 4). However, close inspection of the tablet shows that there is no horizontal ruling here. There is a small and very lightly inscribed horizontal line extending for the length of a single cuneiform sign, but this looks like either a guideline or just the result of the scribe resting his stylus on the clay rather than a ruling (there is a similar short and faint horizontal line between lines 5 and 6). Thus, I am confident that lines 1–4 present a single section.
4 The first preserved sign at the beginning of the line was read as the number 27 by Kugler and Britton. However, the sign is almost certainly 17 not 27. The first wedge which would make the sign 27 is spaced too far away from the other wedges which make up this sign and is almost certainly a mark caused by damage at the edge of the break rather than a real wedge.
9 Britton read the first preserved traces at the beginning of this line as a 10 and took this to be a statement of a ten-day error in the six-year period for Mercury (Kugler had merely marked damage for this sign). However, the single preserved wedge looks too small and in the wrong place to be read as a 10. Furthermore, in Britton's reading we would have to assume that the word 'days' (written either UD or u_4-*mu*.MEŠ) was mistakenly omitted by the scribe (it is correctly added in every other preserved instance on the tablet). I believe that the single wedge is instead the end of the sign MEŠ and we are to restore x u_4-*mu*.MEŠ 'x days' at the beginning of the line.
10 The number 57 is written clearly, but is certainly a scribal error for 47. Kugler and Britton read the number as 47 without comment.

Rev.

1 The two lower wedges and traces of the three upper wedges of the 50 part of the sign 59 are preserved.
3 I restore a DIŠ sign at the beginning of this line, and in parallel at the beginning of each section, because by comparison with the following line there is clearly space for at least one sign in the broken part. A DIŠ sign is the most obvious restoration.
5 The *šá* sign is written to overlap the end of the DUMU sign on both occasions.
6 Britton's reading of this line is erroneous.
7 The surface of this line is almost completely lost, presumably due to general damage, although it is possible that the line was intentionally erased by the scribe. Unfortunately, I cannot make out any of the signs which constitute the king's name.

3.4 Commentary

The text is divided into six sections plus a colophon. Each section presents a short procedure for the computation of the synodic phenomena of a planet (called 'appearances') by taking the observed date of the phenomena and, for some planets at least, applying a correction of a few days. The first section must concern Jupiter although the planet's name is lost at the beginning of the first line. It was believed by both Kugler and Britton that the first section was divided into two sections: one section, corresponding to lines 1–3, which they did not identify followed by a single-line section in line 4 which they took to concern the moon. Kugler and Britton misread the number 17 in this latter line as 27 and took this to be a value for the length of the sidereal month and hence a period after which the moon would repeat conjunctions with stars. However, collation shows that the number is 17 and that line 4 is part of the same section as lines 1–3. The following sections concern respectively Venus, Mercury, Mars, and Saturn, followed by a section for the star Sirius. The order of the planets—Jupiter, Venus, Mercury, Mars, Saturn—follows the common trend in cuneiform texts of placing astrologically benefic planets before malefic ones (Rochberg-Halton 1988), although the order differs slightly from the standard order (Jupiter, Venus, Mercury, Saturn, Mars) which is found in texts of the Seleucid period.

The basic structure of each section is the same. First, it is stated that you go back a certain number of years which is the period for that planet. If there is an error in the period, the size of this error in days is stated. Finally, we are told either to apply a correction for this error in order to 'see' the predicted date of the phenomenon, or that 'day to day' we will 'see' meaning that no correction is necessary. The section for Venus in lines 5–6 illustrates the style of these procedures:

> [Appearances of] Venus: go back 8 years to your rear (and) [4 days you subtract.] 4 days you subtract and see it.

First, we are told that in order to predict the phenomena of Venus we must go back eight years earlier and subtract 4 days from the observed dates. Secondly it reiterates that we must subtract 4 days and says that once we have done this subtraction we will find the predicted dates of the phenomena in the current year. The section for Jupiter is somewhat longer than the other sections, perhaps because it is the first section and so explains the procedure in greater detail. Unfortunately, this section is quite fragmentary and I am unable to make full sense of it.

Table 2 summarizes the periods and the corrections given in BM 45728. Also given in this table are modern estimates of the corrections for these periods when applied to the computation of synodic phenomena. These estimates are taken from Gray (2009). Slightly different corrections are needed when computing passages of the planet past Normal Stars.

The Jupiter section preserves only the magnitude of the day correction, not its sign or the length of the period. Three periods are commonly referenced for Jupiter in other texts: a 12-year period, a 71-year period, and an 83-year period. The 71-year and 83-year periods are used respectively in the Goal-Year Texts for Jupiter's

Table 2 Planetary periods with their day corrections given in BM 45728 with modern values given for comparison

Planet	Period	Correction (text)	Correction (modern)
Jupiter	[12 years]	[+]17 days	+17 days
Venus	8 years	–4 days	–4 days
Mercury	6? years	[+x]	+16 days
Mars	47 years	+12 days	+2 days
Saturn	59 years	0 days	–6 days
Sirius	27 years	0 days	0 days

synodic phenomena and its passages past Normal Stars. The 71-year period is more or less exact, with a mean date correction of zero days. Both the 12- and the 83-year periods require date corrections of about +17 days. However, evidence for the use of the 83-year period is restricted to the computation of passages of Jupiter past Normal Stars and is found only from the Seleucid period. Jupiter's 12-year period is used as the basis of the compilation text BM 36824 (Hunger 2001 No. 54) which covers the end of the sixth and the beginning of the fifth century BC, perhaps contemporary with the composition of our text. Furthermore, the tablet BM 76488 (Ossendrijver 2017), which concerns planetary periods and their application to computing planetary conjunctions, refers to Jupiter's 12-year period and mentions the same +17 day correction. Thus, it seems certain that it is Jupiter's 12-year period that appears in BM 45728.

The 8-year Venus period is used to compute both synodic phenomena and Normal Star passages in goal-year astronomy. It is also used as the basis for compiling Venus observations in the text BM 32299 (+) 42083 (+) 45674 (Hunger 2001 No. 56) which covers most of the fifth and the very beginning of the fourth century BC. Venus's 8-year period is mentioned also in BM 41004 (Neugebauer and Sachs 1967 Text E) and in BM 34560, both of which also state a −4 day correction.

The 6-year Mercury period is not directly attested elsewhere; BM 41004 gives a 13-year period with a −3 day correction and a 46-year period with a −1 day correction and BM 34560 gives the 46-year period stating that it is exact. The 46-year period is used in the Goal-Year Texts. It is tempting, therefore, to see the 6 years given in BM 45728 as a scribal error for 46 years. However, Britton (2002) has noted that 6 years is a valid period for Mercury and has elsewhere argued that it was used to compute the Mercury phenomena in BM 33066 (Britton 2008). In slight support I note that BM 45728 gives a positive day correction for the period, although Britton's claim that the correction was +10 days has been shown to be erroneous. Modern computation suggests a +16 day correction for the 6-year period and a zero day correction for the 46-year period. However, without further evidence it is not possible to decide whether BM 45728 attests to a 6-year period for Mercury or whether this is a scribal error for 46 years.

The 47-year Mars period is used in the Goal-Year Texts for Normal Star passages but synodic phenomena use the 79-year period. However, Gray (2009) has noted that the 79-year period is not necessarily a better period for synodic phenomena: it requires a large day correction, +7 days as opposed to +2 days for the 47-year period, although the standard distribution around these corrections is smaller for the 79-year than the 47-year period. The 47-year period is mentioned on BM 76488 but the day correction is broken. It is also mentioned on BM 41004 in the context of Normal Star passages where a +4 day correction is given.

The 59-year Saturn period is used in the Goal-Year Texts for both synodic phenomena and passages past Normal Stars. The period is cited in BM 34560 where it is said to require a −5 day correction for the computation of synodic phenomena and in BM 41004 where it is said to require a −6 day correction for the computation of Normal Star Passages.

Following the procedures for the five planets we find a procedure for the synodic phenomena of Sirius which cites a 27-year period as exact. Since Sirius phenomena depend purely on the position of the sun (at least for the Babylonians who had no knowledge of precession or the proper motion of stars), this period is effectively equating 27 Babylonian calendar years with 27 solar years. It therefore assumes a 27-year intercalation cycle. The 27-year period for Sirius is implicit in a series of computed dates of Sirius's synodic phenomena covering the late seventh and early sixth century BC found on BM 36731 (Britton 2002).

The tablet ends with a colophon. The tablet names the scribe as Labaši, son of Bel-šar-ibni of the Kānik-bābi family, who is otherwise unattested. Following the name of the scribe is a warning against the theft of the tablet of the type which is found quite commonly on scholarly tablets from Babylonia in the first millennium BC. Finally, the tablet ends with the date on which it was written. Unfortunately, only the day number of the date is preserved; the month, year and king's name are broken away or illegible.

4 BM 35402: An Astrological Text with a Section Containing Several Long and Short Planetary Periods

BM 35402 (= Sp II, 985) was purchased from the dealer Spartali and arrived in the British Museum in the summer of 1879. The tablet is a substantial fragment of a single column tablet. The left and right edges are preserved but both the upper and lower edges are lost. The tablet is divided into at least four sections: the first section of the obverse contains birth omens which predict the number and gender of children born according to various astrological factors, the second section of the obverse concerns two numerical schemes used in astral medicine, the first section of the reverse predicts the life and death of individuals based upon their birth in the 'region' of zodiacal signs, and the final section of the reverse presents a list of planetary periods. Kugler (1907: 48–53) published an edition and analysis of the planetary periods along with Strassmaier's copy of this section. A copy of the complete tablet by Pinches was published as LBAT No. 1593. Reiner (2000) published an edition and study of the first three sections of the tablet but refers simply to Kugler for the final section.

Kugler's discussion of the planetary periods treats them as a piece of astronomical knowledge, separate from the astrological material on the tablet. Kugler's discussion was subsequently used by Neugebauer and Sachs (1967: 206–207) and Hunger and Pingree (1999: 203); both of these later discussions follow Kugler in treating the section as independent of the astrological material on the tablet. I present below an edition of this part of the tablet. For an edition of the other sections of the tablet, see Reiner (2000).

4.1 Transliteration

Rev.

12′ UD.DA ^{d}SAG.ME.GAR GÍD.DA 3 ME 44 MU.MEŠ : UD.DA ^{d}SAG.ME.GAR LUGÚD.DA 1,3 ITU 10 u_4-*mu*
13′ UD.DA ^{d}*dele-bat* GÍD.DA 1,4 ME MU.MEŠ : UD.DA ^{d}*dele-bat* LUGÚD.DA 1,4 ITU 20 u_4-*mu*
14′ UD.DA ^{d}*dele-bat* ⸢LUGÚD⸣.DA 7 UD.MEŠ : *a-na* 14 UD.MEŠ : *ana* 21 UD.MEŠ
15′ UD.DA ^{d}*ṣal-bat-a-nu* 1,5 ITU.MEŠ *a-na* 6 ITU 20 UD.MEŠ
16′ [U]D.DA ^{d}*ṣal-bat-a-nu* GÍD.DA *a-na* 2 ME 1,24 MU.AN.NA [x x]
17′ [UD.DA] ⸢^{d}⸣GENNA *a-na* 10 ITU : UD.DA ^{d}GENNA GÍD.DA *a-na* 5 ME ⸢1⸣ U[Š$^{?}$ MU.MEŠ]
18′ [UD.DA AN.K]U_{10} *sin* GÍD.DA *a-na* 6 ME 1,24 MU.⸢AN⸣.[NA.MEŠ ...]
19′ [UD.DA AN.KU_{10}] ⸢*šamaš*$^{?}$⸣ KI.2 MA [...]
(remainder lost)

4.2 Translation

Rev.
12′ Long period of Jupiter: 344 years. Short period of Jupiter: 63 months 10 days.
13′ Long period of Venus: 6400 years. Short period of Venus: 64 months 20 days.
14′ Short period of Venus: 7 days (alternatively) 14 days (alternatively) 21 days.
15′ (Short) period of Mars: 65 months (alternatively) 6 months 20 days.
16′ Long [p]eriod of Mars: 284 years [...]
17′ (Short) [period] of Saturn: 10 months. Long period of Saturn: 560 [years.]
18′ Long [period of eclip]ses of the moon: 684 yea[rs ...]
19′ [(Long) period of eclipses of] the sun: ditto [...]
(remainder lost)

4.3 Critical apparatus

Rev.
16′ The 2 before the ME sign for hundreds is clear on the tablet. BM 40113 has a 3 here. See further the discussion in the next section.
17′ Kugler read the traces at the end of the line as an 'U' sign for the first wedge of the number 29. BM 40113 has a clear UŠ here, indicating that the previous '1' has the magnitude 60.

4.4 Commentary

As already noted by Neugebauer and Sachs (1967), the list of planetary periods is partially duplicated by the obverse of BM 40113, which will be edited below. Despite there being some small differences between the two sources, it makes sense to discuss both together here.

The basic format of the list of planetary periods is as follows. For each of the planets one 'long' (GÍD.DA) period and one or more 'short' (LÚGUD.DA) periods are given. The long periods are always given in whole numbers of years; the preserved entries range in length from a few hundred years to several thousand years. The short periods are always given in terms of months and/or days, even when the period is longer than one year. As a consequence, these short periods are precisely defined, whereas the long periods are slightly ambiguous in length; because of intercalation the length of a given number of Babylonian years depends upon which year is the starting point.

The list of planetary periods begins with Jupiter, followed by Venus, Mars, and Saturn. Mercury is omitted for reasons that are not evident. Except for the omission of Mercury, the order of the planets is the same as in BM 45728 discussed above.

Table 3 lists the long and short periods given for each planet on BM 35402 and BM 40113. I first consider the long periods. As explained by Kugler (1907), the

Table 3 'Long' and 'short' planetary periods given on BM 35302 and BM 40113

Planet	Long Period		Short Period	
	BM 35302	BM 40113	BM 35302	BM 40113
Jupiter	344y	[…]	63 m 10d	[…]
Venus	6400y	[…]	64 m 20d 7d 14d 21d	7 d […]
Mars	284y	284y	65 m 6 m 20d	5 m 6d
Saturn	560y	560y	10 m	10 m
Lunar Eclipse	684y	684y	–	–
Solar Eclipse	684y	684y	–	–

344-year period for Jupiter can be constructed by combining the 12-year period with the 71- and 83-year periods for Jupiter used in the Goal-Year Texts: 344 = 4 × 71 + 5 × 12 = 4 × 81 + 12. The 6400-year Venus period is simply the common 8-year period multiplied by 800, although why this is done is unknown. The Mars period of 284 years may be constructed by combining the 79- and 47-year periods used in the Goal-Year Texts: 284 = 79 × 3 + 47. This 284-year period is used in both System A and System B for Mars. Hunger and Pingree (1999: 203) suggest that the Saturn period might be constructed by combining the Goal-Year period of 59 years with a 29-year period: 560 = 59 × 9 + 29. They note that the period used in System A and System B for Saturn, 265 years, is also constructed from these periods: 265 = 59 × 4 + 29.

The origins of the short periods are harder to explain. Some may have been constructed from the intervals of visibility and invisibility found in MUL.APIN II i 44–67 (Hunger and Steele 2019: 206). For example, MUL.APIN gives a length of 1 year or 12 schematic months for Jupiter's period of visibility and two lengths of 20 days and 1 month for its period of invisibility. The period of 63 months and 10 days given on BM 35302 is equal to 5 times 12 months plus 5 times 20 days. This may simply be chance, however. The 7-, 14- and 21-day periods for Venus may relate to statements in MUL.APIN that Venus's invisibility in the west can last for zero days, 3 days, 7 days, or 14 days. MUL.APIN gives three options for the duration of Mars's invisibility: 2 months, 3 months and 10 days, or 6 months and 20 days. This latter interval is the same as one of the short periods given on BM 35302. The other short period on BM 35302, 65 months, and the period given on BM 40113 cannot be related to MUL.APIN, however. Finally, the 10-month period for Saturn cannot be related to MUL.APIN.

A period of 684 years is given for eclipses of both the sun and moon. The period can be constructed from the 18-year Saros period either simply as 684 = 38 × 18 or 684 = 19 × 2 × 18. 19-years is the so-called Metonic cycle, used as an intercalation cycle from the early fifth century BC onwards. Thus, 684 years is a combination of the Saros and Metonic cycles. As noted already by Kugler (1907), however, the same 684-year eclipse period is mentioned in BM 34085 (= LBAT No. 1599), an astrological text which relates eclipses to hostilities in the land. BM 34085 Obv. II 9 reads:

[57] A.RÁ 12 11,24 6 ME 1,24 MU.ME ^{d}*sin u* ^{d}*šamaš* AN.KU$_{10}$ [...]
[57] times 12 is 11,26 (=) 684 years (of) solar and lunar eclipses [...]

This passage suggests an alternative origin of the 684-year period whose astronomical—if it is indeed astronomical and not numerological—meaning I do not understand.

5 BM 40113: A Second Astrological Text with the Same Section Containing Several Long and Short Planetary Periods

BM 40113 (= 1881–2–1, 79) was purchased from the dealer Spartali and arrived in the British Museum in November 1880. The tablet preserves most of the width but probably less than half the height of the original tablet. The right edge and part of the lower edge are preserved. The obverse contains the end of a section dealing with planetary periods. The reverse presents a scheme associating the months of the year with short omen apodoses. BM 40113 was identified by Sachs as containing a partial duplicate of BM 35402. Several entries from the list of planetary periods were quoted by Neugebauer and Sachs (1967: 206–207) and mentioned by Hunger and Pingree (1999: 203), but the tablet has not previously been published.

5.1 Transliteration

Obv.

1′	[...] x [...]	
2′	[UD.DA ...] ⸢GÍD.DA⸣ *a-na* [...]	
3′	[UD.DA ... LU]GÚD.DA *a-na* [...]	
4′	[UD.D]A ^{d}*dele-bat* ⸢LUGÚD$^{?}$.DA$^{?}$⸣ *a-na* 7 UD.MEŠ [...]	
5′	[UD.D]A ^{d}*ṣal-bat-a-nu* GÍD.DA *a-na* 2$^{?}$ ME ⸢1,24⸣ [MU.MEŠ]	
6′	[UD.D]A ^{d}*ṣal-bat-a-nu* LUGÚD.DA *a-na* 5 ITU 6 UD.MEŠ	
7′	[UD.D]A ^{d}SAG.UŠ GÍD.DA *a-na* 5 ME 1 UŠ MU.MEŠ	
8′	[UD.D]A ^{d}SAG.UŠ LUGÚD.⸢DA⸣ *a-na* 10 ITU.MEŠ	
9′	[UD.D]A AN.KU$_{10}$ *sin* GÍD.DA *a-na* 6 ME 1,24 MU.MEŠ	
10′	[UD.D]A AN.KU$_{10}$ *šamaš* GÍD.DA *a-na* 6 ME 1,24 MU.MEŠ	
11′	[x] SAG.MEŠ	KUR URI^{ki}
12′	[x] MURUB$_4$	KUR NIM.MA^{ki}
13′	[x] ⸢x⸣.MEŠ	KUR MAR.^{ki}
14′	[BAR GU$_4$ SIG] ŠU IZI KIN DU$_6$ APIN GAN AB ⸢ZÍZ ŠE⸣	

Rev.

1	[BAR x]	⸢x⸣ *u* ŠÈG$^{?}$	⸢x⸣ [...]
2	[GU$_4$ x]	[x] RA	⸢x⸣
3	[SIG x]	KÚR	DU [...]
4	[ŠU x]	ÍD$^{?}$	UD
5	[IZI] ⸢UM⸣	ZI	KÚR
6	[KIN] ⸢UM⸣	ŠE NU	SI
7	[DU$_6$] ⸢UM⸣	EN.TE.⸢NA$^{?}$⸣	*dan-nu*
8	[APIN] UM	ZI ŠU^{ki}	
9	[G]AN UM	⸢GIG$^{?}$⸣	GÁL
10	AB UM	⸢ZI⸣ KUR	NIM.MA^{ki}
11	⸢ZÍZ UM	x NU⸣	⸢x⸣

6 Translation

Obv.
1′ [...]
2′ Long [period of ...]: [...]
3′ [Sh]ort [period of ...]: [...]
4′ Short [peri]od of Venus: 7 days [...]
5′ Long [peri]od of Mars: 284 [years.]
6′ Short [peri]od of Mars: 5 months 6 days [...]
7′ Long [peri]od of Saturn: 560 years.
8′ Short [peri]od of Saturn: 10 months.
9′ Long [peri]od of eclipses of the moon: 684 years.
10′ Long [peri]od of eclipses of the sun: 684 years.
11′ [...] heads?: Land of Akkad.
12′ [...] middle: Land of Elam.
13′ [... feet?]: Land of Amurru.
14′ [Month I, Month II, Month III,] Month IV, Month V, Month VI, Month VII, Month VIII, Month IX, Month X, Month XI, Month XII.

Rev.
1 [Month I ...:] ... and rain ... [...]
2 [Month II ...:] ...
3 [Month III ...:] The enemy will come forth.
4 [Month IV ...:] ...
5 [Month V] ...: Attack of the enemy.
6 [Month VI] ...: Barley will not prosper.
7 [Month VII] ...: Strong cold.
8 [Month VIII] ...: Attack of Subartu.
9 [Month I]X ...: Disease?.
10 Month X ...: Attack of the land of Elam.
11 Month XI ...: ...

6.1 Critical Apparatus

Obv.
5′ The initial 2 of the number 2 ME 1,24 could be a 3, but the duplicate BM 35302 has a clear 2 and a 3 would make no sense. Only 21+x remains of the 24.
11′ I do not understand why 'heads' is plural.
13′ The previous two lines refer to the 'head' (i.e. top/beginning) and 'middle' and therefore we would expect this line to begin with a reference to the 'feet' (i.e. bottom/end). Groups of three entries like this are quite common in astrological texts (e.g. BM 32339+32407+32645 and BM 37361 (Ossendrijver 2018). However, the traces of the first preserved sign do not look like the expected GÌR.

6.2 Commentary

The obverse of the tablet contains the end of a list of planetary periods which partially duplicates that found on BM 35402 Rev. 12–19 and is discussed in the commentary to that tablet above. The list is followed immediately by three statements which associate the beginning/top ('head'), middle and bottom/end (probably 'feet') of something with the three lands of Akkad, Elam, and Amurru. There is no ruling to indicate that this is a different section of the text to the list of planetary periods. Finally, the last line of the obverse presents a list of the standard Late Babylonian logograms for the twelve months of the year. Again, there is no ruling to indicate a section division. The reverse of the tablet contains a scheme which associates months of the year with various omen apodoses.

7 Final Remarks

As I noted in the introduction, statements concerning planetary periods appear in texts that are both primary astronomical and those that are primarily astrological. The three tablets edited above illustrate this point: BM 45728 is clearly an astronomical tablet, presenting procedures for predicting astronomical phenomena from earlier observations whereas BM 35402 and BM 40113 are astrological tablets. This raises the question of whether the statements about planetary periods served the same purposes in astronomical and astrological texts. In particular, do these statements of planetary periods in astrological texts serve a direct astrological purpose, or are they simply astronomical knowledge that happened to be included in an astrological tablet.

Several tablets are known which bring together apparently unrelated material on the same tablet. For example, two of the tablets which preserve uranology (constellation description) texts on one side have completely unrelated material on the other side: one, BM 66958, contains a series of observations of Mercury on the other side, and the second, MLC 1884, contains a non-astral text referring to the topography of the city of Uruk (Beaulieu et al. 2018). Other tablets are known which contain sections which are not directly related but instead share a common theme or are founded on a similar body of knowledge. For example, BM 34713 (= LBAT No. 1499) contains a copy of a so-called 'Three Stars Each' text which assigns one star to each of the three celestial paths of Anu, Enlil, and Ea for each month of the year, and a copy of a zodiac-based rising time scheme (Steele 2017). At first sight these two texts seem to be unrelated, but on closer inspection we find that the rising time scheme quotes from the 'Three Stars Each' text. Thus, we have three possibilities: material on the same tablet may be directly related, it may be indirectly but thematically related, or it may be unrelated. In the latter case, however, the material is almost always (there only a very small number of exceptions) separated into different sections and often written on different sides of the tablet.

The tablet AO 6455 (= TU 11) falls into the third group. It contains several sections containing astrological material alongside sections giving procedures for computing the length of the month, eclipse times, and the duration of visibility of the moon (Brack-Bernsen and Hunger 2002). The astrological material bears no thematic or other relation to the astronomical procedures on the tablet and it always presented in separate sections. Thus, it seems likely that the scribes copied completely unrelated material onto the same tablet. The same does not seem to be true of BM 35302 and BM 40113, however. Immediately following the statements of planetary and lunar periods we have statements which refer to the lands of Akkad, Elam, and Amurru. These statements are clearly presenting astrological associations between countries and something astronomical (probably parts of constellations or signs of the zodiac). Unfortunately, the beginning of these lines is lost so we do not know what the association is, but these statements are similar to those found on many other tablets which associate countries and cities with various things such as the signs of the zodiac, constellations, the watch of the night when an eclipse takes place, the quadrant of the moon which is eclipsed, and so on (Steele 2015). Crucially, however, these statements follow directly the statements of planetary and lunar periods, without a horizontal ruling indicating the beginning of a new section. Unfortunately, the corresponding lines of the duplicate section on BM 35302 are lost to damage, but there is no reason to suppose that this tablet did not contain the same material giving astrological associations with countries. Given that all of the other material on BM 35302 and BM 40113 is astrological, it seems safe to conclude that the statements of planetary periods are to be considered within the context of astrology rather than predictive astronomy.

This conclusion is also supported by looking at the statements of the planetary periods themselves which are clearly different in both style and content to those found in texts which provide procedures for how to predict future astronomical phenomena (BM 45728, BM 41004, and BM 76488). These latter texts present planetary periods accompanied by statements of the day and/or position corrections which need to be applied when they are being used and are written in the form of second person instructions. Planetary periods given in astrological texts, however, are always given as simple third-person statements of apparent fact and the periods themselves are usually stated in whole numbers of years with no indication of any day corrections. The only exception to this is the ‘short’ periods on BM 35402 and BM 40113, which are given in months and days, but these are not day or position corrections that can be used in making predictions of astronomical phenomena. Furthermore, several of the periods given on BM 35402 and BM 40113 would not work for computing phenomena, as would have been obvious to the Babylonian scholars. Thus, it seems likely that the planetary periods are astrological or numerological constructions rather than statements of astronomical knowledge. Exactly how these periods were constructed, and why, remains unknown, however.

This brings us to a final question: why do planetary periods appear in astrological texts? Ossendrijver (2019) has suggested that they were used to make advance astrological predictions in the same way that they were used in goal-year astronomy to make astronomical predictions. The idea here is that if, for example, rain, flood, or a change in the value of commodities in the market is associated with the phenomena of the planets, then it would be possible to predict future rain, flood, or changes in the value of commodities by predicting the phenomena of the planet, which can be done using the period. Thus, one could instead simply apply the period to the weather or the market directly without having to first predict the planetary phenomena. This explanation, while plausible in itself, need not be the only use of planetary periods, however, and does not account for those periods which do not work as astronomical periods. Further research is needed into astrological texts such as BM 34085 (= LBAT No. 1599), which manipulates the 18-year eclipse period for astrological reasons that I do not fully understand, and other astrological texts which perform numerical manipulations on numbers which seem to relate to planetary periods.

Bibliography

Aaboe, A., 1964, 'On Period Relations in Babylonian Astronomy', *Centaurus* 10, 213–231.

Beaulieu, P.-A., Frahm, E., Horowitz, W., and Steele, J. M., 2018, *The Cuneiform Uranology Texts: Drawing the Constellations* (Philadelphia: American Philosophical Society).

Brack-Bernsen, L., 1999, 'Goal-Year Tablets: Lunar Data and Predictions', in N. M. Swerdlow (ed.), *Ancient Astronomy and Celestial Divination* (Cambridge, MA: The MIT Press), 149–177.

Brack-Bernsen, L. and Hunger, H., 2002, 'TU 11: A Collection of Rules for the Prediction of Lunar Phases and of Month Lengths', *SCIAMVS* 3, 3–90.

Britton, J. P., 2002, 'Treatments of Annual Phenomena in Cuneiform Sources', in J. M. Steele and A. Imhausen (eds.), *Under One Sky: Astronomy and Mathematics in the Ancient Near East* (Münster: Ugarit-Verlag), 21–78.

Britton, J. P., 2008, 'Remarks on Strassmaier Cambyses 400', in M. Ross (ed.), *From the Banks of the Euphrates: Studies in Honor of Alice Louise Slotsky* (Winona Lake: Eisenbrauns), 7–34.

Gray, J. M. K., 2009, 'A Study of Babylonian Goal-Year Planetary Astronomy', PhD dissertation, Durham University.

Gray, J. M. K. and Steele, J. M., 2008, 'Studies on Babylonian Goal-Year Astronomy I: A Comparison Between Planetary Data in Goal-Year Texts, Almanacs and Normal Star Almanacs', *Archive for History of Exact Sciences* 62, 553–600.

Gray, J. M. K. and Steele, J. M., 2009, 'Studies on Babylonian Goal-Year Astronomy II: The Babylonian Calendar and Goal-Year Methods of Prediction', *Archive for History of Exact Sciences* 63, 611–633.

Heeßel, N., 2008, 'Astrological Medicine in Babylonia', in A. Akasoy, C. Burnett and R. Yoel-Tlalim (eds.), *Astro-Medicine: Astrology and Medicine, East and West* (Firenze: Sismel), 1–16.

Hunger, H., 1976, 'Astrologische Wettervorhersagen', *Zeitschrift für Assyriologie und Vorderasiatische Archäologie* 66, 234–260.

Hunger, H., 2006, *Astronomical Diaries and Related Texts from Babylonia. Volume VI: Goal Year Texts* (Vienna: Österreichishe Akademie der Wissenschaften).

Hunger, H., and Pingree, D., 1999, *Astral Sciences in Mesopotamia* (Leiden: Brill).

Hunger, H., and Steele, J. M., 2019, *The Babylonian Astronomical Compendium MUL.APIN* (Abingdon: Routledge).

de Jong, T., 2019, 'A Study of Babylonian Planetary Theory I. The Outer Planets', *Archive for History of Exact Sciences* 73, 1–37.

Kugler, F. X., 1907, *Sternkunde und Sterndienst in Babel. I* (Münster in Westfalen: Aschendorffsche Verlagsbuchhandlung).

Neugebauer, O., 1955, *Astronomical Cuneiform Texts* (London: Lund Humphries).
Neugebauer, O. and Sachs, A., 1967, 'Some Atypical Astronomical Cuneiform Texts. I', *Journal of Cuneiform Studies* 21, 183–218.
Ossendrijver, M., 2012, *Babylonian Mathematical Astronomy: Procedure Texts* (New York: Springer).
Ossendrijver, M., 2017, 'BM 76488 – A Babylonian Compendium about Conjunctiond and Other Planetary Phenomena', in J. M. Steele and M. Ossendrijver (eds.), *Studies on the Ancient Exact Sciences in Honor of Lis Brack-Bernsen* (Berlin: Edition Topoi), 127–146.
Ossendrijver, M., 2018, 'BM 32339+32407+32645: New Evidence for Late Babylonian Astrology', in S. V. Panayotov and L. Vacín (eds.), *Mesopotamian Medicine and Magic: Studies in Honor of Markham J. Geller* (Leiden: Brill), 401–420.
Ossendrijver, M. 2019, 'Babylonian Market Predictions', in J. Haubold, J. M. Steele and K. Stevens (eds.), *Keeping Watch in Babylon: The Astronomical Diaries in Context* (Leiden: Brill), 53–78.
Reiner, E., 2000, 'Early Zodiologia and Related Matters', in A. R. George and I. L. Finkel (eds.), *Wisdom, Gods, and Literature: Studies in Assyriology in Honour of W. G. Lambert* (Winona Lake: Eisenbrauns), 421–427.
Rochberg-Halton, F., 1988, 'Benefic and Malefic Planets in Babylonian Astrology', in E. Leichty (eds.), *A Scientific Humanist: Studies in Memory of Abraham Sachs* (Philadelphia: University Museum), 323–328.
Rochberg, F., 1998, *Babylonian Horoscopes* (Philadelphia: American Philosophical Society).
Rochberg, F., 2014, 'Periodicities and Period Relations in Babylonian Celestial Sciences', in M. Kozuh, W. F. M. Henkelman, C. E. Jones and C. Woods (eds.), *Extraction and Control: Studies in Honor of Matthew W. Stolper* (Chicago: Oriental Institute), 211–218.
Roughton, N. A., 2002, 'A Study of Babylonian Normal-Star Almanacs and Observational Texts', in J. M. Steele and A. Imhausen (eds.), *Under One Sky: Astronomy and Mathematics in the Ancient Near East* (Münster: Ugarit-Verlag), 367–378.
Sachs, A., 1974, 'Babylonian Observational Astronomy', *Philosophical Transactions of the Royal Society of London*, A 276, 43–50.
Steele, J. M., 2000, 'Eclipse Prediction in Mesopotamia', *Archive for History of Exact Sciences* 54, 421–454.
Steele, J. M., 2011, 'Goal-Year Periods and Their Use in Predicting Planetary Phenomena', in G. Selz and K. Wagensonner (eds.), *The Empirical Dimension of Ancient Near Eastern Studies – Die empirische Dimension altorientalischer Forschungen* (Wien: LIT Verlag), 101–110.
Steele, J. M., 2015, 'Mesopotamian Astrological Geography', in P. Barthel and G. van Kooten (eds.), *The Star of Bethlehem and the Magi* (Leiden: Brill), 201–216.
Steele, J. M., 2017, *Rising Time Schemes in Babylonian Astronomy* (Dordrecht: Springer).
Steele, J. M., 2021, 'The Continued Relevance of MUL.APIN in Late Babylonian Astronomy', *Journal of Ancient Near Eastern History*, 9, 259–277.

The Research About the Antikythera Mechanism: Old and New Questions

Efthymios Nicolaidis

Abstract This article presents the history of research on the Antikythera Mechanism from its discovery in 1902 in the eponymous shipwreck kept at the Archaeological Museum of Athens to the current day in 2025. The first theories and dates were formulated in 1902 by the archaeologists Ioannis Svoronos and Adolph Wilhelm on the basis of some of the inscriptions. At the same time, the naval officer Periklis Rediadis tried to interpret the gearing. In 1905, the philologist Albert Rehm was the first to suggest that the Mechanism was a planetarium. The admiral Ioannis Theophanides constructed the first model in 1934. Working on the Mechanism from 1958 onwards, the historian of science Derek de Solla Price obtained permission to take X-rays of it in 1971, performed by the physicist Charalampos Karakalos, from which Price constructed a new model displaying the motions of the sun and the moon. In the late 1980s and early 1990s, Michael Wright and Alan Bromley conducted the first 3D X-ray imaging, and Wright made a new model comprising hypothetical planetary gearing and spiral dials on the back, identifying the upper back dial as a calendar. In 2005, an interdisciplinary team created the Antikythera Mechanism Research Project and obtained new imagery using the latest IT and photo technology. Based on this imagery, much more of the inscriptions were revealed, the lower back dial was identified as an eclipse dial, and the planetary hypothesis was confirmed by the reading of the inscriptions. Since then, the planetary gearing has been challenged by other scholars, and a diving campaign has been initiated in order to find the lost gears.

Keywords Antikythera mechanism · Planetarium · Greek astronomy

This article is dedicated to the memory of John Hugh Seiradakis, eminent member of the Antikythera Mechanism Research Project team, who passed away on the third of May 2020.

E. Nicolaidis (✉)
National Hellenic Research Foundation, Athens, Greece

SYRTE, Observatory of Paris, PSL, Paris, France
e-mail: efnicol@eie.gr

D. P. Morgan, T. Quan (eds.), *Mathematics and Astronomy in the Ancient World*,
https://doi.org/10.1007/978-3-031-93809-2_12

1 The First Investigations

The famous "Antikythera shipwreck," a big Greek merchant boat that transported antiquities and other valuable objects from the Aegean sea to the West *circa* 60 B.C., was found in 1900 by Greek sponge divers off Symi Island (then belonging to the Ottoman Empire). In the Eastern season of that year, on their way to the coasts of North Africa, the two small ships of captain Dimitrios Kontos and his crew of eight divers and eleven sailors stopped due to bad weather a few meters away from the rocky shore of South-eastern Antikythera island. There, by incredible chance, they discovered the wreck at a depth of about 50 meters. From late November 1900 to September 1901, the same sponge divers, at the service of the Greek Ministry of Education and Religious Affairs and with the support of the Greek Royal Navy, conducted the first organised underwater archaeological campaign ever, attempting to salvage the precious cargo of the ship. About forty statutes, precious glass items, luxurious beds and other expensive objects were discovered together with pieces of the ship and other things not worthy to be noticed by the archaeologists of the time, who were mainly interested in art. The Greek and international press followed the salvage operation day by day, and the most praised finding was the beautiful statute called the Antikythera ephebe. All the findings were transported at the National Archaeological Museum in Athens.

On Tuesday May 21, 1902,[1] the Athenian journal *Asty* published a small front-page article titled "Important Find at the National Archaeological Museum: the Discovery of a Bronze Plate with Inscriptions—the Things from Antikythera." The content of the article is as follows:

> Yesterday morning,[2] the former minister of Education Mr S. Stais, together with his wife and sister-in-law, paid a long visit to the Archaeological Museum. During his visit to the room of the Antikythera find, he carefully examined the useless pieces placed in a special location.
>
> Among these useless pieces he noticed three fragments of a bronze plate covered by a thick layer of bronze oxides and other material. On these pieces appear a series of interlocking gears belonging to a single system. On this plate he observed a long, carefully-written inscription of which only the verso was visible, the recto being covered by foreign material.
>
> This inscription will probably solve the question of the date of the shipwreck, about which much has been written in Greece and abroad, always based on suppositions. The fact that this was found by Mr S. Stais is a remarkable coincidence, since it was during his time as minister that the inestimably important antiquities of Antikythera that now adorn our Archaeological Museum were found.[3]

Two days later, on Tuesday May 23, 1902, the same journal published another article titled "The Inscribed Plate of Antikythera: What It Is. Yesterday's Studies." In

[1] Dates are given in the Julian calendar. Greece reformed its calendar as late as 1923, when February 16 became March 1.

[2] In fact Stais visited the museum on Saturday May 19 (see the article in the journal *Asty*, dated May 23, 1902).

[3] https://srv-web1.parliament.gr/display_doc.asp?item=46228&seg=37229

this article were mentioned the efforts of two archaeologists to decipher the visible writings on the fragments of the mysterious plate: the specialist of Ancient Greek coins, Ioannis Svoronos, and the director of the Austrian Archaeological School in Athens, Adolph Wilhelm, a specialist of Ancient Greek inscriptions,. The two specialists were able to decipher the following letters:

ΠΡΟΕΧΟΝ
ΦΕΡΕΙΟΝ
ΤΟΣΤΟΔΕ
ΤΗΣ ΑΦΡΟ (of Aphro..)
ΓΝΩΜΩ (gnomo..)
ΗΛΙΟΥ ΑΚΤΙΝΑ (Solar ray)

We read among other things in the article that:

> This inscription is written on a copper plate, very damaged, and it belongs to a meteoroscopical[4] device,—an astrolabe following Svoronos—which was used in the so-called Great Navigation during ancient times. This instrument, of a complicated mechanism, comprises gears of various dimensions, and it seems to be contained in a box, which when opened presents in one part the mechanism and, in the other, the inscription that is the user's manual. This is certified by the aforementioned words that have been read and are related to the instrument and to the zodiac circle.
>
> From the shape of the carefully-incised letters, Mr Wilhelm dates the plate to no earlier than the 2nd c. B.C., but from the shape of their lines they could be written as late as the 1st or 2nd century A.D. …

This first study of the fragments of the plate later called "the Antikythera Mechanism" would inaugurate more than a century of research drawing scholars from all over the world with the aim of understanding the Mechanism's functions and producing a reliable reconstruction of the device.

The first publications that presented the Mechanism to the scholarly public were those of Svoronos and of the naval officer Periklis Rediadis, who was on the ship *Mykali* that participated to the diving campaign of 1900–1901. Svoronos focused on the inscriptions, and Rediadis on the gears. From Svoronos' description, we know that the four main fragments (named by Svoronos A to D) were already identified. The common belief of Svoronos and Rediadis was that this "slab" was an astrolabe used for navigation.[5]

The next important step in research on the Mechanism was the visit to the Archaeological Museum of Athens by the German philologist and amateur astronomer, Albert Rehm. Intrigued by Rediadis' description of the slab, Rehm stopped in Athens in September 1905 to examine the object on his way from Germany to Miletus. He spent only four hours at the Museum and took photographs of the fragments.

---

[4] Meteoroscopical: "to watch the phenomena of the sky."

[5] Rediadis (1910) and Svoronos (1903 [1908]). The history of research on the Mechanism is presented in detail in Alexander Jones (2017: 16–46). Jones also gives the relevant bibliography in what is today the most complete and up-to-date book presenting the Antikythera Mechanism.

The fragments Rehm saw in 1905 were quite different than those described by Svoronos in his 1903 publication. The museum's conservators had cleaned the fragments, several of which were fragmented into further pieces. Much more of the gearing and letters were visible than in 1902. Moreover, Rehm being familiar with a kind of astronomical inscriptions called *parapegma*, he was able to identify similar inscriptions on fragment C. The *parapegma*, found carved in stones and written on papyrus, list stars and constellations that become visible on a certain day of the solar year, often together with meteorological "predictions." What is more, on a small fragment that resulted from the restoration procedure (now fragment 19), Rehm noticed three numbers, 19, 76 and 223, which he was able to identify with the Meton, Kalippus and Saros astronomical cycles, respectively.

Rehm never formally published his findings but instead communicated them in lectures and correspondence. He was convinced that the fragments were part of a planetarium. In his unpublished notebook he even draws a part of this planetarium comporting an epicycle (see Fig. 1).[6] Being a philologist, he knew Cicero's text about ancient planetaria and believed that the Mechanism belonged to this category.[7] Moreover, he had deciphered the word *στηριγμός*, meaning the turning point of a planet on the zodiac, which confirmed his hypothesis.

Rehm's planetarium hypothesis was followed by the next to study the fragments, the admiral of the Greek Navy, Ioannis Teophanides. After his retirement from the Navy in 1923, Theophanides worked on historical subjects, and in 1929 he contributed to the Greek *Great Military and Naval Encyclopaedia* with an article on Greek maritime technology wherein he presented the Mechanism as a navigational instrument.[8] In 1934 he published his interpretation of the Mechanism in French in the *Proceedings of the Academy of Athens*, and he constructed the first model of it. Disassembled, this model is now on display at the National Archaeological Museum in Athens alongside other historical and functional models. Theophanides believed that the Mechanism displayed the motions of the Sun, the Moon and four of the five planets known in antiquity. He also believed that this planetarium used stereographic projection to allow the user to read off the positions of the planets both with respect to the ecliptic and with respect to the celestial equator. In this, he was probably influenced by the contemporary beliefs that this device was an astrolabe (consisting of the stereographic projection of the celestial sphere) and that it was used as a navigation instrument.

World War II and the ensuing Greek Civil War of 1946–1949 brought research to a halt. When Greece was attacked, first by Italy and then by the Third Reich, the exhibits of the Archaeological Museum in Athens were boxed and buried to avoid destruction. In the 1950s, before their return to the rooms of the Museum, the fragments of the Mechanism that suffered from this hasty transportation were again

[6] A. Rehm, "Notizbuch" (unpublished notebook), Bayerische Staatsbibliothek, Rehmiana III/7, 1906.

[7] Cicero, *De Re Publica*, 1.23.

[8] Theophanides (1929).

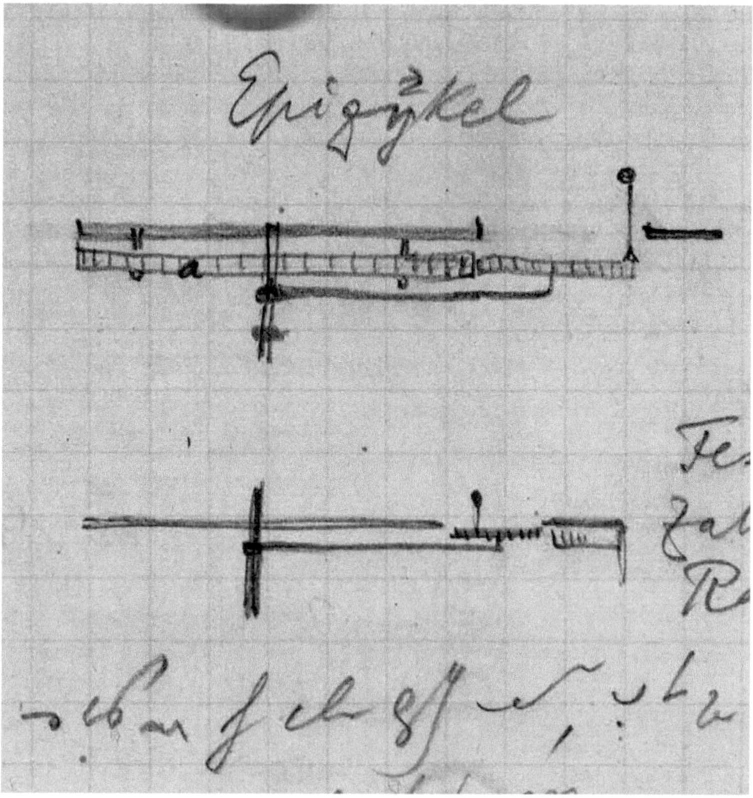

Fig. 1 Albert Rehm's notebook page with his drawing of a planetarium

restored. By then, the Mechanism was more fragmented and more gears were visible.

2 The First Radiographs

Derek de Solla Price (1922–1983) was an English-Argentinian physicist, and a specialist of metals and of the history of astronomical instruments. He was notably familiar with late medieval astronomical calculating instruments. He was made aware of the Antikythera Mechanism by the publications of Rediadis and Theophanides, and in order to study it further he succeeded in obtaining high-quality photographs of the restored fragments. Price realised the necessity of making an autopsy, and in 1958 he managed to obtain a grant from the American Philosophical Society in order to visit the Archaeological Museum in Athens. At that time, he held a temporary position at the School of Historical Studies at the Institute for Advanced Study in Princeton.

Price spent ten days during the summer of 1958 at the Museum, where he studied the fragments assisted by Georgios Stamires, a Greek epigraphist. Price realised that the Mechanism has a lot to do with calendars, and he was the first to understand that it was a computer. Back in the US, he raised public awareness about the device, describing his encounter with it as "like opening a pyramid and finding an atomic bomb," and stating that the Greeks who built the Mechanism "were not far behind where we are now."[9] In 1959 he published his findings in a front-page article in *Scientific American*, which bolstered interest in the Mechanism around the world.[10] The fact that it was the science fiction writer Arthur C. Clarke who encouraged the editor to invite Price to write an article is an example of the level to which Price brought the Mechanism into the public eye.[11]

Price's main contribution to the study of the Mechanism came later, when in 1971 he successfully petitioned the Greek Atomic Energy Commission to perform a radiographic analysis of the fragments. The fact that the Museum had not yet taken X-rays of the Mechanism by that point was due to the prohibitive amount of electricity required to penetrate such a thick piece of metal. By contrast, gamma-ray radiography, as applied to archaeological material in the late 1960s, could do the job with the electrical power available on site. Charalampos Karakalos, a physicist at the Greek National Center of Scientific Research "Demokritos" (founded in 1961 to conduct research on atomic energy), undertook the radiography of the major fragments, initially using gamma radiation from a weak source of the isotope Thulium-170. The resulting imagery was very poor until Karakalos managed to obtain X-ray images with the electric power available. The radiography of fragment A revealed 27 gears, of which six were completely unknown, and two only guessed at. Karakalos and his wife Emilia also estimated the number of the teeth on the gears.

Even with this imagery, Price's task to reconstruct the Mechanism was not easy. The radiograms did not fully resolve the question of which gears shared the same axle, nor which were interconnected, let alone the matter of missing gears. Moreover, extrapolation from gear fragments to determine the total number of their teeth was somehow uncertain. To enhance the results, Karakalos moved the X-ray source to obtain images from various angles, but even then the result was not absolutely clear.

Based on Karakalos' results, Price published the book *Gears from the Greeks* in 1974,[12] and he constructed a model now on display at the National Archaeological Museum in Athens (see Fig. 2). On the front, Price's model features a pointer showing the zodiacal (sidereal) motion of the Sun and another showing the zodiacal motion of the Moon. To arrive at this, he modified Karakalos' estimation of the number of teeth on the big gear of fragment A (which he rightly believed to be the

[9] Interview in *The Science News-Letter*, 75.3 (January 17, 1959), p. 36.

[10] Price (1959).

[11] Jones (2017: 27).

[12] Price (1974).

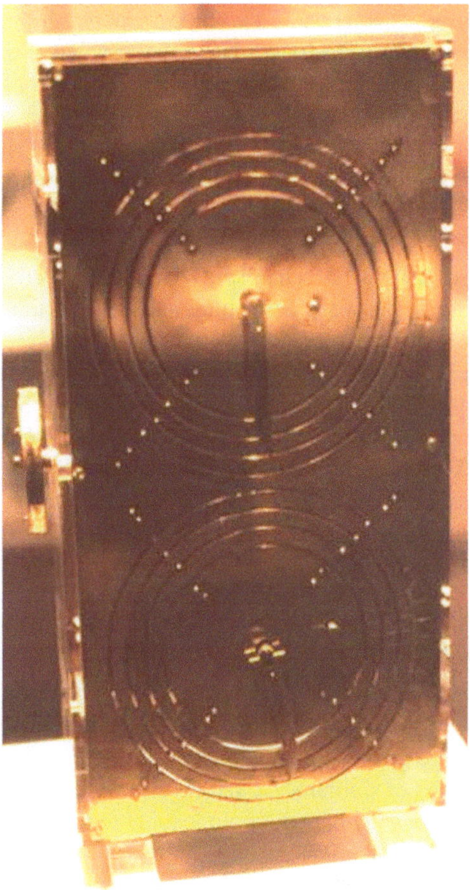

Fig. 2 Derek de Solla Price's model showing its interpretation of the back dials as concentric circles

Sun motion gear) from 128 to 127, obtaining 19 revolutions of the Sun (Metonic cycle) in 254 revolutions of the final gear of the train, which represents the total sidereal revolutions of the Moon during 19 sidereal revolutions of the Sun. On the back, Price understood that they were two dials resulting from two different gear trains. The challenge was to interpret the large gear visible on the reverse side of fragment A, which has two tiers of teeth and carries other, smaller gears. Price interpreted this gear as a "differential turntable," and his reconstruction of the back of the Mechanism comports two dials with concentric circles that display the lunar months and the lunar years. In fact, Price's reconstruction shows a Mechanism with a back much less sophisticated than its front.

Price's model was subsequently contested for its problematic gear ratio. From a mechanical point of view, if the small gears turn too fast compared to the big ones, the Mechanism would quickly break. The English mathematician Eric Christopher Zeeman criticized Price's ratio of Sun/Moon gearing, believing that it would be difficult to construct a working model that runs smoothly with such a high output ratio using the triangular teeth of ancient Greek gears.[13]

Derek de Solla Price was also known for his theory of the exponential growth of science, and this theory seems to apply to studies of the Antikythera Mechanism. The next important step in research after Price's reconstruction came from the collaboration between two physicists, the Englishman Michael Wright, the first curator at the Science Museum (London) and a specialist of ancient clockwork mechanisms, and the Australian Alan Bromley, the second historian of computing. Before studying the Mechanism, Wright had already encountered Greek astronomical gearing in his work describing and constructing a model of a Byzantine sundial with calendrical gearing.[14] Wright and Bromley spent a considerable amount of time examining the fragments at the National Archaeological Museum in Athens from 1988 to 1993. In order to reconstruct a 3D model of the interior gears, Wright applied the method of linear tomography. This method consists of a linearly moving X-ray source and a film plate moving linearly in the opposite direction. The result, as the plane of the image remains halfway between the source and the plate, is that the plane is clear, and the rest of the image is blurred. Wright and Bromley though that this method would allow them to determine the common axles and the interconnections between gears.

The results were exceptional. Sadly, while Wright continued their work, published their findings and constructing a new model of the Antikythera Mechanism, Alan Bromley passed away in 2002 at the age of 57 due to a chronic illness. Wright was later able to confirm Karakalos' hypothesis that one small gear lying on the large gear visible on the reverse of fragment A was in fact a stacked pair of distinct gears of similar size, and he discovered a second such pair. He also determined the missing gears in the train leading to the upper back dial. Familiar with mechanisms with gears, Wright took into serious consideration the functionality of the gear ratios. He demonstrated that the large back gear is not a differential gear, and he made an accurate count of the number of teeth on every gear.

Wright's greatest modification of Price's model concerned the dials on the back. By precise measurement, Wright demonstrated that the two dials did not comport concentric rings or bands of metal (*annuli*), as previous researchers believed, but that they were spirals, the upper one of five revolutions, and the lower of four. As Price had already noticed, these rings were divided into cells bearing inscriptions. Wright was unable to read the engravings of the cells of the upper dial, but he demonstrated that these cells represented the 235 lunar months of a Metonic cycle and, accordingly, that this dial was a lunisolar calendar.

As for the lower dial, he (mistakenly) hypothesized based on the data available at the time that it displays the latitude of the Moon, which determines the alignment of the Sun, Moon and Earth and, therefore, eclipses. As for the front dial, lastly, Wright later discovered a "pin and slot" device for the motion of the Moon: the axle

[13] Zeeman (1986).

[14] Field and Wright (1985).

of the small gear of the Moon is not fixed on the larger gear but can move in a slot, thus rendering the inequality of lunar motion with less gears.[15]

Up to this point, Wright's model was based solely on extant pieces and the parts obviously missing from the extant gear trains. However, Wright's complete reconstruction would include hypothetical gearing for all five planets known in antiquity based on the principle of a bearing circle and an epicycle. Wright based this hypothesis on the fact that the word Venus (Aphrodite) appears on the inscriptions, that the mean Sun motion wheel (the large gear of fragment A, recto) is too big for the needs of the existing gearing, that all the gears are behind this wheel, and that there are protuberances on the wheel that obviously bore something else. Wright's planetary hypothesis is also based on the fact that they are mentions of *planetaria* in Ancient literature, but not of devices displaying only the Sun and Moon.

Wright has written extensively on the Mechanism and constructed a working model using the same tools than those used in antiquity.[16] His only anachronism was intentional: he assembled the mechanism with screws in order to disassemble it for demonstration and modification. Modified according to the findings of the Antikythera Mechanism Research Project in 2005, this model is now on display at the National Archaeological Museum in Athens (see Fig. 3).

Fig. 3 Michael Wright's model with the hypothetic planetary pointers

[15] Wright (2005).

[16] Some publications of Wright (2002, 2003, 2005, 2013).

3 The Latest Research: Bladerunner and Sophisticated Surface Imaging

The exponential growth of interest in the Mechanism and in forming an interdisciplinary team to conduct further investigation lead in 2005 to the creation of the Antikyhera Mechanism Research Project (AMRP). The AMRP was the brainchild of Mike Edmunds, an astrophysicist at Cardiff University, Tony Freeth, a mathematician and filmmaker, and John Seiradakis, an astronomer at Thessaloniki University. In the year 2000, Edmunds had already published an article with the pithy title "The Antikythera Mechanism: Still a Mystery of Greek Astronomy,"[17] and Freeth had published two such articles in 2002: "The Antikythera Mechanism: 1. Challenging the Classic Research" and "The Antikythera Mechanism: 2. Is it Posidonius' Orrery?"[18] The other founding members of the initial AMRP academic team were all from Greece: the astrophysicist Xenophon Moussas (University of Athens), the physicist and historian of science Yanis Bitsakis (National Hellenic Research Foundation) and the philologist and palaeographer Agamemnon Tselikas (National Bank of Greece Cultural Foundation). The academic team worked closely with the team at the National Archaeological Museum, the chemist Dr. Eleni Magou, the archeologist-museologist Mary Zafeiropoulou and a number of conservators led by Gerassimos Makris.[19] The project was conducted under the aegis of the Greek Ministry of Culture and was initially funded by the Leverhulme Trust and, subsequently, by Greek foundations.

The AMRP initiated the last archaeometric analysis of the Mechanism, which took place in September–October 2005 and involved unpreceded technical means.

In September 2005, Hewlett-Packard's Mobile and Media Systems Laboratory sent a special imaging system operated by three specialized scientists (Tom Malzbender, Dan Gelb and Bill Ambrisco), the machine consisting of a dome that surrounds the sample under examination and takes a series of still photos to analyse the three-dimensional structure of the surface.[20] The next step was the most important: in October 2005, the Museum welcomed another team of specialists from X-Tek Systems and led by the company's owner, Roger Hadland. They used the prototype of a very powerful new X-ray machine, the eight-tonne "Bladerunner," which was originally designed to search for minute cracks in turbine blades. This machine gave astonishingly detailed three-dimensional images of the Mechanism at a resolution better than a tenth of a millimetre, using the latest "microfocus" X-ray techniques.

Together with the surface 3D imaging, the new X-ray tomography doubled the number of letters that could be read off the Mechanism, especially the mirror

[17] Edmunds and Morgan (2000).

[18] Freeth (2002a, 2002b).

[19] http://www.antikythera-mechanism.gr/project/team

[20] Images of this procedure can be found at: http://www.antikythera-mechanism.gr/data/ptm/full-resolution-ptm

writing noted by Svoronos in 1902 (i.e., the imprint of text from plates that had not survived). The AMRP team published a first collaborative article presenting the initial archaeometric results in *Nature*.[21] The team confirmed Wright's reproduction of the upper back dial (235 Metonic cycle months) and that the Mechanism likewise provided the Calippus cycle (Meton's cycle multiplied by four, which gives a better approximation of the correspondence between the solar and lunar calendar every 76 years). However, the main breakthrough for the reconstruction of the back of the Mechanism was the discovery that the lower back dial was divided in 223 cells, representing the Saros eclipse cycle. The AMRP team provided astonishingly high-quality images of the writing in these cells indicating the kind of the eclipse (solar or lunar), its date and also the hour. It has been established that the lower back dial comprised a smaller one displaying the 54-year exeligmos cycle, used to predict successive eclipses with similar properties and location. As to the front, the AMRP confirmed the existence of a Lunar and a Solar pointer, that the ring of the calendar could be manually turned to accommodate the gradual shift of the Egyptian calendar year relative to the seasons, as well as the existence of the pin and slot device. As for the back dial pointers, AMRP imaging revealed the record-player-like shape of their heads, which allows them to follow the engraved path of the spiral.

In 2006, the AMRP began collaborating with Alexander Jones (Institute for the Study of the Ancient World, New York) and John Steele (Brown University). It also published its definitive reconstruction of the gearing (see Fig. 4), which has won the consensus of specialists such as Michael Wright, who has since modified his own model to demonstrate the functionality of the new gearing. In 2007, the team was able to decipher the inscriptions on a small dial of about 2 cm in diameter contained in the upper back spiral dial, which was believed by Wright to convert from the

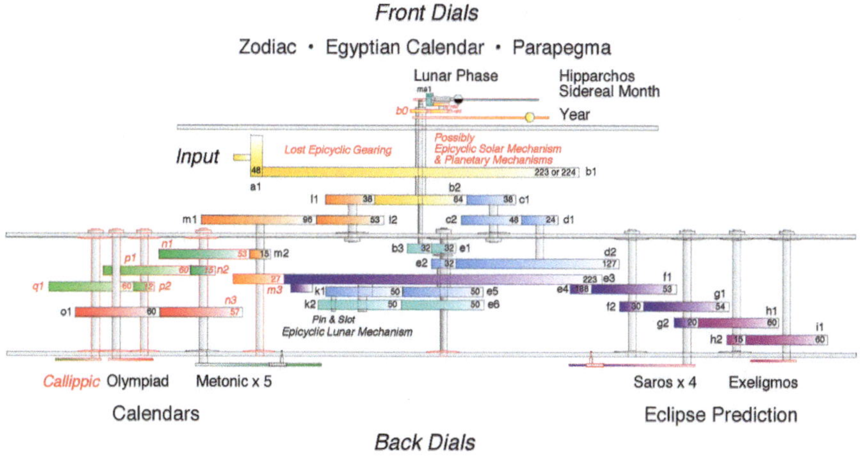

Fig. 4 The gearing as defined by AMRP team

[21] Freeth et al. (2006).

Metonic to the Calippus cycle. To their surprise, they found the names of the four main Hellenic games: the Olympics, the Nemeans, the Pytheans and the Isthmians. This four-year cycle was thus identified as the "games dial" and understood to indicate the year that these games took place. In addition, the researchers found the names of another, minor games—the games of Naa in Epirus—and the classicist Paul Iversen (Case Western Reserve University) later found mention of Halieia, the Rhodian games.[22] Alexander Jones has argued that the inclusion of these two minor games may solve the question of the maker and the owner of the Mechanism. According to Jones (2017: 87–94), the former inscribed the games of his native town, Rhodes, and the latter asked that the games of his own region be likewised inscribed upon the Mechanism (that is, of Dodona in Epirus). The fact that the month names on the Metonic calendar match those used in Epirus corroborates Jones' hypothesis.

The next important presentation of new findings came in 2016 with the publication of a special issue of the journal *Almagest* devoted to the decipherment of the Mechanism's inscriptions.[23] The new photos and transcriptions revealed not only text mentioning the Sun, Moon and their motions, but also the planets and their motions. The individual, specialised articles provide detailed information and interpretations of inscriptions that concern, in order, the front dial and *parapegma*, the back dial and back plate, the front cover, and the back cover. These results are summarised in the issue's preface, where the nineteen members of the enlarged AMRP project affirm the following:

> *Front dial:*
> Most of the square plate was occupied by a large circular dial surrounded by two graduated scale rings. The Greek names of the signs of the zodiac were inscribed on the inner Zodiac Scale and the Greek names for Egyptian months on the outer Egyptian Calendar Scale, which was manually moveable to accommodate the gradual shift of the Egyptian calendar year relative to the seasons. Pointers revolving around the dial represented the motions around the zodiac of the Sun, the Moon, and the five planets known in antiquity, as well as the date in the Egyptian calendar year. Short texts (single words and letters) were inscribed on the dial rings. The remains of these texts, collectively designated *Front Dial Inscriptions*, are all in Fragment C.[24]
>
> *Front cover:*
> A long text, the *Front Cover Inscription* (FCI), was inscribed on the Front Cover Plate. Its remains, which were attached to Fragment C as it was discovered in 1902, now comprise Fragment G and numerous small fragments (21, 23, 26, 27, 29, 37–44, 49, 54–56, and 60). The text describes the synodic cycles of motion of the five planets relative to the Sun and around the zodiac.[25]
>
> *Back cover:*
> The inscriptions served several distinct functions. The dial inscriptions gave the positions of the pointers immediate meaning, in terms of astronomy and systems of time-reckoning.

---

[22] Iversen (2013).

[23] "The Inscriptions of the Antikythera Mechanism," *Almagest* 7–1 (2016).

[24] Allen et al. (2016: 21).

[25] Allen et al. (2016: 31).

> The pointer positions predicted recurring astronomical and terrestrial events, e.g. eclipses, planetary positions and athletic competitions. Back Plate Inscription and Parapegma Inscription augment the information on astronomical phenomena predicted by the dial displays. The Front Cover Inscription is a more general description of the behavior of the planets displayed by pointers on the front dial, while the Back Cover Inscription helps the viewer to identify the meaning of the displays on both faces of the Mechanism.[26]

Therefore, the readings confirm Wright's earlier hypothesis of the existence of a planetary gearing on the Mechanism. In 2016, the only question that seemed unresolved was the dating of the instrument.

The discovery of the Saros dial and its glyphs indicating the dates of eclipses may solve this question if we compare these dates with historical eclipses. Of all possible matches, the best is between the dial's first cell and the eclipse of May 12, 205 B.C. If we take this as the Mechanism's computational starting date, the Saros dial would cover the period from 205 to 187 B.C.[27] However, this dating seems too early if we compare the many similarities between the Mechanism with Geminos' (first cent. B.C.) astronomy, as pointed out by Alexander Jones.[28]

4 New Questions and New Diving Campaign

The new data and their publication have breathed surprising new life into research on the Mechanism. Thought to have been definitively settled, the details of the gearing are once again up for debate; the dating has yet to be resolved; the maker and the owner continue to be subjects of investigation; and most importantly, despite the affirmations of the AMRP team, the existence of the planet gears is being challenged.

In an article in *Nature Astronomy* in 2018 summarizing our present knowledge about the Mechanism, John Hugh Seiradakis (1947–2020) and Mike Edmunds present an important unsolved problem:

> A particular fragment of the Mechanism (fragment D) remains a mystery. It contains a well-preserved gear with 63 teeth—perhaps suitable for a Mercury or other planetary mechanism—and a strange, bean-shaped bronze piece. No reconstruction has included this in a satisfactory way, although remarkably the bronze piece has a shape that would enable it to act as a cam in a mechanical arrangement that would display the effects of 'the equation of time'—the difference between time as measured by a sundial and a mean solar time, essentially the solar anomaly recognized by Ptolemy.[29]

The discussion about the fragment D had begun in 2012, in an article co-authored by Alexander Jones and Tony Freeth, which suggested that the mechanical elements contained in the fragment could belong to an epicyclic system for representing the

---

[26] Allen et al. (2016: 23).

[27] Carman and Evans (2014) and Freeth (2014).

[28] Jones (2017: 238ff.).

[29] Seiradakis and Edmunds (2018).

Fig. 5 Freeth et al. hypothesis of the display of the planets

motion of Venus.[30] This hypothesis reappeared in 2021, in a paper by Tony Freeth et his collaborators.[31] This paper proposed a reconstruction of the front display surface of the Mechanism with moving concentric rings, each of which bears a globe symbolizing a planet. This fascinating hypothesis was made in 2011 by Alexander Jones and rejected by Tony Freeth.[32] Since then, Freeth has apparently revised his original view, so it is likely that this system of concentric rings, which closely resemble the representation of the Cosmos in old texts (see Fig. 5), will be accepted for the time being by the majority of Mechanism researchers.

According to a common belief, the planetary gearing would have been placed over the mean Sun wheel of fragment A. The axle of this wheel might have born the necessary gears for the planets as in Wright's reconstruction.[33] In 2018, however, an article published in the *Journal for the History of Astronomy* by Aristeidis Voulgaris (who is constructing working models of the Mechanism), Christophoros Mouratidis and Andreas Vossinakis claimed that this axle cannot extend more than 10 mm beyond the wheel, while it should extend at least 70 mm in order to bear the additional planetary gearing.[34] Based on the CT images of the AMRP, the authors detected a conical hole on the top of the mean Sun wheel's axle that served to fix the

[30] http://dlib.nyu.edu/awdl/isaw/isaw-papers/4/

[31] Freeth et al. (2021)

[32] See A. Jones' collection of research notes 'The "Cosmos" of the Antikythera Mechanism' at https://archive.nyu.edu/handle/2451/61729

[33] See the virtual reconstruction by Mogi Vicentini at: http://www.antikythera-mechanism.gr/data/models/computer-models

[34] Voulgaris et al. (2018).

original metal piece on a bow lathe in order to construct the axle. Indeed, ancient artisans used to make a hole with a conical drill on both sides of a piece of metal to fix it on a bow lathe. By moving the bow back and forth with the leather cord with this device, the craftsman rotated the piece, and he shaped it with an abrasion tool. By measuring the shape and dimensions of the conical hole, one can determine the maximum length of the missing metal above the hole (the base of the cone cannot exceed the diameter of the axle). In fact, this conical hole had already been noticed by M. Wright and A. Bromley. Wright believes that it is due either to corrosion or to an earlier investigator who touched a pointed instrument to the corroded and friable fragments (for instance, someone might have sought to plant the point of a pair of compasses on each of the axles in question in an attempt to take measurements, as in gauging centre-distances).[35] On the other hand, concerning the fragment D discussed above, Aristides Voulgaris' group believes 'that this fragment is part of a mechanism which shows the Moon's movement in latitude at the side of the mechanism. In this case the mechanism is moving directly from the front gear and not from the side as the majority of researchers believe.

In 2010, the renewed interest in the Mechanism resulted in the first important exhibition on the Mechanism at the National Hellenic Research Foundation in Athens and, in 2012, to the first exhibition of all the objects from the Antikythera wreck at the National Archaeological Museum. It also led to the organisation of a new diving campaign.

After the campaign of 1900–1901, Jacques Cousteau and Frédéric Dumas made brief dives at the site of the shipwreck in 1953, and Cousteau returned in 1976 in a collaboration with the Greek Institute of Marine Archaeology. Cousteau found coins, a statuette and other small objects. The new diving campaign of 2014–2019 involved a multidisciplinary team composed of specialised divers, archaeologists, engineers and historians of science and technology under the supervision of the Greek Ephorate of Underwater Antiquities (which belongs to the Ministry of Culture) and the sponsorship of the Swiss watchmaker Hublot and the Aikaterini Laskaridis Foundation. The first aim of this campaign was to relocate and produce a complete map the site using a remote-controlled underwater camera. Following this mapping, the area was divided into a grid, and teams of divers operating in threes used metal detectors and ultraviolet lighting to begin a systematic search of each square within the grid. They then used a sand pump to collect small pieces hidden under the accumulation of material now covering most of the shipwreck.

Though the teams relied on modern diving techniques, one notes that due to the depth of the site at approximately 45–55 meters, the divers had to use a special gas in their tanks and could not work for more than 45 minutes per day. Another obstacle to the research of what remains of the shipwreck is the short periods of the yearly diving campaigns. During the five years from 2014 to 2019, it was impossible to organise campaigns any longer than fifteen days per year, and none could take

[35] Private communication by M.T. Wright, January 12, 2020.

place in all of 2018. This was due to various factors such as funding, autorisation, and weather conditions in this part of the island. Last but not least is that parts of the shipwreck are now under huge rocks that have fallen there over the last two millennia because of earthquakes and other geological phenomena. This is why Mathias Buttet, Research and Development Director of the project's sponsor, Hublot, decided to develop an underwater drone, to be operational in 2020, which will assist the divers in the digging phase and provide various media and analytical tools for the research and documentation of the shipwreck.

Despite these challenges, we have recovered a number of small objects as well as part of a dress and an arm belonging to one or more bronze statutes. Moreover, we seem to have spotted the general location of the passenger cabins. If the theory that the Mechanism would have been transported in a cabin is correct, this is the place where the divers should concentrate their efforts in a future campaign.

From all the small pieces of metal found in this five-year campaign, one is of particular interest as concerns the Mechanism. On Sunday September 10, 2017, the divers found a round metal object matching the dimensions of the mean Sun wheel of fragment A (see Fig. 6). The X-ray image of this object reveals that it is decorative and bears the figure of a bull, which usually symbolizes Jupiter. Immediately, the question arose as to whether this piece is a decorative cover meant to fit over the mean Sun wheel on the front of the Mechanism, and whether it might explain the protuberances on that wheel. If this turns out to be true, then we can be certain that the Mechanism did not feature any planetary gearing, displaying only the zodiacal motions of the Sun and Moon. In all likelihood, however, this piece probably has nothing to do with the Mechanism and is instead part of another object carried by the ship.

In 2021, the excavation permit was renewed for five years. The archaeologist Angeliki Simosis is in charge of the excavation and the diver-archaeologist

Fig. 6 Left: Fragment A. Right: Piece found September 10, 2017

Alexandros Sotiriou is in charge of the divers. There have already been two missions, in May–June 2021 and 2022. In 2021, after the removal of a large rock that had covered part of the wreck, several objects were recovered, including the head of a statue that had been found in 1901 and a base of a currently unknown statue. Dating the shells that crashed with the fall of the rock, this rock must have fallen on the wreck around AD 100. In 2022, advanced archaeometric methods were used to determine the level of the wreck in relation to the various sediments that have since been deposited. Investigations continue.

Bibliography

Allen, M., et al. 2016. General Preface to the Inscriptions. *Almagest* 7(1): 5–35.

Carman, C. C., and Evans, J. 2014. On the Epoch of the Antikythera Mechanism and its Eclipse Predictor. *Archive for History of Exact Sciences* 68: 693–774.

Edmunds, M. G., and Morgan, P. 2000. The Antikythera Mechanism: Still a Mystery of Greek Astronomy. *Astronomy & Geophysics* 41: 10-17.

Field, J. V., & Wright, M. T. 1985. Gears from the Byzantines: A Portable Sundial with Calendrical Gearing. *Annals of Science* 42(2): 87–138.

Freeth, T. 2002a. The Antikythera Mechanism: 1. Challenging the Classic Research. *Mediterranean Archaeology & Archaeometry* 2(1): 21–35.

Freeth, T. 2002b. The Antikythera Mechanism: 2. Is it Posidonius' Orrery? *Mediterranean Archaeology & Archaeometry* 2(2): 45–58.

Freeth, T. 2014. Eclipse Prediction on the Ancient Greek Astronomical Calculating Machine Known as the Antikythera Mechanism. *PLOS ONE* 9(7): e103275.

Freeth, T., et al. 2006. Decoding the Ancient Greek Astronomical Calculator Known as the Antikythera Mechanism. *Nature* 444(7119): 587–591.

Freeth et al. 2021, A Model of the Cosmos in the Ancient Greek Antikythera Mechanism, *Scientific Reports*, v. 11, no aet. 5821.

Iversen, P. 2013. The Antikythera Mechanism and Rhodes. Paper presented at the workshop *The Antikythera Mechanism: Science and Innovation in the Ancient World*, Lorentz Center, Leiden, June 17–21, 2013.

Jones, A. 2017. *A Portable Cosmos: Revealing the Antikythera Mechanism, Scientific Wonder of the Ancient World.* Oxford: Oxford UP.

Price, D. J. de Solla. 1959. An Ancient Greek Computer. *Scientific American* 201(6): 60–67.

Price, D. J. de Solla. 1974. *Gears from the Greeks*. Transactions of the American Philosophical Society N.S. 64.7. Philadelphia: American Philosophical Society.

Rediadis, P. 1910. Τὸ ἐξ Ἀντικυθήρων στρολάβον (The Astrolabe of Antikythera). *Ἀρχαιολογική εφημερίς* 10: 158–172

Seiradakis, J. H., and Edmunds, M. G. 2018. Our Current Knowledge of the Antikythera Mechanism. *Nature Astronomy* 2: 35–42.

Svoronos, I. 1903 [1908]. *Die Funde von Antikythera*. Athens: Beck & Barth.

Theophanides, I. 1929. Ἀγίου Παύλου (πλοῦς) (Navigation of Saint Paul). In *Μεγάλη Στρατιωτικὴ καὶ Ναυτικὴ Ἐγκυκλοπαίδεια* (Great Military and Naval Encyclopedia), 83–96 (pp. 89–96 are misnumbered as 97–104). Athens: Megali Stratiotiki kai Nautiki Egkyklopaideia.

Voulgaris, A., Mouratidis, C., and Vossinakis, A. 2018. Conclusions from the Functional Reconstruction of the Antikythera Mechanism. *Journal for the History of Astronomy* 49(2): 216–238.

Wright, M. T. 2002. A Planetarium Display for the Antikythera Mechanism. *Horological Journal* 144(5): 169–173.
Wright, M. T. 2003. Epicyclic Gearing and the Antikythera Mechanism: Part I. *Antiquarian Horology* 27(3): 270–279.
Wright, M. T. 2005. Epicyclic Gearing and the Antikythera Mechanism: Part II. *Antiquarian Horology* 29(1): 51–63.
Wright, M. T. 2013. The Antikythera Mechanism: Compound Gear Trains for Planetary Indications. *Almagest* 4(2): 4–31.
Zeeman, E. C. 1986. Gears from the Greeks. *Proceedings of the Royal Institution* 58: 137–156.

Thābit ibn Qurra's Handling of Indian Trigonometry

Mimura Taro

Abstract In this paper, I point out Thābit ibn Qurra's usage of Indian astronomical and mathematical knowledge in his works, showing that he learned very well, not only Greek mathematics and astronomy, but also Indian astronomy and mathematics. Then, I elucidate, with recourse to his biographical information, why he knew Indian astronomy and mathematics. In al-Ma'mūn's days, scholars in the court became interested in Greek science, and owing to their growing interest, those who were engaged in it competed with each other in their knowledge for positions at court. As a result, they tried to expand their research groups by recruiting people in order to increase their knowledge of Greek science. This expansion gave promising persons the opportunity to build connections with prominent scholars at court, and, for Thābit, the chance to learn Greek science at the Banū Mūsā's home. In the course of working with the Banū Mūsā family, he rapidly amplified his knowledge of Greek science, using the abundant resources that they had collected, eventually becoming the top scholar in the field. Since Indian astronomy and mathematics were also necessary knowledge for scholars engaging in Greek science in Baghdad, Thābit learned them also with the family. Therefore, there is no surprise that he mastered not only Greek astronomy and mathematics but also their Indian counterparts.

Keywords Thābit ibn Qurra · Indian astronomy · Bīrūnī

1 Introduction

Thābit ibn Qurra (d. 901) was one of the notable scholars studying Greek mathematics and astronomy in the Abbasid dynasty.[1] He is especially known for his translations of important Greek scientific works. For example, he revised Isḥāq ibn

[1] On the life and times of Thābit ibn Qurra, see Rashed (2012: 113–122).

Mimura Taro (✉)
University of Tokyo, Meguro-ku, Tokyo, Japan
e-mail: taromimura@g.ecc.u-tokyo.ac.jp

D. P. Morgan, T. Quan (eds.), *Mathematics and Astronomy in the Ancient World*,
https://doi.org/10.1007/978-3-031-93809-2_13

Ḥunayn's (ca. 830–910/911) Arabic translations of Euclid's *Elements* and Ptolemy's *Almagest*, and his would become the standard Arabic versions of these texts in the Islamicate world.[2] In addition, Thābit also wrote summaries of such Greek works as the *Almagest* and Nicomachus' *Introduction to Arithmetic*, and he composed original works based on his knowledge of Greek science, including *On the Measurement of the Conic Section Called Parabola* and *Treatise on the Motion of Two Luminous Bodies [i.e., the Sun and the Moon]*.

As we know from a variety of historical documents, Thābit ibn Qurra was writing at a time of growing interest in Greek science among Abbasid scholars that had begun under the seventh caliph, al-Ma'mūn (r. 813–833). Consider Ḥabash al-Ḥāsib (d. after 869), an astrologer active in al-Ma'mūn's time and known to have composed several works on astronomy.[3] Among these, in the introduction to what is one of the earliest extant *zīj*es in Arabic, Ḥabash recounts the history of astronomy to his day and the place of Ptolemy at al-Ma'mūn's court[4]:

> Al-Ma'mūn had outstanding knowledge and exhibited a deep interest in the subtleties of things and the mysteries of sciences, and he had especially a passionate devotion to the science of stars. So, he compared that which was found by the Greeks' books in the Canon [i.e., the *Almagest*, see below] and others, that which was found by the Indians in the book of the *Sindhind*[5] and the *Arkand*,[6] and that which was found by the Persians in the *Shāh Zīj*[7] and others. Then, he saw that they were not in agreement with each other, and he found that each one of them sometimes agreed with what is correct, but sometimes diverged from the method of the truth. Since he became aware of that [matter], he ordered Yahyā ibn Abī Manṣūr al-Ḥāsib to go back to the principle of the books on stars, and to gather the scholars on this art and the sages of his time in order to cooperate with each other for investigating the principles of this science and for attempting to correct it, since Claudius[8] Ptolemy already demonstrated that the comprehension of the art on stars whose knowledge was sought was not impossible. Then, Yahyā obeyed that which he ordered him [to do], and he gathered the scholars of this art and the sages who were esteemed at that time. Next, he and they went back to the principles of these books, and examined them carefully, and compared what they found described in them [i.e., these books]. As a result, they saw that among all the books, there was nothing more correct than Claudius Ptolemy's book entitled *Almagest*; for Ptolemy already showed in it the attainment to the truth of the correctness by clear reasoning and geometrical proofs, and he said that he had

[2] Note that Grupe (2012) identifies another Arabic version composed by Thābit.

[3] On the life and times of Ḥabash al-Ḥāsib, see Kennedy et al. (1999: 8–11).

[4] A *zīj* is an astronomical handbook containing solutions to astronomical problems as well as tables. Ḥabash's *zīj* is preserved in two manuscripts: Istanbul, Süleymaniye Library, Yeni Cami MS 784 and Berlin, Staatsbibliothek zu Berlin, Wetzstein I MS 90. Of the two, the Istanbul MS has a longer introduction, only the first half of which is found in the Berlin MS. For the Arabic text and an English translation of the Istanbul MS introduction, see Sayılı (1995).

[5] The *Sindhind* (from the Sanskrit word *Siddhānta*) is a genre of Indian astronomical books collecting solutions to astronomical problems; on the *Siddhānta*s, see Plofker (2009: 66–70). The contents of the *Sindhind* are discussed below.

[6] According to Bīrūnī, the *Arkand* was a translation of Brahmagupta's *Khandakhādyaka*; see Haddad et al. (1981: 207–211).

[7] On this Persian *zīj*, see Van Bladel (2014: 262).

[8] Here and below, "Claudius" is read *qalūdhī*, versus *falūdhī* in Sayılı's text.

observed the motions of the Sun, the Moon, and the planets in their positions of the celestial sphere, and examined them in all the conditions of them, and [he said] that the observations and examinations led him to an error and a defect happening in the observations by the people who adjusted the planets before him, and [he said] that he corrected this error completely and defect of which the observations and examinations informed him, and next he established the positions of the planets in this book of his [i.e., the *Almagest*] according to what was found by the observations and measurements in his time after correcting and improving that. Therefore, they placed that book in the position of a canon for themselves, and next they used the instruments with which observations were made, such as the armillary sphere, and so on, and followed what Ptolemy described concerning the observations, and examined the motions of the Sun, the Moon, and the planets on different occasions in Baghdad.[9]

In short, al-Ma'mūn probed the very principles upon which different astronomical traditions were built, and the result was that he and his scholars accepted the *Almagest* as their astronomical research model (i.e., "the canon").

The scholars at al-Ma'mūn's court were equally interested in Greek mathematics, medicine, and philosophy, out of which came the Arabic translations of such works as Euclid's *Elements*, the *Almagest*, and Galen's books.[10] The most prominent members of this movement were three brother known by the Banū Mūsā brothers (Muḥammad [d. 873], Aḥmad, and Ḥasan) and al-Kindī (ca. 800-after 670), whose flurry of translations lend to the impression that they were competing with one another for authority over the Greek tradition.

The sons of al-Ma'mūn's astrologer (*munajjim*) and childhood companion, Mūsā ibn Shākir, the Banū Mūsā were among the highest-ranking and most prominent scholars at the seventh caliph's court. Best known for their Arabic translation of Apollonius' *Conics*, they contributed to a host of Greek sciences including mathematics, astronomy, and mechanics.[11] That theirs was the quest to come in first is seen in an episode reported by Ibn Abī Uṣaybi'a (1203–1270) concerning a power game between them and al-Kindī.

Abū Yūsuf Ya'qūb ibn Isḥāq al-Kindī, the first Arab philosopher, was an author on a variety of subjects also including mathematics, astronomy, astrology, and medicine.[12] Al-Kindī is known to have organized a translation group comprised principally of Syriac Christians, called the "Kindī Circle," which tackled scientific and philosophical works including the so-called *Theology of Aristotle*. As recorded in Ibn Abī Uṣaybi'a's *Sources of Information about the Classes of Physician* (*Uyūn al-anbā' fī ṭabaqāt al-aṭibbā'*), the Banū Mūsā tricked al-Kindī into losing the trust of the tenth Abbasid caliph, al-Mutawakkil (rt. 822–861), and they tried to take his books away from him.[13] This would suggest that the Banū Mūsā and the Kindī

[9] Sayılı (1955: 148–9). Here and elsewhere all translations are the author's unless otherwise noted.

[10] See Gutas (1998).

[11] On the life and times of the Banū Mūsā, see Hill (1979: 3–6).

[12] On the life and times of al-Kindī, see Adamson (2007).

[13] On this episode, see Adamson and Pormann (2012: lxv–lxviii).

Circle were in aggressive competition and that the stakes were as much political as academic. It also reminds us that, since none of the power players could read Greek, their success depended as much upon language assistants as their access to original works.

Where al-Kindī hired Syriacs to this end, the Banū Mūsā brothers had Thābit ibn Qurra. As we read in his biography (below), Thābit was recruited by the Banū Mūsā for his language skills, to assist them in their study of Greek science, and he would go on from there to make his own name as an independent scholar. Although Thābit's achievement concerning Greek science in the Abbasid period is well known, we lack direct evidence for how he came to learn Greek scientific knowledge. In this paper, I will focus on his trigonometrical works, trying to elucidate how he learned the subject and how he utilized his knowledge of Greek science in his works.

2 Thābit ibn Qurra's Treatise on the Sector-Figure

One of Thābit ibn Qurra's most popular works on Greek mathematics is his *Treatise on the Sector-Figure* (*Risāla fī al-shakl al-qattā'*),[14] the "sector-figure" comprising spherical arcs for applying Menelaus' theorem as in Fig. 1. The *Treatise on Sector-Figure* would become the go-to handbook on Menelaus' theorem in the Islamicate world, seeing extensive use, for example, in Ibn al-Haytham's (d. ca. 1040) commentary on the *Almagest* and, later, multiple translations into Latin.[15]

At the beginning of the *Treatise on the Sector-Figure*, Thābit introduces two relationships of spherical chords on the sector-figure given by Ptolemy in the *Almagest* (Fig. 1; here "Cd 2 arc AE" denotes the chord of double the arc AE)[16]: that is,

Fig. 1 Diagram of the "sector-figure"

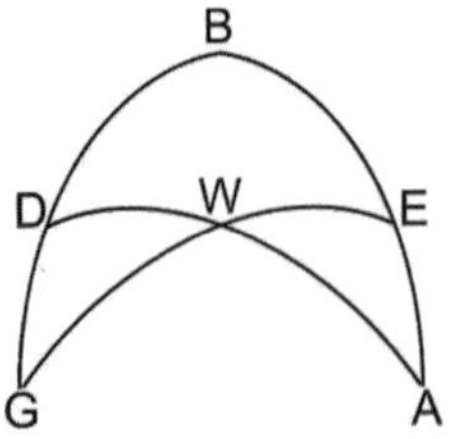

[14] See the critical edition with English translation in Lorch (2001: 42–113) and the French translation of Bellosta (2009: 365–390).

[15] Ibn al-Haytham's commentary to the *Almagest* is preserved in the manuscript Ahmet III (Istanbul) MS 3329, on which see Sabra (1998: 33–38). On the life and times of Ibn al-Haytham, see Langermann (2007). On the Latin translations of Thābit's *Treatise on the Sector-Figure*, see Lorch (2001: 30–37).

[16] Lorch (2001: 42–47, 154); cf. Bellosta (2009: 365–366).

$$Cd\,2\,arc\,AE\,/\,Cd\,2\,arc\,EB = Cd\,2\,arc\,AW\,/\,Cd\,2\,arc\,WD \cdot Cd\,2\,arc\,DG\,/\,Cd\,arc\,2\,GB, \text{and}$$
$$Cd\,2\,arc\,AB\,/\,Cd\,2\,arc\,BE = Cd\,2\,arc\,AD\,/\,Cd\,2\,arc\,DW \cdot Cd\,2\,arc\,WG\,/\,Cd\,2\,arc\,GE.$$

One notes that while Greek astronomical works like the *Almagest* used chords for calculating arc (e.g., chord AB for arc ACB, i.e., double the arc AC, in Fig. 2), post-transmission Indian scholars instead developed and opted for the sine (e.g., R sin arc AC = AH) in the interest of simplicity.[17] In addition to the sine, they produced the cosine (e.g., R cos arc AC = OH) and the versed-sine (e.g., R version arc AC = HC).[18] The fact that Thābit uses chords thus illustrates that he is well acquainted with the Greek way of expressing arcs.

After presenting these two relationships, Thābit notes that Ptolemy gives only part of their geometrical proofs.[19] As such, he devotes the first half of the *Treatise on the Sector-Figure* to extending Ptolemy's proofs. In its second half, he then goes on to explore all modes of expressing compound ratios concerning Menelaus' theorem. Where Thābit enumerates the modes, he does so in a table following a discussion of the theory of compound ratios.[20]

Fig. 2 Relationship between chords and sines

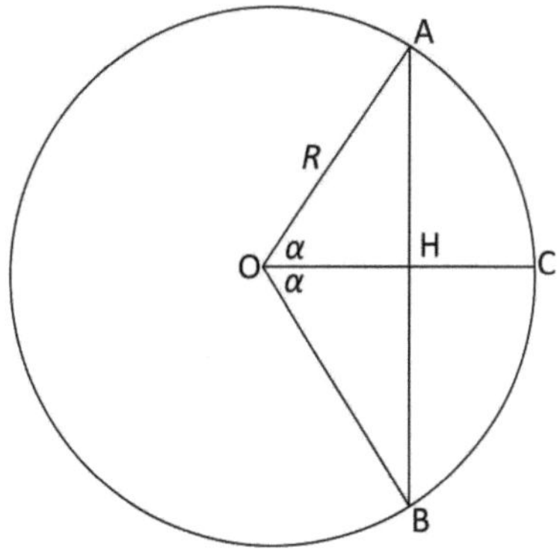

[17] Pre-modern trigonometrical quantities were determined in a particular circle called "the base circle." In this paper, I use "R" for the radius of the base circle. Thus, "the sine of arc AB" in pre-modern texts means R sin arc AB.

[18] See Van Brummelen (2009: 94–134). Note that, in India, spherical trigonometry only first saw use with the introduction of Islamic astronomy in the Mughal period.

[19] Lorch (2001: 46–47); cf. Bellosta (2009: 366).

[20] Lorch (2001: 86–111); cf. Bellosta (2009: 382–390). One notes that Thābit's interest here in the composition of rations through the sector-figure is probably the inspiration behind the separate work *Treatise on the Composition of Ratios* (*Kitāb fī ta'līf al-nisab*), edited and translated in Lorch (2001: 168–307) and Crozet (2009: 428–535).

Most of what Thābit introduces by way of propositions to complete Ptolemy's proofs for the two cases of the sector-figure are extensions of the geometrical proofs of the *Almagest* itself, treating the relationship between arcs on the surface of a sphere.[21] One notices, however, that the proof of one of the lemmas stands out in that it focuses on triangles inscribed in a sphere[22]:

> [If] from one of any two great circles occurring on the surface of a sphere, two arcs less than semicircles are cut off, which are adjacent to one of the two points of the intersection of the two [circles], and [if] from the ends of the two arcs two perpendiculars are drawn to the plane of the other circle, then the ratio of the chord of the double of one of the arcs to the chord of the double of the other arc is as the ratio of the perpendicular drawn from its end to the perpendicular drawn from the end of the other arc, whether the two arcs are together in one [and the same] direction or in two different directions.
>
> Example: two circles ABGD and AEGW [see Fig. 3],[23] which are great circles occurring on the surface of the sphere, intersect at A and G; and two arcs are cut off from circle AEGW, each of which are less than semicircle, namely [arcs] AE and AZ; and from two points E and Z two perpendiculars are drawn to the plane of circle ABGD.
>
> I say: the ratio of the chord of double the arc AE to the chord of double the arc AZ is as the ratio of the perpendicular that is drawn from point E to the perpendicular that is drawn from point Z.

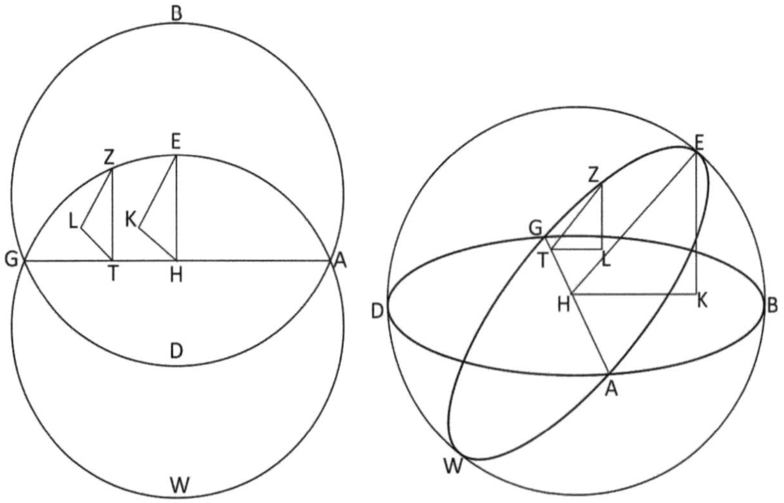

Fig. 3 Diagram drawn by Thābit (left) and the three-dimensional representation of it (right)

[21] On Ptolemy's proofs, see Toomer (1984: 64–69).

[22] Lorch (2001: 62–67); cf. Bellosta (2009: 372–373). In this paper, when I translate letters denoting points, I use Sabra's transcription system as described in Hogendijk (1985: 131): ا = A, ب = B, ج = G, د = D, ه = E, و = W, ز = Z, ح = H, ط = T, ي = I, ك = K, ل = L, م = M, ن = N, س = S, ع = O, ف = F, ص = C, ق = Q, ر = R, ش = J, ت = t, ث = h, خ = X, ذ = V, ض = d, ظ = Y, غ = g.

[23] Figure 3 (left) is the diagram drawn by Thābit. I add Figure 3 (right) for modern readers.

Proof: the common section of two circles ABGD and AEGW is a diameter of the two [circle], so let it be diameter AG. We draw from two points E and Z two perpendiculars to AG, which are EH and ZT. Then, if the two [perpendiculars] are also perpendicular to the plane of circle ABGD, what we wanted [to prove] is clear, because they are the sines (*jayb*) of two arcs AE and AZ; if they are not like that, then we draw from two points E and Z two perpendiculars to the plane of circle ABGD, which are EK and ZL, so the two [perpendiculars] are parallel to each other, and two lines EH and ZT are also parallel to each other. When two lines containing an angle are parallel to two other lines containing another angle, the two angles are equal to each other. So, angle HEK is equal to angle TZL; and two angle EKH and ZLT are right. So, triangles EHK ZTL are similar to each other. So, the ratio of EH to ZT is as the ratio of EK to ZL. But, the ratio of EH to ZT is as the ratio of the chord of double the arc AE to the chord of double the arc AZ, because they [i.e., lines EH and ZT] are the sines of them [i.e., arcs AE and AZ]. So, the ratio of the chord of double the arc AE to the chord of double the arc AZ is as the ratio of perpendicular EK to perpendicular ZL. Likewise, what we said will also be clear, if one of two arcs AE and AZ is in the direction of AW. That is what we wanted to prove.

Here, to prove the proposition that Cd 2 arc AE/Cd 2 arc AZ = EK/ZL, Thābit draws two lines EH and ZT perpendicular to AG, and then he considers the following two cases.

Case 1: When ET and ZT are perpendicular to circle ABGD (see Fig. 4),

$$\mathrm{ET} = \mathrm{EK} = \mathrm{R}\sin \mathrm{arc}\,\mathrm{AE}, \mathrm{and}\,\mathrm{ZT} = \mathrm{ZL} = \mathrm{R}\sin \mathrm{arc}\,\mathrm{AZ},$$

where we suppose the radius of a great circle = R, so the proposition is clear.

Case 2: When ET and ZT are not perpendicular to circle ABGD, we draw EK and ZT perpendicular to circle ABGD (see Fig. 3b). Since EK is parallel to ZL, and EH is parallel to ZT, angle HEK is equal to angle TZL. Given that angles EKH and ZLT are right, triangle EHK is similar to triangle ZTL. But, since EH = R sin arc AE, and ZT = R sin arc AZ,

$$\mathrm{EH} / \mathrm{ZT} = \mathrm{Cd}\,2\,\mathrm{arc}\,\mathrm{AE} / \mathrm{Cd}\,2\,\mathrm{arc}\,\mathrm{AZ}.$$

Fig. 4 Diagram for Case 1

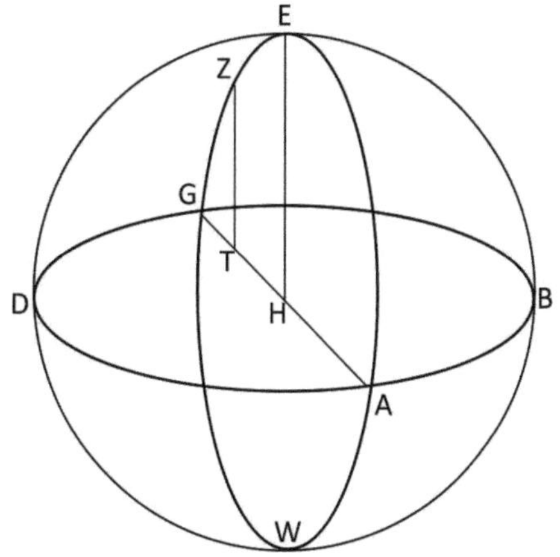

Therefore, Cd 2 arc AE/Cd 2 arc AZ = EH/ZT = EK/ZT. At the end, he notes that this proposition can be proved in the case when one of arcs AE and AZ is on semicircle AW.

Clearly, in this proof, Thābit uses triangles inscribed in the sphere. Like sines, this is alien to the *Almagest*, which handles arcs using only a chord attached to the surface. That Thābit uses both suggests that he is borrowing from Indian astronomy.

3 Inscribed Triangles in Indian Astronomy

The first work of Indian astronomy to contain such elements of Greek astronomy as the celestial sphere, the epicycle-eccentric theory, and plane trigonometry is Āryabhata's *Āryabhatīya* of 499 CE.[24] Marked by the geometrical representation of phenomena, this Greco-Indian hybrid then saw further development in the *Siddhānta*s, of which we will consider Brahmagupta's (598–670) *Brāhmasphuṭasiddhānta* and its method of determining the length of daylight.[25]

Let us first say a word about what we are reading. All *Siddhānta*s share the same basic table of contents: a chapter on planetary positions followed by a chapter on the celestial sphere.[26] They share the same practical orientation: the provision of concrete solutions to concrete astronomical problems (without a word to their rationale). Most strikingly, they also share the same rather unique form of verse (*sūtra*s). One notes that the reason for this lies in traditional Indian education. In India, knowledge was transmitted orally from master to disciple, the master intervening to explain a text only after the students had memorized its basic contents. In astronomy, the *Siddhānta*s are thus written the way they are for the sake of memorization, and the reasoning behind the solutions existed beyond the text, to be provided orally at a later date.[27]

Second, as illustrated in Fig. 5, to determine the length of daylight on a given day the *Siddhānta*s use the "day circle" that the Sun describes over said day and the "day arc" (arc BAC in Fig. 5) it makes from rise to set. The length of the day arc varies according to the Sun's position on the ecliptic. Regarding the length of the day arc, Indian scholars focused on the difference between half the day arc and a quarter of a circle (arc AB in Fig. 5), and they named the sine of this difference "the earth-sine."[28] One notes that scholars in the Islamicate world called the arc on the equator corresponding to this difference "the equation of daylight" (arc EF in Fig. 5).[29]

---

[24] See Pingree (1976).

[25] On the *Brāhmasphuṭasiddhānta*, see Plofker (2009: 319).

[26] See Ôhashi (1994).

[27] See Yano (2006).

[28] See Plofker (2009: 91–97).

[29] See Van Brummelen (2009: 176).

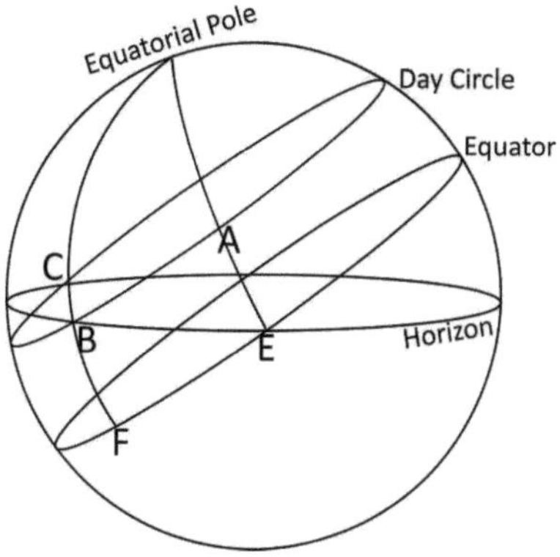

Fig. 5 Diagram for the day-circle

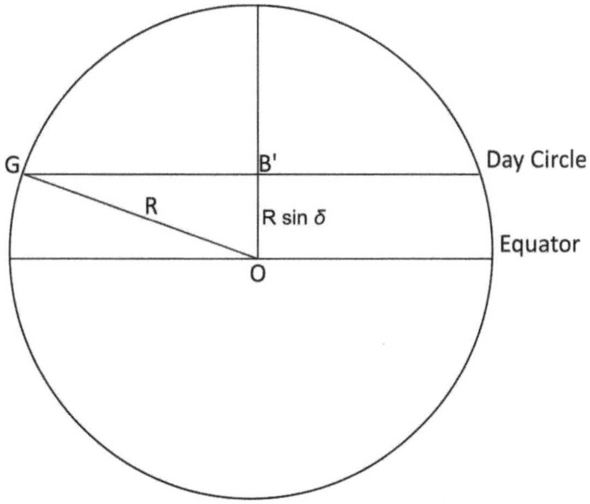

Fig. 6 Diagram for the day-circle and the equator

The following is Brahmagupta's solution for determining the length of daylight in *Brāhmasphuṭasiddhānta* 2.56–58:

> After the square of [the sine of] the given declination is subtracted from the square of the radius, the square root of the remainder is the radius of its own day circle north or south of the equinox. The sine of the declination, multiplied by the [noon] equinoctial shadow, divided by twelve, is the earth-sine, constant on its own day-circle. [The earth-sine] multiplied by the radius [and] divided by the half [-diameter] of its own day circle is the sine of decrease or increase [in day length].[30]

In this solution, Brahmagupta first calculates the radius of the day circle (see Fig. 6, which shows the day circle and equator projecting on the meridian):

[30] Translation modified from Plofker (2009: 93–94).

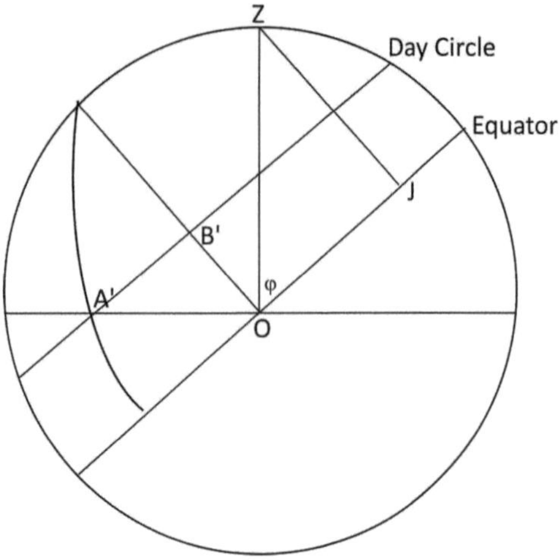

Fig. 7 Diagram for the day-circle and the equator

$$R_{day}\left(= GB' = \sqrt{GO^2 - OB'^2}\right) = \sqrt{R^2 - R\sin^2\delta},$$

where R_{day} denotes the radius of the day circle, and δ denotes the given declination of the day circle.

He then determines the earth-sine by

$$R\sin\delta \cdot s/12,$$

where *s* denotes the length of the noon equinoctial shadow of the Sun.

This determination can be explained as follows. As Fig. 7 shows,

$$\text{the earth sine} = A'B' = OB' \cdot ZJ/OJ = R\sin\delta \cdot \sin\varphi/\cos\varphi,$$

where φ denotes the terrestrial latitude. From the relationship between the gnomon and the Sun at noon on the day of equinox (see Fig. 8),

$$R\sin\varphi/R\cos\varphi = s/g,$$

where g denotes the gnomon length. When we suppose that the gnomon length be the Indian standard twelve digits, we have his solution, namely, the earth-sine = R sin $\delta \cdot$ s/12.

Finally, Brahmagupta converts the earth-sine into the sine of the equation of daylight by multiplying R/R_{day}, because, as we see in Fig. 5,

$$\text{the sine of the equation of daylight} = R\sin\text{arc}\,EF = R\sin AB\,R/R_{day} = \text{the earth sine} \cdot R/R_{day}.$$

This example should give the reader an idea the *Siddhānta*s' concision. This and the use of astronomical-geometrical terms like "the sine of the declination" would have certainly aided in memorizing this solution, but they hardly reveal the reasoning behind it. For that, in the absence of a contemporary master, we must turn to later texts.

We know from later texts that such solutions taking recourse to the concept of the celestial sphere were constructed on plane trigonometry and similar triangles inscribed in the celestial sphere. In the *Siddhānta Siromanī*, for example, Bhāskara II (1114–1185) emphasizes the importance of triangles in such matters thus:

> Those who have a clear knowledge of the sphere having thus immediately formed thousands of triangles should explain the doctrine of the sphere to their pupils.[31]

Interestingly, the Persian polymath Abū al-Rayḥān Muḥammad ibn Aḥmad al-Bīrūnī (973–1048) recorded a proof in Arabic for Brahmagupta's solution in his *Exhaustive Treatise on Shadows* (*Ifrād al-maqāl fī amr al-ẓilāl*).

Bīrūnī wrote on a range of subjects including mathematics, astronomy, astrology, and medicine, though he was particularly famous for his discussion of Indian scholarship in *India* (*Kitāb fī taḥqīq mā li'l-Hind min maqūla maqbūla fi'l-ʿaql aw mardhūla,* i.e. *Book on the Verification of the Categories Pertaining to the Indians Rationally Accepted or Rejected*).[32] In addition to *India*, Bīrūnī also showcases his knowledge of Indian astronomy in such monographs as his *Exhaustive Treatise on Shadows*.[33]

In chapter 22 of his *Exhaustive Treatise on Shadows*, entitled "On the Amounts of the Day and Night, and the Differences of the Ascensions," Bīrūnī discusses how to determine day length and rising times. He opens with a quote of the above passage from the *Brāhmasphuṭasiddhānta*:

> Brahmagupta said in[34] the *Brāhmasiddhānta*: To know the equation of daylight, multiply each of the sine of the declination of a planet and the total sine[35] by itself, and take the root of the difference between them, so it [i.e., the root] is a radius of the day circle of the planet. Then, multiply the sine of the declination of the planet by the shadow at the equatorial region [= the noon equinoctial shadow], and divide what comes by twelve. Multiply what results by the total sine, and divide the result by the radius of the day circle of the planet. What results is the sine, which you should make the arc, namely, the *prāṇa*s[36] of the equation of daylight.[37]

---

[31] *Siddhānta Siromanī*, Gola-adyāya, 7.49; tr. Sastri and Wilkinson (1974: 176). On the life and times of Bhāskara II, see Plofker (2009: 318–319).

[32] On *India*, see Sachau (1888). On the life and times of Bīrūnī, see Kennedy (1970) and Yano (2007).

[33] For the Arabic text, see Bīrūnī (1948: the second tract). In this paper, I call this edition "the Hyderabad edition." For an annotated English translation, see Kennedy (1976).

[34] Read *fī*; *min* in the Hyderabad edition.

[35] The total sine is R sin 90° = R.

[36] A *prāṇa* is a basic unit for calculating arc in Indian astronomy. See Kennedy (1976: Vol. II, 99).

[37] Bīrūnī (1948: 126); cf. the English translation of Kennedy (1976: Vol. I, 174).

Fig. 8 Gnomon and the Sun at noon on the day of equinox

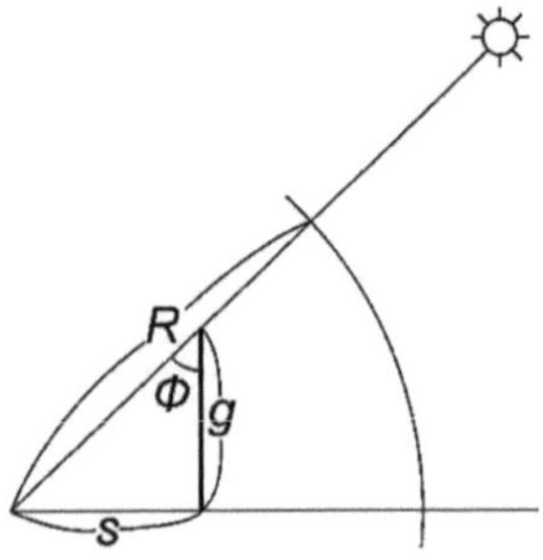

Bīrūnī changes the wording, and adds information like "the equation of daylight," but the passage is otherwise clearly the same as that quoted above. Notably, the radius of the day circle is determined by $\mathrm{R}_{\mathrm{day}} = \sqrt{R\sin 90^{\circ 2} - R\sin^2\delta}$; the earth-sine is calculated by R sin δ · s/12; and the sine of the equation of daylight is obtained by the earth sine · R sin 90°/$\mathrm{R}_{\mathrm{day}}$.

Curiously, while Bīrūnī adds the Islamic term "equation of daylight," he omits the Indic term "earth-sine" in the original. Why? Owing to the aforementioned characteristics of the *Siddhānta*s and contemporary Indian education, one imagines that Bīrūnī's writings on Indian astronomy must have relied on the oral explanations of an informant.[38] If so, this is probably the result of his informant adapting his conceptual vocabulary to one familiar to Islamic scholars, since Bīrūnī's works on Indian scholarship otherwise take delight in explaining all things unique to India.[39]

After quoting Brahmagupta's solution, Bīrūnī then provides the following proof:

> Proof: Let ABG be the orb of the meridian [see Fig. 9],[40] and AG the common part between the plane of it [i.e., the meridian] and the plane of the horizon, and BEH the common part between it [i.e., the plane of the meridian] and the plane of the equator with pole T, and let the common part between the planes of the orb of the meridian and of the day circle of the planet be KM. We draw TE, so SM is the sine of the equation of daylight in the day circle [i.e., the earth-sine]. We make EZ, a perpendicular to AG, which is equal to the gnomon, and we draw ZH parallel to AG, so it is the shadow of the equatorial region. We drop perpendicular GL[41] to TE. Then, because [arc] AB is the complement of [arc] GT,[42] the two triangles BED and ELG are similar and equal to each other, and triangle MSE is similar to both of them. So, the ratio of MS to SE is as the ratio of HZ to ZE. SE is the sine of the declination, and EK is the total sine, which is equal in square (*yaqwā*) to [the square of] it

[38] On this point, see Pingree (1975).

[39] On Bīrūnī's approach to introducing Indian scholarship, see Sachau (1888: xvii–xxvii).

[40] Figure 9 (left) is the diagram drawn by Bīrūnī. I add Figure 9 (right) for modern readers.

[41] Read *GL*; *HL* in the Hyderabad edition.

[42] Read *GT*; *HT* in the Hyderabad edition. Here, arc AB = π/2 – arc GT.

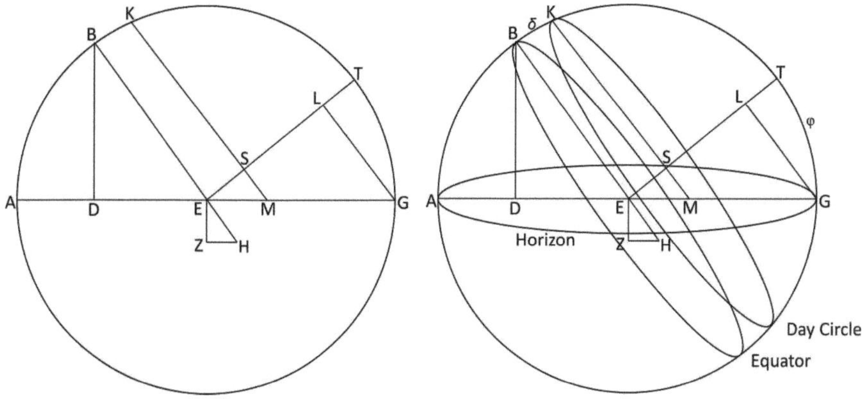

Fig. 9 Diagram drawn by Bīrūnī (left) and the three-dimensional representation of it (right)

> [i.e., SE] and [the square of] SK,[43] so SK[44] is the sine of the complement of the declination [= the cosine of the declination], which is what is called the radius of the day circle of the planet, and the ratio of it [i.e., SK] to SM is as the ratio of the total sine, I mean BE, to what corresponds to it [i.e., SK] along the radius of the sphere. This is the transformation; I mean the transformation of SM in terms of the value in which SK is the total sine.[45] Thus, the sine of the equation of daylight is known.[46]

In this proof, Bīrūnī constructs triangles inscribed in the sphere, and he proves Brahmagupta's solution by using their similarities. First, from the similarity of triangles BED and MSE (and the similarity of triangles BED and EZH), we have MS/SE = HZ/ZE, so

$$\text{the earth}\quad \text{sine} / \text{R} \sin \delta = \text{s} / \text{g}.$$

Thus, the earth-sine = R sin δ · s/12, when g = 12. Next, from the relationship $EK^2 = SE^2 + SK^2$, we have $R \sin^2 90° = R \sin^2 \delta + SK^2$, so $R_{day} = SK = R \cos \delta$. Finally, from

$$\text{SK} / \text{SM} = \text{R} / \text{the sine of the equation of daylight},$$

the equation of daylight is known, because

$$\text{the sine of the equation of daylight} = \text{the earth}\quad \text{sine} \cdot \text{R} / \text{R}_{\text{day}}.$$

[43] That is, $EK^2 = SE^2 + SK^2$.

[44] Read *fa-SK*; *wa-SK* in the Hyderabad edition.

[45] I.e., the radius of the base circle for the earth-sine, which is $R_{day} = SK$ (= R cos δ), is transformed into R in order to obtain the sine of the equation of daylight.

[46] Bīrūnī (1948: 127); cf. the English translation of Kennedy (1976: Vol. I, 174–175).

In concluding, Bīrūnī notes that this relationship is used to transform the radius of the base circle of the earth-sine from R_{day} to R, since the sine of the equation of daylight is the sine corresponding to the earth-sine on the equator circle.

Bīrūnī's use of the term "be equal in square (*yaqwā*)," from the Greek δύνασθαι,[47] suggests that his aim was to compose a Greek-style proof, but his use of the similarities of the inscribed triangles obviously follows the Indian-style reasoning presumably learned from his informant. This is exactly what we see in Thābit ibn Qurra's *Treatise on Sector-Figure*, as discussed in the previous section, and it is thus reasonable to conclude that there too it was a result of the author's double-exposure to both Greek and Indian astronomy. Indeed, taking a step back, we must remark that Thābit seems to be conscious that he is importing an Indian method into this Greek-style proof, because in every other Greek-style proof he consistently uses the Greek chord. This brings us to a question: Why did Thābit know Indian astronomy?

4 The Abbasid Transmission of Indian Astronomy

The reception of foreign knowledge in the Abbasid dynasty began under the second caliph, al-Manṣūr (rt. 754–775).[48] In *Meadows of Gold and Mines of Gems* (*Muruj al-dhahab wa-ma'ādin al-jawhar*), Mas'ūdī (ca. 896–956) records the following testimony by a contemporary scholar, Akhbārī:

> [Akhbārī said:] He [al-Manṣūr] was the first caliph who approached astrologers (*munajjim*) and used astrological judgements. There were with him the astrologer Nawbakht the Zoroastrian, who converted to Islam under him [al-Manṣūr] and was the father of this family of the Nawbakhts; the astrologer Ibrāhīm al-Fazārī, the author of a verse on stars as well as other [works on] astrological and astronomical sciences; and the astrologer Alī ibn 'Isā, the Astrolabist.
>
> He was the first caliph who had books translated from foreign languages into Arabic, including the *Book of Kalīla and Dimna*[49] and the *Book of Sindhind*. Books by Aristotle on logic and other [subjects] were translated for him. The *Book of the Almagest* by Ptolemy, the *Book of the Arithmetic* [by Nicomachus of Gerasa], the book by Euclid, and other ancient books were translated from classical Greek, Byzantine Greek, Pahlavi, Persian, and Syriac.[50]

[47] On this term, see Decorps-Foulquier and Federspie (2008: 220n35).

[48] See Gutas (1998: 28–60).

[49] The *Book of Kalīla and Dimna* was a collection of fables explaining how nobles ought to behave; see De Blois (1990).

[50] For the Arabic text with French translation, see De Meynard and De Courteille (1861–1917: Vol. VIII, 290–291); cf. the English translation in Gutas (1998: 30).

Sindhind is an Arabic transliteration of the Sanskrit word *Siddhānta*, signaling that the translation of the *Siddhānta*s had begun at al-Manṣūr's court alongside works of Greek astronomy.

Of the two, it is in fact the translation of "the *Book of Sindhind*" that is the better documented. For example, Ibn al-Qifṭī's (d. 1248) *History of Learned Men* (*Ta'rīkh al- ḥukamā'*) quotes the following story from Ādamī's (fl. ca. 925) *zīj*[51]:

> In the year 156 [= 772/3 CE], there came to the caliph al-Manṣūr, a man from India who was a master of the calculation known by the [name] *Sindhind* concerning the motions of the planets accompanied with equations made according to *kardajas*[52] calculated for each half degree, together with topics on operations of the celestial sphere concerning the two [i.e., solar and lunar] eclipses and the ascensions of the zodiacal signs as well as other things [found] in a book consisting of a number of chapters…

> Then, al-Manṣūr ordered that this book should be translated into Arabic and that there should be composed from it a book which the Arabs might use as a basis for the motions of the planets. So, Muḥammad ibn Ibrāhīm al-Fazārī was in charge of this, and he wrote from it the book the astrologers (*munajjims*) called the *Great Sindhind* (*al-Sindhind al-kabīr*).[53]

At first glance, it is unclear whether *Sindhind* was the name of a set of astronomical calculations or the title of an astronomical book.[54] However, given that the *Siddhānta*s were memorized and orally transmitted, we can surmise that the Arabic *Sindhind* was probably more the product of dictation than the translation of some one written work bequeathed by the Indian master.

Just as the *Sindhind* inspired Fazārī's *Great Sindhind*, it also inspired the *zīj*es such as Ādamī's, which would quickly become one of the most popular genres of astronomical writing in the Islamicate world.[55] Indeed, the success of *zīj* literature illustrates that Indian astronomy had a greater impact on the Arabic tradition than did its Greek counterpart. The reason of this, one imagines, is probably the practical, astrological orientation of the scholars at al-Manṣūr's court, for whom the mathematical inventions and trigonometrical devices of the *Siddhānta*s would have presumably been better adapted.[56]

To come back to our question, it would seem that Thābit knew Indian astronomy because it was simply part of the larger background in the early Abbasid period. In fact, outside of the *Treatise on the Sector-Figure*, we find traces of Indian mathematics and astronomy in Thābit's larger oeuvre.

---

[51] On the life and times of Ādamī, see Ragep and Bolt (2007).

[52] The term *kardaja* is an Arabic transliteration of the Sanskrit word *kramajyā*, meaning "sine." However, this term was used in the Islamicate world for denoting arguments of sines and equations; see Haddad et al. (1981: 214).

[53] For the Arabic text, see Lippert (1903: 270–271); cf. the English translation in Pingree (1970: 105–106).

[54] See Van Bladel (2014: 262).

[55] For a survey of the *zīj* genre, see Kennedy (1956) and King et al. (2001).

[56] See Gutas (1998: 45–52).

5 Traces of Indian Mathematics and Astronomy in Thābit ibn Qurra's Work

Let us turn to Thābit ibn Qurra's book on sundials, entitled the *Book on the Horary Instruments Called Plane Sundials* (*Kitāb fī ālāt al-sāʿāt allatī tusammā rukhāmāt*).[57] In this work, Thābit classifies the locations of the plane where the hour-lines of the sundial are drawn, and for each location, he explains how to calculate and construct the hour-lines.

Thābit opens with an introductory section entitled "On Knowing the Sine Called 'Versed' from Every Known Arc as well as [Knowing] the Arc Called 'Versed' from Every Known Versed-sine." In this section, he explains how to calculate these elements as follows:

> If an arc is known, and it is smaller than a quarter of a circle, subtract it from a quarter of a circle, take the sine of what is left, and then subtract it from the maximum sine. As a result, what is left is the versed sine of this arc.[58]

As previously mentioned, the sine and versed-sine are both Indian in origin, so this shows that Thābit is equally as familiar with them as is with the Greek-origin chords that predominate in his *Treatise on the Sector-Figure*. In fact, one notes that he tends to use Greek-origin chords in presenting Greek-style geometrical proposition in the *Treatise on the Sector-Figure*, and Indian-origin sines in *Siddhānta*-style calculations, as we have here. In other words, he would seem to distinguish between the two traditions and apply their conceptual vocabularies to different ends.

In addition to these mathematical tools, one also finds several Indian-origin astronomical concepts in Thābit's other works. For example, in *The Almagest Simplified* (*Tashīl al-majisṭī*), he mentions the concept *jawzahar* in explaining planetary motion:

> None of the seven planets except for the Sun is always moving on the plane of the zodiacal orb; they are moving only on circles cutting the zodiacal orb at two points. These circles are called the inclined orbs, and these two points at which the two circles are cutting each other are named "nodes": one of these two points is named "head of *jawzahar*," namely the [point] which when a planet passes, it inclines to the north of the zodiacal orb; the other point is named "tail," namely the [point] which when a planet passes, it inclines to the south of the zodiacal orb. However, the [points] characterized by the name "head of *jawzahar*" or "tail" are the two points of the Moon's orb.[59]

From Middle Persian *gozihr*, *jawzahar* refers to the "dragon" whose head and tail (*rāhu* and *ketu*), in India, represent the nodes of the lunar orb. The concept was first introduced into Arabic under the Abbasid dynasty, where it saw use in astrological and astronomical works.[60] Interestingly, Thābit specifies that *jawzahar* only applies

---

[57] See the critical edition and French translation in Morelon (1987: 130–168).

[58] For the Arabic text with French translation, see Morelon (1987: 132–133).

[59] For the Arabic text with French translation, see Morelon (1987: 4).

[60] On the notion *jawzahar*, see Pingree (1997: 39).

to the Moon, suggesting that his contemporaries erroneously applied it to all inclined orb.[61]

6 Thābit ibn Qurra's Early Career

As previously mentioned, several Arabic historical records underline that Thābit ibn Qurra began his studies in Greek science with the Banū Mūsā. Their first contact is recorded in Ibn al-Nadīm's (ca. 932–990) *Catalogue* (*Fihrist*):

> He [Thābit] was a money-changer in Ḥarrān[62]; Muḥammad ibn Mūsā [the eldest of the Banū Mūsā] took him as a companion, when he came back from the Roman region, because he recognized that he had language ability. It is said that he studied under Muḥammad ibn Mūsā, so he became learned at his house.[63]

Although this does not mention Thābit's age when they met, it does show that Muḥammad trained Thābit, suggesting that Thābit had not received sufficient education (including Greek science) in his homeland of Ḥarrān. Thus, we can conclude that since Thābit presented multilingual talent, Muḥammad decided to teach Thābit Greek science from a beginner's level to make him a language assistant for the Banū Mūsā's research thereupon.

Soon, Thābit's training paid off, and he became an invaluable assistant to his benefactors. As we read in the preface to their Arabic version of Apollonius' *Conics*, the middle brother, Aḥmad, completed his translation with the help of Hilāl ibn Abī Hilāl al-Ḥimṣī and Thābit ibn Qurra.[64] The case of the Arabic version of the *Conics* impresses on us that in the course of assisting the Banū Mūsā, Thābit had precious opportunities to learn quite a few Greek scientific works with them. As a result, he became a proficient scholar on Greek science, so that he was regarded as a colleague, not a mere assistant, of them by later scholars, such as Bīrūnī in his *Chronology of Ancient Nations* (*Al-Āthār al-bāqīya 'an al-qurūn al-khālīya, i.e., The Remaining Signs of Past Generations*).[65]

In the chapter entitled "On the natures of the months used in the preceding eras" of his *Chronology of Ancient Nations*, Bīrūnī mentioned an astronomical book written by Muḥammad ibn Mūsā ibn Shākir, the eldest of the Banū Mūsā, with the following note:

[61] For example, 'Alī ibn Sulaymān al-Hāshimī (fl. 890) uses the *jawzahar* for the nodes of all the planetary orbs; see Haddad et al. (1981: 10).

[62] Ḥarrān was one of the central cities in Upper Mesopotamia, now in Turkey. On the myth about this city, see Van Bladel (2009: 64–118).

[63] The Arabic text is in Flügel (1871: 272); cf. the English translation in Dodge (1998: 647).

[64] For the Arabic text with English translation of this preface, see Toomer (1990: 620–629); cf. the French translation of Rashed (2008: 500–507).

[65] The Arabic text is in Sachau (1878). Sachau also publishes an English translation of this work: Sachau (1879).

> As for the book referred to, it is that linked to Thābit ibn Qurra, since he [Thābit] was a protégé (*ṣanīʿa*) of these people and was one among them[66] and revised their knowledge for them.[67]

From this note, we know that although this work was attributed only to Muḥammad ibn Mūsā and did not have the name of Thābit, it was obvious for Bīrūnī that it was jointly composed by Muḥammad and Thābit. Moreover, the last part of it, namely, "[Thābit] revised their knowledge for them" reveals his judgment that Thābit was as competent in Greek science as the Banū Mūsā brothers were, showing that through his works with them he became comparable to them concerning the study on Greek science.

What is remarkable in this note is that when Bīrūnī explained his opinion, he declared that Thābit was a protégé (*ṣanīʿa*) of the Banū Mūsā and he belonged to their family. Here, the Arabic word *ṣanīʿa* is closely related to the word *iṣṭanaʿa* which means "to foster someone's career," and someone called a *ṣanīʿa* of a person was one who was reared, educated and trained by this person.[68] Thus, this reasoning by al-Bīrūnī confirms that it was publicly acknowledged that Thābit's connection to Muḥammad was very strong like the bond between a foster son and the father.

This note illustrates that because Thābit was fostered by Muḥammad and established a strong tie to him, he assisted the Banū Mūsā family for their study on Greek science. With the help of Thābit, they aimed to increase their knowledge on Greek science as a family, and they endeavored to retain the place of the authority concerning Greek science in the court.

The above analysis of Thābit's early career suggests that he learned science in Baghdad around al-Ma'mūn's time. As I have mentioned before, in al-Ma'mūn's day, when Greek astronomy became dominant, scholars continued to utilize calculation techniques coming from Indian astronomy. Hence, it is only natural that Thābit studied both the Greek and Indian traditions of astronomy and mathematics at the Banū Mūsā family's home.

7 Conclusion

In this paper, I have highlighted Thābit ibn Qurra's usage of Indian astronomical and mathematical knowledge in his works, showing that mastered both the Greek and Indian traditions. Then, I elucidated, with recourse to his biographical information, why he knew Indian astronomy and mathematics. In al-Ma'mūn's day, court scholars became interested in Greek science, and owing to their growing interest, those who were engaged in it competed with one other in their knowledge for positions at court. As a result, they tried to expand their research groups by recruiting

[66] Read *wa-man bayna-hum*; *wa-min bayni-nm* in Sachau's text.

[67] The Arabic text is in Sachau (1878: 52); cf. Sachau's English translation in Sachau (1879: 61–62).

[68] See Mottahede (1980: 82–95).

people in order to increase their knowledge of Greek science. This expansion gave promising persons such as Thābit the opportunity to build connections with prominent scholars at court.

As for the case of Thābit, this opportunity provided him the chance to learn Greek science at Banū Mūsā's home. In the course of working with the Banū Mūsā family, Thābit rapidly amplified his knowledge of Greek science, using the abundant resources that they had collected to eventually become the top scholar in the field. Since Indian astronomy and mathematics were also necessary knowledge for scholars engaging in Greek science in Baghdad, he also learned them with the family. Therefore, there is no surprise that he came to master not only Greek astronomy and mathematics but also their Indian counterparts.

Bibliography

Adamson, Peter. 2007. *Al-Kindī*. New York: Oxford University Press.

Adamson, Peter and Pormann, Peter. 2012. *The Philosophical Works of Al-Kindī*. Karachi: Oxford University Press.

Bellosta, Hélène. 2009. Le traité de Thābit ibn Qurra sur *La figure secteur*. In *Thābit ibn Qurra: Science and Philosophy in Ninth-Century Baghdad*, ed. Roshdi Rashed, 225–390. Berlin: Walter de Gruyter.

Bīrūnī. 1948. *Rasā'il al-Bīrūnī*. Hyderabad-Dn.: The Dāiratu'l-Ma'ārifi'l-'Osmania (Osmania Oriental Publications Bureau).

Crozet, Pascal. 2009. Thābit ibn qurra et la composition des rapports. In *Thābit ibn Qurra: Science and Philosophy in Ninth-Century Baghdad*, ed. Roshdi Rashed, 391–426. Berlin: Walter de Gruyter.

Decorps-Foulquier, Micheline and Federspie, Michel. 2008. *Apollonius de Perge: Coniques*, vol. 1.2, Berlin; New York: Walter de Gruyter.

Dodge, Bayard. 1998. *The Fihrist: a 10th Century AD Survey of Islamic Culture*. Chicago: KAZI Publications.

Flügel, Gustav. 1871. *Kitāb al-Fihrist*. Leipzig: Vogel.

De Blois, François. 1990. *Burzōy's Voyage to India and the Origin of the Book of Kalīlah wa Dimnah*. London: Royal Asiatic Society.

De Meynard, C. B. and De Courteille, P. 1861–1917. *Les prairies d'or* 9 vols. Paris: Imprimerie impériale.

Grupe, Dirk. 2012. The "Thābit-Version" of Ptolemy's *Almagest* in MS Dresden Db.87. *Suhayl*, 11: 147–153.

Gutas, Dimitri. 1998. *Greek Thought, Arabic Culture: The Graeco-Arabic Translation Movement in Baghdad and Early Abbasid Society (2nd–4th/8th–10th Centuries)*. London and New York: Routledge.

Haddad, F.I., Kennedy, E.S. and Pingree, D. 1981. *The Book of the Reasons behind Astronomical Tables*. Delmar, N.Y.: Scholars' Facsimiles & Reprints.

Hill, Donald R. 1979. *The Book of Ingenious Devices (Kitāb al-ḥiyal) by the Banū (sons of) Mūsà bin Shākir*. Dordrecht; Boston: D. Reidel Pub. Co.

Hogendijk, J.P. 1985. *Ibn al-Haytham's Completion of the Conics*. New York: Springer.

Kennedy, E.S. 1956. A Survey of Islamic Astronomical Tables. *Transactions of the American Philosophical Society*, 46: 123–177.

Kennedy, E.S. 1970. Al-Bīrūnī. In *Dictionary of Scientific Biography*, ed. Charles C. Gillispie, vol. 2, 147–158. New York: Charles Scribner's Sons.

Kennedy, E.S. 1976. *The Exhaustive Treatise on Shadows by Abu al-Rayḥān Muḥammad b. Aḥmad al-Bīrūnī* 2 vols. Aleppo: University of Aleppo, Institute for the History of Arabic Science.
Kennedy, E.S., Kunitzsch, P. and Lorch, R.P. 1999. *The Melon-Shaped Astrolabe in Arabic Astronomy*. Stuttgart: F. Steiner.
King, D.A., Samsó, J. and Goldstein, B.R. 2001. Astronomical Handbooks and Tables from the Islamic World (750–1900): an Interim Report. *Suhayl*, 2: 9–105.
Langermann, Tzvi. 2007. Ibn al-Haytham: Abū ʿAlī al-Ḥasan ibn al-Ḥasan. In *The Biographical Encyclopedia of Astronomers*, eds. Thomas Hockey et al., 556–567. New York: Springer.
Lippert, Julius. 1903. *Ibn al-Qifṭī's Ta'rīkh-i Ḥukamā', auf Grund der Vorarbeiten Aug. Müllers, herausgegeben von Julius Lippert*. Leipzig: Dieterich'sche Verlagsbuchhandlung.
Lorch, Richard. 2001. *Thābit ibn Qurra: On the Sector-figure and Related Texts*. Frankfurt am Main: Institute for the History of Arabic-Islamic Science at the Johann Wolfgang Goethe University.
Morelon, Régis. 1987. *Thābit ibn Qurra: Œuvres d'astronomie*. Paris: Belles lettres.
Mottahede, Roy. 1980. *Loyalty and Leadership in an Early Islamic Society*. Princeton, N.J.: Princeton University Press.
Ôhashi, Yukio. 1994. Astronomical Instruments in Classical Siddhāntas. *Indian Journal of History of Science*, 29: 155–313.
Pingree, David. 1970. The Fragments of the Works of Al-Fazārī. *Journal of Near Eastern Studies*, 29: 103–123.
Pingree, David. 1975. Al-Bīrūnī's Knowledge of Sanskrit Astronomical Texts. In *The Scholar and the Saint: Studies in Commemoration of Abu'l-Rayhan al-Bīrūni and Jalal al-Din al-Rūmī*, ed. Peter J. Chelkowski, 67-81. New York: Hagop Kevorkian Center for Near Eastern Studies, New York University Press.
Pingree, David. 1976. The Recovery of Early Greek Astronomy from India. *Journal for the History of Astronomy*, 7: 109–123.
Pingree, David. 1997. *From Astral Omens to Astrology from Babylon to Bīkāner*. Roma Ist. Italiano per l'Africa e l'Oriente.
Plofker, Kim. 2009. *Mathematics in India*. Princeton: Princeton University Press.
Ragep, F. Jamil and Bolt, Marvin. 2007. Ādamī: Abū ʿAlī al-Ḥusayn ibn Muḥammad al-Ādamī. In *The Biographical Encyclopedia of Astronomers*, eds. Thomas Hockey et al., 12. New York: Springer.
Rashed, Roshdi. 2008. *Apollonius de Perge: Coniques* vol. 1. Berlin; New York: Walter de Gruyter.
Rashed, Roshdi. 2012. *Founding Figures and Commentators in Arabic Mathematics*. London: Routledge.
Sabra, A.I. 1998. One Ibn al-Haytham or Two? An Exercise in Reading the Bio-bibliographical Sources. *Zeitschrift für Geschichte der Arabisch-Islamischen Wissenschaften* 12: 1–50.
Sachau, E. 1878. *Chronologie orientalischer Völker*. Leipzig: In commission bei F.A. Brockhaus.
Sachau, E. 1879. *The Chronology of Ancient Nations*. London: W.H. Allen & Co.
Sachau, E. 1888. *Alberuni's India: An Account of the Religion, Philosophy, Literature, Geography, Chronology, Astronomy, Customs, Laws and Astrology of India About A. D. 1030* 2 vols. London: Trübner.
Sastri, B.D. and Wilkinson, L. 1974. *The Súrya siddhánta, or an Ancient System of Hindu Astronomy: Followed by the Siddhánta siromani*. Amsterdam: Philo Press.
Sayılı, Aydın. 1995. The Introductory Section of Ḥabash's Astronomical Tables Known as the 'Damascene' Zīj. *Dil ve Tarih-Coğrafya Fakultesi dergisi*, 13: 133–151.
Toomer, G., 1984. Ptolemy's Almagest. London: Duckworth.
Toomer, G. 1990. *Apollonius: Conics Books V to VII: The Arabic Translation of the Lost Greek Original in the Version of the Banū Mūsā*. New York: Springer.
Van Bladel, Kevin. 2009. *The Arabic Hermes: from Pagan Sage to Prophet of Science*. New York: Oxford University Press.

Van Bladel, Kevin. 2014. Eighth-Century Indian Astronomy in the Two Cities of Peace. In *Islamic Cultures, Islamic Contexts: Essays in Honor of Patricia Crone*, eds. Asad Ahmed et al., 257–294. Leiden: Brill.

Van Brummelen, Glen. 2009. *The Mathematics of the Heavens and the Earth: the Early History of Trigonometry*. Princeton, N.J.: Princeton University Press.

Yano, Micho. 2006. Oral and Written Transmission of Exact Sciences in Sanskrit. *Journal of Indian Philosophy*, 34: 143–160.

Yano, Michio. 2007. Bīrūnī: Abū al-Rayḥān Muḥammad ibn Aḥmad al-Bīrūnī. In *The Biographical Encyclopedia of Astronomers*, eds. Thomas Hockey et al., 131–133. New York: Springer.

Estimating the Gregorian Arrival Date of the Māori to Aotearoa-New Zealand

Ockie Simmonds and Tanikawa Kiyotaka

Abstract At the beginning of the twentieth century, western historians working from traditional oral genealogies and narratives determined that a 'Great Fleet' carrying the Polynesian ancestors of the Māori left the Central Eastern Pacific to settle New Zealand in the year AD 1350. Over the past century, though the scholarship behind it has been thoroughly refuted, this date has become entrenched in New Zealand's national consciousness and has remained largely unchallenged. Focusing on the crew of the *Tainui* and *Te Arawa*, and combining the sort of traditional sources used in previous studies with archaeo-astronomy, this paper argues from the mythologised description of what can only be a total solar eclipse that the celebrated 'Great Fleet' in fact arrived in December 1408, almost sixty years later than previously thought.

Keywords Luni-solar calendar · Archeo-astronomy · Māori

1 Introduction

The proto-Polynesians, descendants of their Austronesian ancestors, set off from continental Asia at least 6000 years ago, travelling through Taiwan and the Philippines (King 2003: 31; Evans 1998: 15). Following the rising sun, they continued from there to sail across one of the widest stretches of the Pacific. Skirting around Papua and Melanesia,[1] they island-hopped through Micronesia on their way

[1] Eastern Polynesians (including Māori) are morphologically, socially, and culturally distinct from the peoples of Papua and Melanesia, and as European explorers found out later, they would have

O. Simmonds (✉)
Society of Māori Astronomy Research and Traditions, Wellington, New Zealand
e-mail: ockie.simmonds@smsa.nz

Tanikawa Kiyotaka
National Astronomical Observatory of Japan, Mitaka, Tokyo, Japan

D. P. Morgan, T. Quan (eds.), *Mathematics and Astronomy in the Ancient World*,
https://doi.org/10.1007/978-3-031-93809-2_14

to the central Pacific (Buck 1938: 41–45; Lewis 1977: 13), and their culture and social order sustained the journey.

One of the greatest indicators of this cultural continuity, and of the path of their migration, is the myth of the proto-Polynesian ancestral God Māui-tikitiki. In Micronesia, 3000 kilometres from the shores of mainland China, the proto-Polynesian's ancestral God Māui-tikitiki is said to have fished up the islands of Fai and Yap in the Caroline Islands (Lewis 1978: 59). The Polynesian god Māui-tikitiki is said to have fished up the islands of Tonga, Rakahanga (Cook Islands) and Hawai'i (Buck 1938: 53). Later still, Māori legends record that their god, Māui-tikitiki, fished up the North Island of New Zealand (Grey 1885 [1988]: 26). Indeed, the Māori name for the North Island is Te Ika-a-Māui, or 'the Fish of Māui'.

In settling the wide expanses of the Pacific Ocean, the Polynesians mastered the art of navigating by the stars long before Eurasian sailors began using the sextant in the mid-eighteenth century. Their mastery of this skill allowed them to colonise every habitable island in the Polynesian triangle about 500 years before Europeans entered the Pacific, and it led the ancestors of the Māori to New Zealand, the last major land mass on earth to be inhabited by man (Lee 2018: 18; Buck 1938: 60, 259; Evans 1998: 21).

Famously, working from Polynesian traditions and *whakapapa* genealogies, the ethnologist S. Percy Smith (1840–1922) argued that the ancestors of the Māori arrived in a 'Great Fleet' of 12 canoes from Tahiti in the year 1350, where they found New Zealand already scattered with earlier Polynesian settlers.[2] In the century since, the archaeological record has confirmed this rough sequence of events. Notably, the earliest evidence of human settlement points to migrants from such Eastern Polynesian islands as Marquesas, the Society Islands and Tuamotu having taken root by the thirteenth century (King 2003: 48). The very first explorers likely arrived during the moa-hunting period, in the eleventh and twelfth centuries (Duff 1950 [1977]: xii), but evidence for the sort of 'Great Fleet' mass migration described by Smith indeed appears around the late thirteenth century and continues into the fifteenth century (King 2003: 50). As to their reason for setting out, traditional narratives stress famine, war and political and religious persecution, and we now know that this may have coincided with significant changes in climatic conditions surrounding the Little Ice Age (Bridgman 1983). Lastly, one notes that differences in inland-origin explain the different traditions, social orders, different tribes, and different dialects in New Zealand (Condliffe 1971: 171, 290; Buck 1938: 273).

Of course, while archaeology has confirmed the broader picture recounted by Smith's traditional Māori sources, the radiocarbon dating of human settlements is not nearly so precise as to date them to a specific year (Burley et al. 2015). Nevertheless, over the past century, Smith's hypothetical date of AD 1350 has been

---

been liable to being eaten if passing through. Rather, one finds Polynesian language, culture, and genes only the outskirts of this central zone between Island Southeast Asia, Micronesia and Polynesia.

[2] Smith (1910: 75–77). On Smith as the origin of this date, see King (2003: 49), Hiroa (1949 [1977]: 7) and Simmons (1976: 43).

given official status and celebrated in the New Zealand educational system and public consciousness by both Pākehā (Europeans) and Māori alike. In this paper, we would like to propose a different date.

Specifically, in this paper we will examine the sort of traditional *whakapapa* (genealogies), narratives, legends and stories at the basis of Smith's work, adding to his sources, critically reflecting on his approach, and anchoring our chronology to the archaeo-astronomical event of what appears to be a total solar eclipse reported therein. In short, the story surrounding the crew of the Great Fleet's first journey inland culminates in the 'great darkness' of a total solar eclipse, which has inspired legends, geographical place names, celebrations and, in our day, public and government sponsored events. Meticulous oral genealogies that can be cross-referenced between different tribes and islands allow us trace back individual Māoris' ancestry a precise number of generations to these figures, and we will discuss how these have been used to calculate the Māori's date of arrival as well as the limitations and contradictions inherent therein. Taking away from these materials a simple upper and lower bounds, we will then survey the historical eclipses that occurred between them, finding only one that matches the location, description and seasonal context provided in the traditional narrative, which will allow us to fix dates to this excursion, their landing, their departure, and the stops the 'Great Fleet' made on their way there.

2 Legends to Eclipses

2.1 Eclipses in Memory

In literate cultures with astronomers and scribes, eclipse events are often recorded for study and historical posterity, those such as China, Korea, Japan, Vietnam, and West Asian and European countries leaving us with observational data in a sufficiently familiar and detailed form as can be used towards modern astronomy. In cultures without writing, most are lost to history, be they forgotten or remembered in some less recognisable form. This does not mean that they have left us with nothing in this regard, but simply that we must learn to recognise the distant echo of what was once the memory of an eclipse in the stories that have been handed down.

Ancient Greece provides a good example of how eclipses were recorded under similar conditions, specifically, the period 600–360 BC, prior to the emergence of astronomy and the systematic observation and recording of eclipses. In their study of this period, Tanikawa and Sôma (2010) considered deep and partial solar eclipses over this region. During this period, there were 114 eclipse events over the Greek mainland, Aegean Sea and parts of Turkey, while the contemporary Greek historians Herodotus (*c*.484–*c*.425 BC), Thucydides (*c*.460–*c*.400 BC) and Xenophon (*c*.430–354 BC) record only six eclipses events (5.3%). Furthermore, between 500 and 350 BC there were 63 eclipse events that were observable over the Greek world, only five of which (7.9%) were recorded. Left to historians, poets and storytellers, it was only a handful of the total number of eclipses that entered into memory.

What did this handful have in common? Tanikawa and Sôma have proposed that that the solar eclipses that tend to enter into the cultural memory of a people without astronomers or writing are key events, for which they propose the following criteria:

(i) when a solar eclipse is either total, annular or very deep;
(ii) when it co-occurs with an important historical event;
(iii) when there is an observer there to record the historical event;

To this we would add a fourth:

(iv) when commemorated, more specifically, in traditional legends, stories, chants and songs.[3]

2.2 *In Japanese Legend*

By way of comparison, Japanese legend describes Amaterasu天照, goddess of the sun, entering a rocky cave, at which the outside world went pitch black, making day and night indiscernible. The legend tells of efforts made by the assembled gods to entice Amaterasu back out of the cave, which, when they succeeded, immediately restored the cosmic order. As recorded in the *Nihon shoki* 日本書紀, the legend uses two phrases that are strongly suggestive that the story may have been about a total solar eclipse: the gods outside the cave 'gathered long-singing birds of the Eternal Land and made them cry' 聚常世之長鳴鳥使互長鳴, possibly referring to roosters; also, inside the cave, Amaterasu 'slightly opened the rock door and peeped out' 細開磐戸窺之, invoking the brilliant diamond ring (coronal) effect just after totality.[4]

That this legend records the experience of ancient people of watching a total solar eclipse is generally accepted. However, narrowing down the exact eclipse it may have been involves weighing additional information, such as eclipse paths and the debated relation between Amaterasu and the historical Queen Himiko卑弥呼 (*c.*170–248), known from contemporary Chinese accounts. With their combined data sets, Tanikawa and Sôma (2010) have estimated that the historical 'stone cave eclipse' occurred on 13 July AD 158, noting that this Japanese legend relates to a similar one in Korea linked to the same eclipse.[5]

[3] See Harris et al. (2013: 325–336) and Simmons (1976: 11).

[4] *Nihon shoki* 日本書紀, 1 ('Shindai' 神代), passage 7.

[5] Cf. Tanikawa et al. (2014).

2.3 In Māori Legend

The story of Amaterasu bears a striking resemblance to a well-known Māori legend called 'Hatupatu and the Witch', one of several stories, gathered in the Appendices 1–3, of the first Māori settlers of the Northern Island as passed down through oral tradition. To put the legend of Hatupatu in context, we will begin here with the arrival of two of the vessels of the 'Great Fleet' and their respective crews' travels across the central volcanic region of the North Island. Their encounters and observations are still known today, and they can be used to corroborate other events of the time.

It is said that the sailing vessels *Te Arawa* and *Tainui* arrived in New Zealand at Whangaparaoa, on the East Coast. The *Te Arawa* people explored the East Coast with some people settling at Maketū, in the Bay of Plenty, and some near Whakatane. One group under the leadership of Tia and his younger brother, Hatupatu, moved inland (west) from Maketū to explore the central volcanic plateau (Grace 1959 [1970]: 59). At the time of their exploration, around the fourteenth or fifteenth century, the Taupō plateau was a vast, bleak volcanic wasteland (King 2003: 51). Under Ngātoroirangi, the spiritual leader and navigator of *Te Arawa,* the other group travelled west from Whakatane to Taupō.[6] Ngātoroirangi and his group moved inland, crossing the Tarawera volcanic plateau and the Kaingāroa plains to arrive at Lake Taupō, a distance of about 120 km from the East Coast.

Tia and Hatupatu travelled via Rotorua and Horohoro and eventually arrived at a place Tia named Atiamuri, about 120 km from Maketū. Tia and his group then travelled up the Waikato River to seek its source and arrived at Lake Taupō, a distance of about 50 km from Atiamuri. Upon their arrival, Tia and his people set off to explore the shores of the lake.

During their travels, Tia and Ngātoroirangi met lakeside whilst they explored the eastern shores of Lake Taupō. Ngātoroirangi warned Tia and his people that he must not be followed, otherwise he would issue a *mākutu*, a curse, or evil spell to stop Tia's group (Grace 1959 [1970]: 62; Stafford 1967: 21). Tia's people were aware of the great authority and spiritual power of Ngātoroirangi, and they would certainly obey his wishes. Tia and his group thus left Ngātoroirangi, travelling north-west towards the Karangāhape cliffs on their way to the mountain of Titiraupengā on the far west side of Lake Taupō. In the opposite direction, Ngātoroirangi travelled to what is now known as Rangipō, just south of Lake Taupō. While in the general area of Lake Taupō, it is said of both the parties of Tia and Ngātoroirangi that 'darkness settled over them', suggesting the totality of a solar eclipse.

Grace and Stafford's research suggests that Hatupatu had left Tia's group earlier and headed back to their settlement at Maketū, in the Bay of Plenty. He followed the Waikato River back, passing through Atiamuri. At this point, as described above, Hatupatu was overtaken by a solar eclipse and apparently took shelter in the rock

[6] Personal communication: Rev. Rahu Katene, 31/03/2018. Rev. Rahu Katene is a descendant of Ngātoroirangi and his wife Kearoa, who travelled on the *Te Arawa*.

now known and still revered as Te Kōwhatu ō Hatupatu, the Rock of Hatupatu. The story goes that Hatupatu was pursued by a witch named Kurangaituku and hid inside a rock-cave until the witch abandoned her search, after which he safely escaped back to his family and village.[7] From the traditional narrative, the day was cold, wet and overcast. However, Hatupatu may well have directly seen part of the eclipse: the legend tells of flames emanating from the witch's fingers, which is suggestive of the brilliant diamond coronal effect and that he may therefore have seen the solar eclipse just before totality.[8]

The story of Hatupatu, Tia and Ngātoroirangi's respective journeys left its mark on local place names, which hold significant importance that local tribal authorities in the central volcanic plateau have retained in legend. Māori commonly name places and people after significant historical events, and these toponyms are suggestive of an eclipse event that occurred at the time. Namely, there is Lake Taupō, or Taupō-nui-ā-Tia, literally 'a great darkness settled over Tia', such as occurs during a total eclipse.[9] There is also Rangipō, plateau south of Lake Taupō where Ngātoroirangi was during eclipse event, whose name means 'darkening sky'.[10]

Oral tradition and the place-names with which they are ascribed would suggest that the three intrepid explorers from the *Te Arawa* found themselves within the path of a total solar eclipse. Figure 1 provides the probable location of the three explorers at the time.[11] Returning to Tanikawa and Sôma (2010) and the four conditions under which solar eclipses enter into the cultural memory of pre-literate societies, we note that this eclipse meets all four: it was total (i), coincided with important historical events (ii), was witnessed (iii), and was commemorated in oral tradition and toponyms (iv).

According to the traditional narrative, the impact of this singular solar eclipse event was principally due to the influence and powers of Ngātoroirangi, the

[7] The legend of Hatupatu has been Westernised for reader consumption over the past 150 years. However, the Māori narrative records that Hatupatu story is traditionally and historically different to the Western version. See Cowan (1925: 120) and Grey (1855 [1988]: 114).

[8] Note that the Rock of Hatupatu is on the very edge of the area of totality during the solar eclipse of AD 1409.

[9] Lake Taupō, or Taupō-nui-ā-Tia, covers 616 km^2 and is the largest lake in New Zealand. Traditional tribal narratives describe the name of Taupō-nui-ā-Tia to describe 'the great enveloping of Tia by night, or darkness, affecting his eyesight'. This is possibly in reference to the *karakia* spell invoked by Ngātoroirangi to stop others from following him to claim the mountains of Tongariro (Reed 1977: 118). However, according to the Ngāti Tūwharetoa people, whose tribal region covers this area, the name Taupō-nui-ā-Tia originated from the blotting out of Tia's view of the surrounding country, or the great covering of Tia by darkness (Grace 1959 [1970]: 67). In the Māori language a *taupō* also used to refer to a rough black and yellow cloak (Williams 1975: 224).

[10] The Rangipō is a desolate plateau near the volcanic mountains south of Lake Taupō. From *rangi* ('sky') and *pō* ('night', 'darkness'), the literal meaning of Rangipō is 'dark sky', though Western scholars render this variously as a 'windy sky' or 'bleak weather' (Stafford 1967: 22). According to the historical narrative, Rangipō was named such by Ngātoroirangi upon encountering winds and sudden darkness (Grace 1959 [1970]: 63), which suggests that the eclipse coincided with a storm.

[11] See Tanikawa et al. (2019).

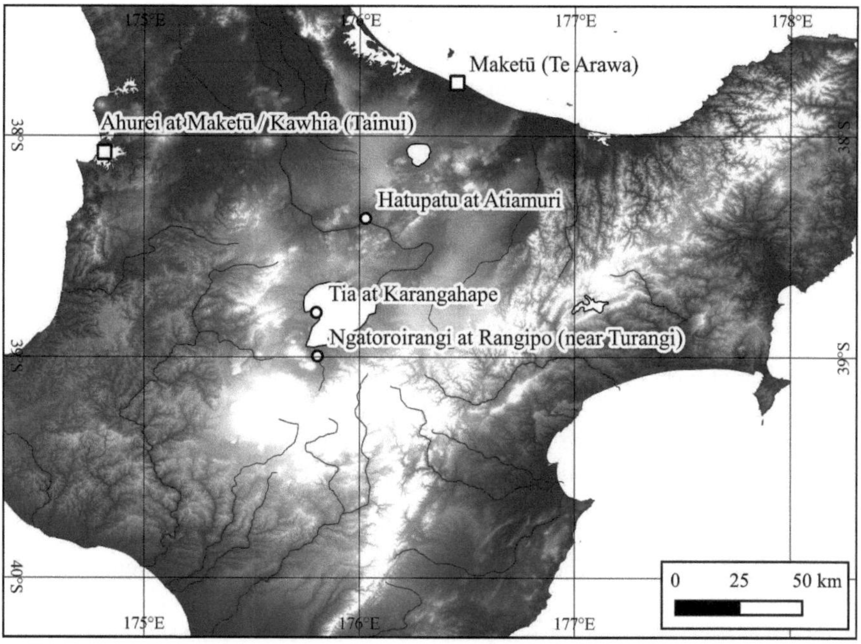

Fig. 1 Locations of explorers during the eclipse. Map generated using the New Zealand National Digital Elevation Model data from Landcare Research New Zealand Limited (https://lris.scinfo.org.nz/layer/48131-nzdem-north-island-25-metre/)

renowned *tohunga*, priest, and navigator of the *Te Arawa*. This event has been etched indelibly into the minds and culture of the Māori. Subsequent retellings have influenced their cultural expectations of the time, ensuring their preservation in an oral society. Even to this day, 600 years later, New Zealanders, both Māori and Pākehā (Europeans), are mindful of what was seen and set in motion, in particular, by that most feared of all *tohunga*, Ngātoroirangi.

By themselves, the solar eclipse event and the spiritual power of Ngātoroirangi, the spiritual leader and navigator of the *Te Arawa*, were not the only factors that impressed this occasion on the newly arrived Polynesian visitors. These were amplified by a mix of natural astronomical events combined with the supernatural.

The perceived supernatural was helped by a large measure of traditional *tapu*, or sacredness, which was entrenched within the culture of the intrepid explorers. *Tapu* literally means 'prohibition by the laws of the native gods' (Mead 1984: 91), and there is no greater *tapu* than the new moon, or *mutuwhenua* ('the whole world is in darkness'), a spiritual day of reflection, quiet contemplation, and sacredness at the end of the lunar month.

To the explorers, a solar eclipse would have been an unexpected and an unnatural event that reinforced the underlying ethos of *tapu* and awe embodied in that day. Adding to the ominousness of this day would have been the weather: again, the traditional narrative emphasises that it was cold, bleak and miserable. Even if the unsuspecting travellers did not directly witness the eclipse of the sun, they would have experienced an unnatural darkening of sky and drop in temperature around them. Psychologically, the culture of *mutuwhenua* would have pre-conditioned their emotions into a state of fear and fright, which the unexpected mid-afternoon solar eclipse would have intensified, helping entrench the occasion into native legends and stories, including fears and superstitions, which still exists today.

3 Genealogical Evidence

Smith's 'Great Fleet' hypothesis would place these events in AD 1350 based on his study of traditional genealogies. Before turning to the eclipse witnessed by Hatupatu, Tia and Ngātoroirangi in the following section, let us then first discuss the genealogical evidence and its limitations, focusing specifically on the tribes descended from the crew of the *Tainui* and *Te Arawa*.

Whakapapa, or genealogy, is key pillar of Polynesian culture, social order and identity (Buck 1938: 21). The term is defined as the ability to recite genealogies in the correct order (Williams 1975: 259), and their recital served as a chronology of historical events associated with key ancestors (Buck 1938: 22; King 2003: 67). In pre-European times, these traditional genealogies were passed down from generation to generation by rote learning and were held in the memory of knowledge-experts, the *tohunga*, or the tribal priests, and were eventually committed to writing starting in the nineteenth century. Between oral and written sources, there exist hundreds of these *whakapapa* that span back to well before the Māori's arrival to New Zealand. Due to the stress laid on exact memorisation and recital, there is furthermore an impressive consistency across different tribal sources, from New Zealand to East Polynesia, as well as across time. Indeed, there are Eastern Polynesian and Māori genealogies that have been found to be fundamentally consistent over a period of 500 years (from the year 1900). New Zealand anthropologist Te Rangi Hiroa (also known as Peter Buck) recognised that there were limitations to their fidelity, as it is impossible to retain detailed memories of people and events over more than a half-millennium, the events prior drifting off into myth, the spiritual world, deities and gods (Buck 1938: 22). Still, that brings us back to the fourteenth century, and cross-referencing independent *whakapapa* allows us to construct fairly reliable genealogies going back to the crew of these ships.

The principal author of this paper, Ockie Simmonds, has gathered extensive, authoritative *whakapapa* in oral and manuscript form concerning the tribes issued from the *Tainui* and *Te Arawa*. He is furthermore a direct descendant of both captains and other members of each vessel through his paternal grandfather, Tiaki Kereti, and his *karanga-kuia*, or tribal grandmother, Rangimahora Reihana-Mete.[12]

[12] Genealogy information sources: Private manuscripts from: Piripi Teimana, Te Hiko, N., Waaka, R., Te Kaponga, *c.*1912, Rangimahora, 1986; Flavell, Te U., 1980.

The vessels *Tainui* and *Te Arawa* departed together from their island homeland in 'Hawaiki' (Rai'ātea, Tahiti). Their crews were related, and the captains were accompanied by their respective wives, children, grandchildren, great-grandchildren, siblings and extended family members. Each vessel had possibly as many as 70–80 passengers aboard. The names and genealogies of the men (and most of the women) on board were retained in memory by traditions, stories, *waiata* (songs) and, then, committed to writing after the Europeans arrived in the nineteenth century (Harris et al. 2013; Simmons 1976: 11).

As illustrated in Fig. 2, there were four generations aboard each vessel when they arrived in New Zealand, and the *Te Arawa* side left two elder living generations behind. Tradition holds that the youngest generation—Mōtai and Uenuku—were born on the voyage to New Zealand, which was estimated to take around 26–30 days.[13] Uenuku's full name is Uenuku-mai-Rarotonga, or Uenuku from Rarotonga (an island located in the Cook Islands). Furthermore, one notes that while Gudgeon (1894: 47) identifies Hoturoa, captain of the *Tainui*, as the elder brother of Tuamatua, the grandfather of Captain Tamatekapua, *Tainui* authorities have the two as cousins.[14]

As illustrated in Table 1, counting from Tiaki Kereti (1896–1958) and Rangimahora Reihana-Mete (1899–1993) in 1900, the *whakapapa* of their tribal

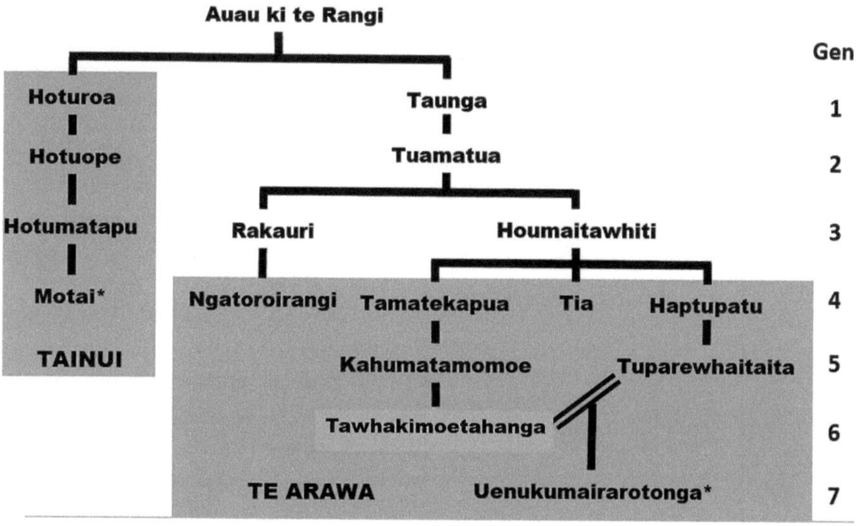

Fig. 2 Genealogy of the crew of the *Tainui* and *Te Arawa*, modified from Gudgeon (1894). The asterisks (*) indicates the two members that were born *en route* to New Zealand in the same year

[13] See Jones (1995: 52) and M. Winiata (ed.), *Tainui Sexcentennial Canoe Celebrations 1350–1950. Souvenir Brochure* (New Zealand: Turangawaewae Māori Adult Education Committee and Tainui Sexcentennial Canoe Celebrations, 1950), p. 5.

[14] Private manuscripts from Te Rangipapanui Te Kapunga, *c.*1912, and Piripi Kereti, 1974.

Table 1 *Whakapapa*—Linear Genealogy from the *Tainui* and *Te Arawa*

Year	*Tainui* 1	*Tainui* 2	*Tainui* 3	*Te Arawa* 1	*Te Arawa* 2	*Te Arawa* 3	Gen
1300			**Hoturoa**				25
		Hoturoa	Hotumatapu				24
1350	**Hoturoa**	Hotumatapu	Hotuope				23
	Hotumatapu	Hotuope					22
1400	Hotuope	M tai	Uetapu				21
	M tai	Uetapu	Rakamamao				20
1450	Uetapu	Rakamamao	Kakati	**Tamatekapua**	**Tamatekapua**	**Tamatekapua**	19
	Rakamamao	Kakati	Tawhao	Kahumatamomoe	Kahumatamomoe	Kahumatamomoe	18
1500	Kakati	Tawhao	Whatihua	Tawhakimoetahanga	Tawhakimoetahanga	Tawhakimoetahanga	17
	Tawhao	Turongo	Uenukuwhangai	Uenukumairarotonga	Uenukumairarotanga	Uenukumairarotonga	16
1550	Turongo	Raukawa	Kotare	Rangitihi	Rangitihi	Rangitihi	15
	Raukawa	Takihiku	Kauwhata	Tuhourangi	Tuhourangi	Tuhourangi	14
1600	Whakatere	Pipito	Wehiwehi	Uenukukopako	Uenukukopako	Uenukukopako	13
	Poututerangi	Tamatewhana	Tutete	Whakauekaipapa	Kiritai (w)	Whakauekaipapa	12
1650	Huriroroa	Maihi	Parekarewa (w)	Kaitapu	Hineaupounamu (w)	Tawakeheimoa	11
	Ahiroa	Pareunuora (w)	Ngatokowaru	Rangipunga	Pareunuora (w)	Rangiwewehe	10
1700	Te Awa	Matau	Matau	Tiori	Matau	Kereru	9
	Te Ngue	Puauea (w)	Puauea (w)	Te Ngakau	Puauea (w)	Wehiwehi	8
1750	Haehaekuku	Parehikanga (w)	Parehikanga(w)	Hoputu	Parehikanga (w)	Whatutapunui	7
	Hakunga	Kainganui (w)	Kainganui (w)	Paneturi (w)	Kainganui (w)	Ngarootu	6
1800	Hinekapuarangi (w)	Pahau	Pahau	Hinekapuarangi (w)	Pahau	Whekiki	5
	Rauti (w)	Rauti (w)	Rauti (w)	Rauti (w)	Rauti (w)	Hikairo	4
1850	Hare Teimana	Hare Teimana	Hare Teimana	Hare Teimana	Hare Teimana	Atareta (w)	3
	Tawhai Kereti	Tawhai Kereti	Tawhai Kereti	Tawhai Kereti	Tawhai Kereti	Heketoro	2
1900	**Tiaki Kereti**	**Tiaki Kereti**	**Tiaki Kereti**	**Tiaki Kereti**	**Tiaki Kereti**	**Rangimahora (w)**	1

Note that each column represents a different, direct line of father-son and father-daughter (W) relations, and that the 'Year' column is approximation based on the rate of 25 years per generation. Dark grey indicates the four generations aboard the respective ships. Sources: private manuscripts from Hare Teimana (dated c.1910), Mitai Kereti (1969–1972), Piripi Teimana (dated 1974), Nigel Te Hiko (dated 2007). Rangipapanui Te Kaponga (dated *c.*1912) and Rangimahora Teimana-Leonard (dated 1982–1992), Hoani Iwikau-Wereta (dated 1984–1987), Iwikatea Nicholson (dated 1985–2014) and Kiripua Te Aomarere (dated 1996)

descendants go back twenty-five generations to Hoturoa, captain of the *Tainui*, and nineteen generations to Tamatekapua, captain of the *Te Arawa*. Specifically, Both Tiaki and Rangimahora have the same grandfather, Te Taute (Hare Teimana) who is a direct descendant of both captains.

Be they Polynesian or Māori, such genealogies cover a rather short historical time frame, by world standards, and as they count back in *generations*, rather than *years*, it is difficult to date historical events thereby with any degree of certainty (Burley et al. 2015). When *whakapapa* were first used by historians to this end in the early twentieth century, it was standard to use the rough estimate of there being 25 years per generation (Simmons 1976: 37; Buck 1938: 23). As to Smith's famous arrival date of AD 1350 for the 'Great Fleet', he 'came up with his date by simply averaging out a great number of unrelated genealogical lines', using the mathematical mean of 22 generations from the captains of various sailing voyagers in the Great Fleet, in this example from the *Tainui* and *Te Arawa* captains to Tiaki Kereti and Rangimahora in 1900 (see Table 1): 1900 − 22 × 25 = 1350 (King 2003: 45; Simmons 1976). Needless to say, there are a number of problems with Smith's methodology.

As shown by the archaeological record, this 25-year rule of thumb provides an adequate method of approximating historical events to within a century or so, but it remains crude and unsatisfactory without supporting evidence. More importantly, European historians like S. Percy Smith failed to understand the half-millennium limitation of reciting genealogies from memory as well as the subtle cultural nuances of Māori and Polynesian genealogies and their inter-generational relationships. When trying to establish a chronological date mechanism a direct linear association of genealogies is fraught with difficulties that early writers did not understand.

There are, for example, a number of important generational disparities across different lines running between the same two people that make a simple conversion from generations into years impossible. In Fig. 2, we see that though Mōtai (*Tainui*) and Uenuku (*Te Arawa*) were both born *en route*, in the same year, they are in fact three generations apart in the family tree, or 75 years by Smith's rule of thumb. Likewise, one notes that the latter's parents, Tawake and Tupare (W) married across generations, making Uenuku both a fourth- and fifth-generation descendant of Tuamatua. In Table 1, we see that, in 1900, the same two descendants of both crews could trace their lineage back to the same ancestor via 11 different lines counting 22–30 generations, this eight-generation disparity amounting to 200 years by the same rule of thumb. Rather than adjusting the mean number of generations or years per generation going back to the *Tainui* and *Te Arawa*, therefore, we should peg our chronology to two facts: that Mōtai and Uenuku were born in the same year—the year of the 'Great Fleet'—and that there were four living generations aboard each vessel when they landed, with two additional living generations on the *Te Arawa* side who had stayed behind in Hawaiki.

This may sound incredible, but six-generation living cohorts were not uncommon until recent times. According to the principal author's discussions with his tribal grandmother Rangimahora Reihana-Mete at age 87 in 1986, for example, Rangimahora's mother, Ranginui Teimana-Leonard (1872–1984), died at 112 in 1984, at which time she was a great-great-great-grandmother. Ranginui was 27 when she had her first child, and Rangimahora 25, but one imagines that their ancestors may well have done so as early as age 13, in which case one may well see great-great-great grandchildren by age 65.[15] The difference is almost enough to fit two additional generations at 25 years a piece, and it highlights why the generational disparity between the voyagers on the *Tainui* and *Te Arawa* can be largely ignored.

On their departure from Hawaiki, the traditional narrative tells us that Tamatekapua's father, Houmaitāwhiti, and his grandfather, Tuamatua, were still alive (Reed 1977: 101; Gudgeon 1894). On the year of arrival—and of Uenuku's birth—this six-generation cohort must have likewise spanned some 65–112 years, with Captain Tamatekapua in his middle age, somewhere around 50, or two full generations old. One notes however that Smith's arrival date of AD 1350 is calculated according to the captains' probable *dates of birth*, ignoring what can be deduced of their *ages at departure* from the number of living generations in their respective families at the time. Therefore, at his own rate of 25 years per generation, we would need to move Smith's arrival date forward to at least AD 1400.

Again, *whakapapa* genealogies may well be consistent and perfectly reliable up to at least several generations from the crew of the *Tainui* and *Te Arawa*, but the

[15] That is, 5 × 13 = 65 years, by which time the fifth generation would be 13, and the sixth generation age 0.

most that we can extract from them in terms of Western calendar years are gross approximations riddled with caveats. Keeping in mind the birth of Mōtai and Uenuku *en route* and the number of living generations aboard, the genealogies in Table 1 suggest an arrival period sometime between the fourteenth and fifteenth century. This is nowhere near as neat as the authoritative date of 'AD 1350', but if we examine this two-century span for eclipses matching that described in traditional narrative, then we may well be able to fix the exact year of arrival.

4 Determining the Time of Arrival and Departure

Dating the arrival of the 'Great Fleet' to New Zealand based on genealogical evidence alone is highly approximate, but if we can identify the historical eclipse witnessed by that the leaders of the *Tainui* and *Te Arawa* expeditions shortly after their arrival, then we can work backwards from an exact date. To this end, the *whakapapa* and archaeological record give us a rough idea of the range of centuries in which we should be looking (14th–15th), and traditional narratives and toponyms help identify the zone which would have fallen in or near the path of totality (the North Island, and the central Taupō volcanic region more specifically). Furthermore, traditional accounts tells us the day was already dark and bleak, heavily overcast, and cold with possible snow and sleet, which eliminates the hot season, in the middle of summer.

The period in question saw four total solar eclipses over New Zealand (all dates are according to the Julian calendar):

- 15 June 1368: early morning, North Island
- 9 October 1409: afternoon, North Island
- 2 January 1424: afternoon, North Island
- 11 December 1433: early morning, North–South Island

As illustrated in Fig. 3, the closest eclipse to iconic date of 1350 is that of 15 June 1368, but it is not a perfect fit: the path of totality fell north of the North Island, and 1368 is too early for a six-generation living cohort. That is, the majority of the genealogical lines on Table 1 above do not go back a sufficient number of generations for Mōtai and Uenuku to have been conceivably been born so early.[16]

We can likewise eliminate the eclipses of 2 January 1424 and 11 December 1433, both of which fall in the hot season. This is not to mention that, in 1424, the path of totality grazes the Southern end of the South Island, and that, in both cases, we would expect the captains and older crew members of the two vessels to be at the extreme end of their predicted life spans at this late date.

The best match by far is the total eclipse of 9 October 1409. As illustrated in the close-up in Fig. 4, the path of totality fell over the central volcanic plateau of the

[16] Note that the columns *Tainui* 5 and *Te Arawa* 6 provide two exceptions.

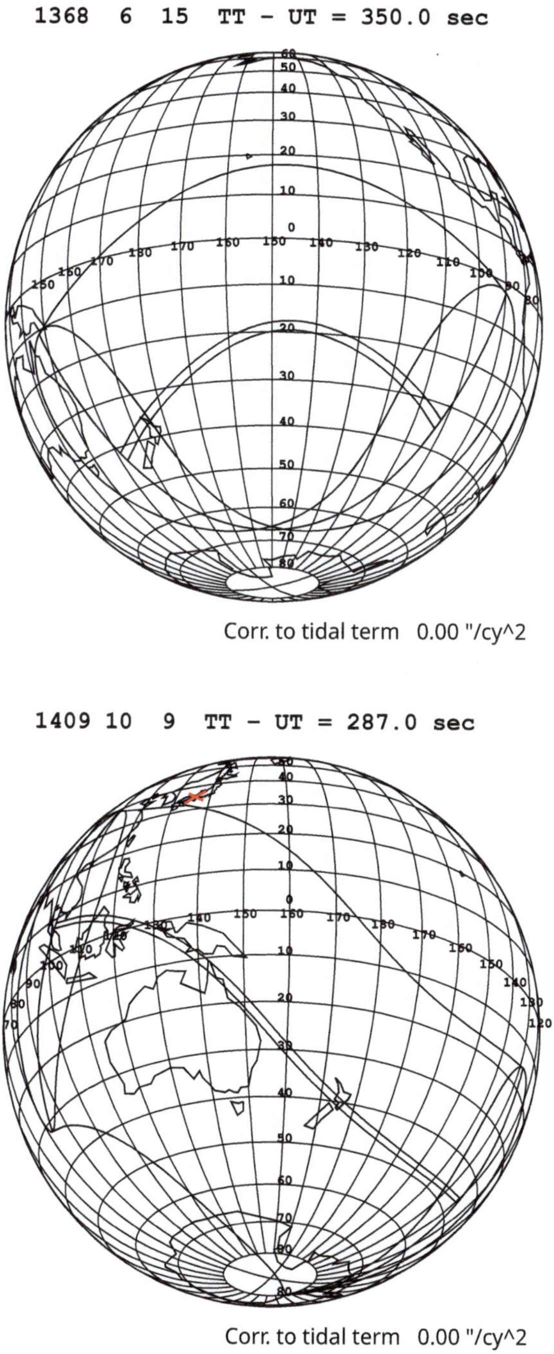

Fig. 3 Candidate eclipses, thirteenth–fourteenth century

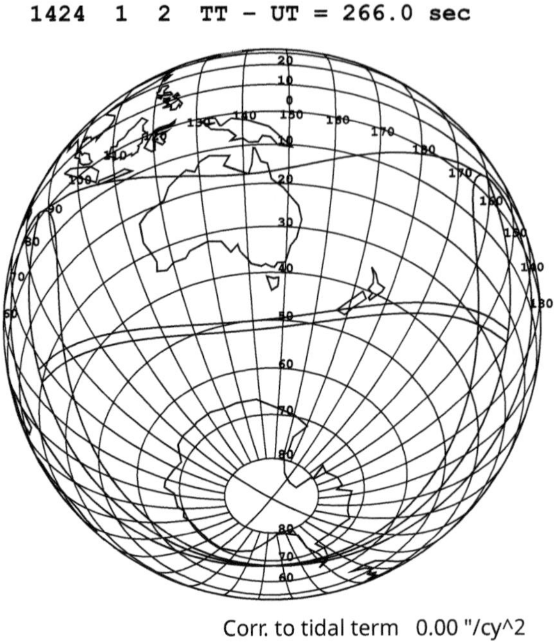

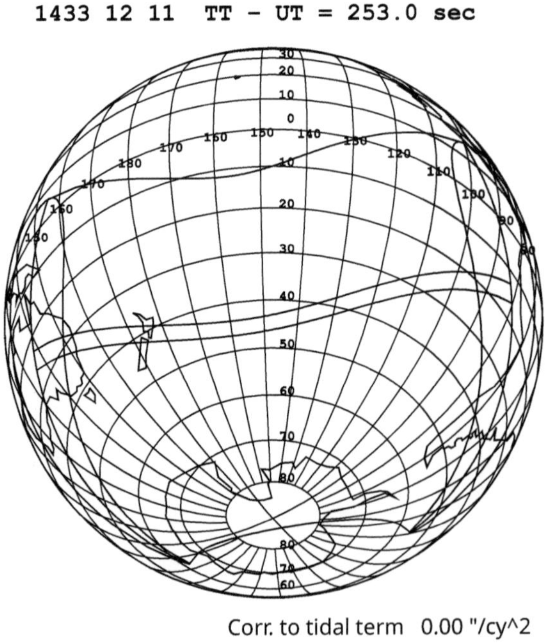

Fig. 3 (continued)

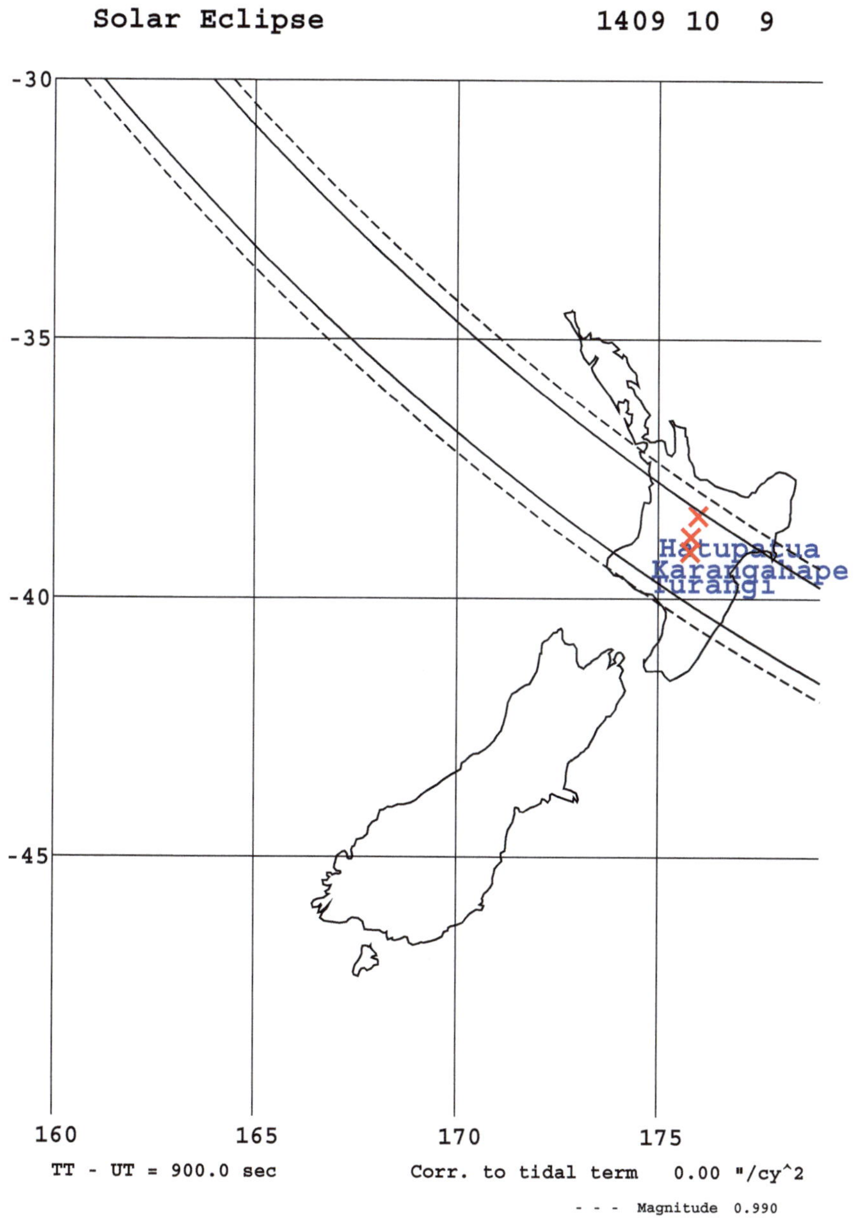

Fig. 4 Total solar eclipse, 9 October 1409. Note that the path of totality would have covered the known locations of Ngātoroirangi and Tia at Rangipo (near Turangi), as well as Karangahape, on the south-west shores of Lake Taupō, placing Hatupatu (at Atiamuri) at the edge of totality

Northern Island, covering the zone of Hatupatu, Ngātoroirangi, and Tia's inland excursion. 1409 is a reasonable date to expect a six-generation living cohort in line with the *whakapapa* in Table 1. Lastly, one notes that October is occasionally more likely to see seasonal cold spells, including snow and sleet, matching the weather conditions described in traditional narrative.

Working back from the date 9 October 1409, when their captains were exploring inland, we can deduce that the *Tainui* and *Te Arawa* had probably arrived the previous summer. According to tradition, the voyagers made landfall near Cape Runaway, on the East Coast, at the time that the *pōhutukawa*, or New Zealand Christmas tree, was in full bloom (Stafford 1967: 428), pointing to a date in mid-December 1408. As to their date of departure, tradition states that the voyage took a total of 30 days and that it began on *orongonui ō te tau ururoa*, or 'the 27th lunar day of the long breadfruit season'.[17]

There are two main bread-fruit seasons in French Polynesia are in July–August and November–April, which, combined with the flowering of the *pōhutukawa* at the end of their journey, suggests a departure date in November. Indeed, this coincides with a period of abundance of fresh fish, breadfruit and other foodstuffs that one might want to stock up on for a long voyage, celebrated in Eastern Polynesia on the last evening rising of Matali'i (the Pleiades), in mid-November.[18] As to an exact date, the previous new moon (*mutuwhenua*) fell on 19 October 1408 (28 October in the proleptic Gregorian calendar),[19] which would place their departure from Hawaiki on the 27th in the traditional Māori lunisolar calendar on 15 November 1408 (24 November in the proleptic Gregorian calendar).

Next, it is said that after about three or four days at sea the *Tainui* and *Te Arawa* arrived at Rarotonga, where they stayed several days (for the birth of Uenuku-mai-Rarotonga, or 'Uenuku, born on Rarotonga') before setting off for their final destination in New Zealand. They left Rarotonga on the day *oo-Uenuku* ('of Uenuku'), day four of the lunar calendar (Jones 1995: 30), Uenuku having apparently been named after the day of their departure for New Zealand. The next new moon falling on 18 November (27 November in the proleptic Gregorian calendar), that would place their departure from Rarotonga on 22 November 1408 (1 December in the proleptic Gregorian calendar).

[17] On the duration of the voyage, see Jones (1995: 52). Traditional Polynesian voyagers were believed to sail from Tahiti to New Zealand in two weeks. In 1769, Captain Cook's *Endeavour* took 49 days; see Rolls, Lala, *Tupaia's Endeavour* (a television documentary), Island Productions 2016, New Zealand. As to the date of departure, Hinano Astrid Drollet, a linguist from Tahiti, has identified the correct spelling and meaning of *te tau ururoa* ('the long breadfruit season'). The Māori had forgotten the word *uru* for 'breadfruit', because it failed to grow in New Zealand, but it is retained in traditional *waiata*, or songs (personal communication, 26/11/2016). See also Buck (1938: 271–272) and Winiata (ed.), *Tainui Sexcentennial Canoe Celebrations 1350–1950. Souvenir Brochure*, p. 5.

[18] Personal communication from H. Drollet, email 18/11/2016.

[19] Lunar phases calculated using the website *WolframAlfa Computational Intelligence* (https://www.wolframalpha.com/)

From there, *Tainui* traditions state that, travelling via Rangitahua (Raoul Island), in the Kermadec Island group (Hiroa 1977: 46), it took them 26 days to arrive in New Zealand, which they did during the dark of the moon, whilst the red *pōhutukawa* flours were in bloom (Jones 1995: 52). This would place the date of their arrival on the subsequent new moon, 18 December 1408 (27 December in the proleptic Gregorian calendar). When the *Te Arawa* arrived at Whangaparāoa, the crew realised that *Tainui* had arrived before them. Tamatekapua, captain of the *Te Arawa*, disputed his defeat by the *Tainui*, and this is still being debated today. According to Stafford (1967: 17), the *Te Arawa* is said to have arrived on the seventh day of the seventh month, placing their arrival on 24 December (2 January 1409 in the proleptic Gregorian calendar).

In a personal communication, Hoturoa Kerr, an expert in traditional sailing, was able to confirm the feasibility of this timeline. Rarotonga, in the Cook Islands, is approximately 500 nm (800 km) from Hawaiki (Ra'iātea, Tahiti), in French Polynesia, which is a modest distance that can be easily sailed within a few days, leaving the *Tainui* and *Te Arawa* with a day or two onshore before once again setting sail. However, it only takes 12 days to sail directly from Rarotonga to New Zealand, which suggests that, to take 26 days, the two ships must indeed have taken a detour through the Kermadec Islands before arriving on New Zealand's East Cape.[20]

5 Conclusion

In summary, the story of 'Hatupatu and the Witch' recounts how, upon their inland expedition to the centre of the North Island shortly after landing in New Zealand, the crew of the *Tainui* and *Te Arawa* saw the daytime sky go dark at new moon—an event, suggestive of a total solar eclipse, so charged with *tapu* as to imprint itself on Māori legend, place names, cultural memory, and customs. Impressively, *whakapapa* genealogies allow us to reliably trace the ancestry of modern Māori generation by generation all the way back to the captains of these vessels and the family that the *Te Arawa* left behind on Hawaiki, allowing S. Percy Smith to estimate their date of arrival as AD 1350 by counting back the mean number of generations by the estimated interval of 25 years per generation. Percy's estimate does not take a number of things into account—the number of living generations reportedly aboard the two vessels, the wide generational discrepancy between *whakapapa*—and in doing so we arrive at an upper and lower bounds within which we can look for historical eclipse paths matching the location, description and seasonal context provided in the aforementioned story. As we discover, the one eclipse that matches these criteria occurred on 9 October 1409, and we can work backwards from there and additional information in traditional sources to establish the *Tainui*'s date of arrival as 18 December 1408, followed by the *Te Arawa* on Christmas Eve, 24 December 1408.

---

[20] Personal communication from H. Kerr, 11/09/2017.

Tentatively proposed at the beginning of the twentieth century, Smith's arrival date for the 'Great Fleet' of AD 1350 was a calculated guess that was quickly embraced by a young colonial country in search of an identity. Being based on both Western chronology and traditional *whakapapa*, moreover, it was a hypothesis that would embed itself with vigour within the social consciousness of both Pākehā and Māori alike. Over the course of the twentieth century, scholars would go on question this date and the very idea of the 'Great fleet' and the planned settlement of New Zealand. It was theorised, for example, that Neolithic Polynesians would have been incapable of navigation, travel, discovery, and settlement across such expanses without modern Western technologies like the sextant and compass. Without offering my by way of an alternative, however, such questioning overlooks the coherency in oral tradition about this period both across different tribes, on different islands, and with what we can calculate backwards through genealogy and celestial mechanics.

Of course, one thing that the present study leaves unresolved is the date of arrival of the very first Eastern Polynesian settlers, the moa hunters. In brief, Smith (1910: 39–77) proposed that New Zealand was first discovered by Kupe, in AD 950, who was followed by his moa hunter descendants in 1100. On the basis of archaeology, Duff (1950 [1977]: 16) would argue that this first people first arrived a century or more earlier, and King (2003:18), by contrast, no earlier than the late thirteenth-century. And today, a number of tribal groups hold fast to the idea that their *whakapapa* proves a line of descent from Kupe and an arrival in 950 or earlier. However, if we can anchor the arrival of the *Tainui* and *Te Arawa* to an astronomical event in 1408, then we may well be able explore links between the genealogies of their crews and those going back to Kupe to situate the latter vis-à-vis the former in terms of generation and revisit the rough date of his arrival. Either way, there is much about the 'Great Fleet' migration and the settlement of New Zealand that remains a mystery. Why did they leave the Eastern Pacific to the cold shores of New Zealand? Was climate change a factor? What was the impact on those who stayed behind? These are questions that will require future investigation and research to be answered.

Acknowledgements We would like to acknowledge the scientific analysis and support of Mitsuru Sôma of the National Astronomical Observatory of Japan. Sôma san's invaluable technical input as an international collaborator has been pivotal to this research. Without his support, scientific direction and patience, this research may not have eventuated. We would thank Pauline Harris, professor at Massey University in Palmerston North (formerly of the University of Victoria of Wellington) and chairperson of the Society of Māori Astronomy, Research and Traditions Inc. for her expert feedback and help reviewing the current paper. Lastly, we would like to thank the many tribal elders of Raukawa whānui for sharing their traditional genealogies, wisdom and knowledge that reach back 600 years to their voyaging ancestors of the 'Great Fleet'. *Nō reira, ka nui te mihi aroha ki a koutou katoa.*

Appendix: Hatupatu and the Witch

The story of Hatupatu says he arrived on the *Te Arawa waka*. One day he went inland bird-hunting, sparing and snaring birds. In the depths of the forests near Atiamuri, by the Waikato river, he was caught by a frightful bird-feathered-like monster called Kurangaituku ('Kura-of-the-Claws') or 'adorned as a flying-bird' as per Cowan (1925: 120). Hatupatu was held by this feathered giantess in her home in a forest cave. Kurangaituku had a kind of aviary of tame birds, and in the crevasses of the cave were lots of small lizards. Hatupatu had to look after the birds while Kurangaituku was out hunting in the forests.

One day Hatupatu escaped and fled from the cave, liberating all the birds before leaving. One of the birds, a small *riroriro*, a little grey warbler, flew off to alert Kurangaituku that Hatupatu had escaped. Kurangaituku returned to the cave to find all the birds gone and the lizards dead. Kurangaituku was furious and set out to capture the fugitive. Kurangaituku was very close to capturing Hatupatu when the latter came upon a great rock. In desperation, he uttered a potent charm to open the rock that was taught to him by the greatest priestly expert, Ngatoroirangi. The rock opened revealing a small cavern. Hatupatu quickly entered the cave, and it closed behind him. Kurangaituku was too late. She clawed and scratched at the rock-cave hoping to capture Hatupatu. Tiring of her efforts, she eventually left, hiding nearby in await for him to emerge. Hatupatu eventually came out from hiding and fled to his homeland over hills and hot boiling thermal pools to the Rotorua district. Kurangaituku set off in pursuit of flying Hatupatu. However, Kurangaituku mistimed her leap over a hot boiling thermal mud-pool, fell in, and was killed, allowing Hatupatu to escape safely back to his family at Lake Rotorua.

The Hatupatu legend is iconic within Māori traditional history. The rock-cave where Hatupatu is said to have taken shelter is located at Atiamuri, next to State Highway One, about 250 km south of Auckland city.[21] Furthermore, Te Kōwhatu o Hatupatu, the Rock of Hatupatu, is the only *wāhi tapu* (sacred place) registered as a Government heritage site in the South Waikato District of the North Island. In the words of Cowan (1925: 119–127), 'The refuge-place of Hatupatu is a peculiarly shaped rhyolite rock, hollowed out as by human agency, which stands at the foot of Ngatuku hill. The hollow interior of the rock is somewhat like an arm-chair seat.' To this day, Māori and Pākehā (Europeans) travellers still lay offerings of *manuka* twigs, ferns, grass, food and money in the cavity of the rock as a form of appeasement to Hatupatu. The practice is supposed to prevent untoward events that may affect a wary traveller's journey.

[21] Known as Hatupatu's Rock, the iconic rock-cave was recognised as a historical landmark by the New Zealand government and, in April 2017, Māori authorities and government officials conducted a ceremony in recognition of the site (Media Release, 20 April 2017: New Zealand Transport Agency, *Restoration of Hatupatu Rock Complete*. https://www.nzta.govt.nz/media-releases/restoration-of-hatupatu-rock-complete/)

Bibliography

Bridgman, Howard A. 1983. Could Climatic Change Have Had an Influence on the Polynesian Migrations? *Palaeogeography, Palaeoclimatology, Palaeoecology* 41(3–4): 193–206.

Buck, Peter H. 1938. *The Vikings of the Sunrise.* Philadelphia: J. B. Lippincott.

Burley, David, Edinborough, Kevan, Weisler, Marshall, and Zhao Jian-xin. 2015. Bayesian Modeling and Chronological Precision for Polynesian Settlement of Tonga. *PLoS ONE* 10(3): e0120795. doi:https://doi.org/10.1371/journal.pone.0120795.

Condliffe, J. B. 1971. *Te Rangi Hiroa, The life of Sir Peter Buck.* Christchurch: Whitcombe & Tombs.

Cowan, James. 1925. *Fairy Folk Tales of the Maori*. Auckland: Whitcombe & Tombs.

Duff, Roger. 1950 [1977]. *The Moa-hunter period of Maori culture*. Wellington: Government Printer.

Evans, Jeff. 1998. *The Discovery of Aotearoa:* Auckland: Reed.

Grace, John Te H. G. 1959 [1970]. *Tūwharetoa, The History of the Māori People of the Taupō District.* Wellington: A. H & A. W. Reed.

Grey, George. 1885 [1988]. *Legends of Aotearoa.* First published 1855 under another name. Hamilton, N.Z.: Silver Fern Books.

Gudgeon, W. E. 1894. Maori Migrations, No II. *Journal of the Polynesian Society* 3(1): 46–51.

Harris, P., Matamua, R., Kerr, H., Smith, T., Waaka, T,. 2013. A Review of Māori Astronomy in Aotearoa-New Zealand. The Journal of Astronomical History and Heritage 16(3), 325–336.

Hiroa, Te Rangi (Peter H. Buck). 1949 [1977]. *The Coming of the Maori (2nd ed.).* Wellington: Whitcoulls Limited.

King, Michael. 2003. *The Penguin History of New Zealand.* Auckland: Penguin Books.

Lee, Michael. 2018. *Navigators & Naturalists: French Exploration of New Zealand and the South Seas (1769–1824)*. Aukland: Bateman Books.

Lewis, David. 1977. *From Maui to Cook: The Discovery and Settlement of the Pacific*. Sydney: Doubleday Australia.

Lewis, David. 1978. *The Voyaging Stars: Secrets of the Pacific Island Navigators*. Sydney: Collins.

Mead, Sidney Moko. 1984. *Customary concepts of the Maori: a Source Book for Maori Studies Students*. Second revised edition. Wellington: Department of Māori Studies, Victoria University of Wellington.

Reed, A. W. 1977. *Treasury of Māori Exploration, Legends Relating to the First Polynesian Explorers of New Zealand*. Wellington: A. H. & A. W. Reed.

Simmons, D. R. 1976. *The Great New Zealand Myth: a Study of the Discovery and Origin Traditions of the Māori*. Wellington: A. H. & A. W. Reed.

Smith, S. Percy. 1910. 'History and traditions of the Maoris of the West Coast, North Island of New Zealand, prior to 1840'. Memoirs of the Polynesian Society v. 1. New Plymouth, N.Z.: the Polynesian Society (first published in *Journal of the Polynesian Society* v. 16–19).

Stafford, D. M. 1967. *Te Arawa: A History of the Arawa People*. Wellington: Auckland Reed.

Tanikawa Kiyotaka 谷川清隆 and Sôma Mitsuru 相馬充. 2010. 'Amana-Iwato' nisshoku kōho ni tsuite『天の磐戸』日食候補について (On Candidate Eclipses of a Supposed Eclipse in Myth 'Amano-Iwato'). *Kokuritsu temmondai hō* 国立天文台報 (Publications of the National Observatory of Japan) 13: 85–99.

Tanikawa Kiyotaka 谷川清隆, Sôma Mitsuru 相馬充 and Ueda Akitoshi 上田暁俊. 2014. Delta T variations from AD 1133 to AD 1267. Paper presented at the Eight International Conference on Oriental Astronomy (ICOA-8), 26–28 March 2014, Hefei, China.

Tanikawa Kiyotaka 谷川清隆, Sôma Mitsuru 相馬充 Simmonds, Ockie, and Iwahashi, K. 2019. Delta T from solar eclipses around 1415 AD. Proceedings of the Fifth Symposium on 'Historical astronomical records and modern science', held 23-24 March 2018. Publication of the National Astronomical Observatory of Japan (131–138).

Jones, Pei Te Hurinui. 1995. *Ngā Iwi o Tainui: nga koorero tuku iho a nga tuupuna* (The traditional history of the Tainui people). Edited by Bruce Biggs. Auckland: Auckland University Press.

Williams, Herbert W. 1975. *A Dictionary of the Māori Language.* 7th ed. Wellington: Government Printer.

MIX
Papier aus verantwortungsvollen Quellen
Paper from responsible sources
FSC® C105338

If you have any concerns about our products,
you can contact us on
ProductSafety@springernature.com

In case Publisher is established outside the EU,
the EU authorized representative is:
Springer Nature Customer Service Center GmbH
Europaplatz 3, 69115 Heidelberg, Germany

Printed by Libri Plureos GmbH
in Hamburg, Germany